Water Splitting

Scrivener Publishing
100 Cummings Center, Suite 541J
Beverly, MA 01915-6106

Publishers at Scrivener
Martin Scrivener (martin@scrivenerpublishing.com)
Phillip Carmical (pcarmical@scrivenerpublishing.com)

Water Splitting

Production of Hydrogen

Edited by

Inamuddin

Tariq Altalhi

Mohammad Luqman

and

Jorddy Neves Cruz

Scrivener Publishing

WILEY

This edition first published 2025 by John Wiley & Sons, Inc., 111 River Street, Hoboken, NJ 07030, USA and Scrivener Publishing LLC, 100 Cummings Center, Suite 541J, Beverly, MA 01915, USA

For more information about Scrivener publications please visit www.scrivenerpublishing.com.

Wiley Global Headquarters
111 River Street, Hoboken, NJ 07030, USA

For details of our global editorial offices, customer services, and more information about Wiley products visit us at www.wiley.com.

The manufacturer's authorized representative according to the EU General Product Safety Regulation is Wiley-VCH GmbH, Boschstr. 12, 69469 Weinheim, Germany, e-mail: Product_Safety@wiley.com

Library of Congress Cataloging-in-Publication Data

ISBN 9781394247622

Cover image: Generated with AI using Adobe Firefly
Cover design by Russell Richardson

Set in size of 11pt and Minion Pro by Manila Typesetting Company, Makati, Philippines

Printed in the USA

10 9 8 7 6 5 4 3 2 1

Contents

Preface

Water splitting for the production of hydrogen is a rapidly evolving field at the forefront of interdisciplinary research and industrial development. It encompasses the integration of multiple scientific disciplines, including chemistry, physics, materials science, engineering, and environmental science, to address the global energy challenge and transition towards a sustainable future. In the context of disciplinary development, water splitting has witnessed significant advancements in understanding the fundamental principles and kinetics of the process. Researchers have explored various methods and catalysts for efficient water splitting, aiming to enhance hydrogen production rates and minimize energy consumption. The field of electrocatalysis has seen remarkable progress, with the development of novel materials, such as metal oxides, metal sulfides, and molecular catalysts, to enhance the performance of electrolyzers. Furthermore, the emergence of photoelectrochemical water splitting has brought together materials science, electrochemistry, and photonics. Researchers have been exploring semiconductor materials, such as metal oxides and metal chalcogenides, to design efficient photoelectrodes capable of harnessing solar energy for water splitting. The field has witnessed breakthroughs in bandgap engineering, surface modification, and tandem cell configurations, enabling improved solar-to-hydrogen conversion efficiency. In terms of industry development, water splitting holds immense potential for sustainable hydrogen production, which can be used as a clean energy carrier. The advancement of water electrolysis technologies, such as alkaline and proton exchange membrane (PEM) electrolyzers, has enabled scalable and cost-effective hydrogen production. These technologies are being increasingly deployed in industries, such as hydrogen fueling stations, energy storage systems, and industrial processes, to meet the growing demand for clean energy solutions. Moreover, the integration of water splitting with other renewable energy sources, such as solar and wind, presents opportunities for the development of integrated systems and the establishment of a hydrogen economy. The ability to store and utilize hydrogen as a versatile energy carrier has the potential to revolutionize sectors like transportation, power generation, and industrial applications, enabling a transition away from fossil fuels and reducing carbon emissions.

Water Splitting: Production of Hydrogen book provides a comprehensive exploration of water splitting, starting with foundational principles on thermodynamics and kinetics, crucial for understanding hydrogen production. It covers diverse methods and catalysts, emphasizing material selection and reaction optimization. Advancing further, the text explores recent innovations in materials and catalysts tailored for water splitting, highlighting synthesis techniques, functional materials, and nanotechnology integration. Detailed insights into electrochemical water splitting, encompassing alkaline and proton exchange membrane (PEM) electrolysis, clarify design principles, operational strategies, and electrolyte enhancements. A significant portion is dedicated to water photoelectrochemistry, analyzing semiconductor materials, photoelectrode design, and solar-to-hydrogen conversion strategies. The book delves into integrated systems, advanced reactors, and the role of artificial intelligence, machine learning, and big data in enhancing water splitting technologies. Concluding with applications, it explores hydrogen's potential as a clean energy carrier in fuel cells, energy storage, and industrial processes. Addressing gaps in current resources, the book focuses on cutting-edge advancements, ensuring researchers stay informed and prepared to contribute to the field's progress. The chapters included in the book are summarized below:

Chapter 1 explores the fundamental thermodynamical principles governing the energy requirements and efficiencies of splitting water molecules into H2 and O2. The thermodynamical parameters discussed include Gibbs free energy, enthalpy, and entropy changes, emphasizing their implications for developing efficient, effective, and sustainable hydrogen production techniques.

Chapter 2 emphasizes the importance of understanding water splitting kinetics for developing efficient catalysts for green energy. It explores the distinct mechanisms of the Oxygen Evolution Reaction (OER) and the Hydrogen Evolution Reaction (HER), discusses various tools for studying their kinetics, and highlights the need for systematic investigation to optimize catalyst design in both alkaline and acidic media.

Chapter 3 presents a comprehensive study of perovskite electrocatalysts for water splitting. Hydrogen fuel from electrochemical water splitting using renewable electricity is a promising future energy source. Effective, stable electrocatalysts are essential for improving water splitting reaction rates. This chapter reviews recent advances in perovskite oxides for catalytic reactions, exploring their mechanisms, composition, and design strategies, including insights from density functional theory.

Chapter 4 explores electrode materials, especially those based on titania nanotubes, that can be used in solar-driven electrochemical water splitting. As water splitting is considered a promising field for the green production of hydrogen, the accurate selection and design of photoelectrodes are of key importance.

Chapter 5 discusses the recent advances of MXenes materials in the electrocatalytic production of hydrogen and oxygen. Strategies include the control of surface terminating groups and structural as well as heteroatom doping, targeting improvements in the catalytic performance of MXenes towards efficient water splitting.

Chapter 6 discusses inorganic photocatalysts for hydrogen production through water splitting, emphasizing the importance of their crystalline structure and semiconductors like TiO2. It covers the fundamental principles of photocatalysis and the physical and chemical properties of photocatalysts. The chapter also explores methods to improve water splitting efficiency, addressing current challenges and prospects.

Chapter 7 surveys contemporary water splitting materials and discusses future requirements in the field. Considering the functional materials used for water separation, the most efficient functional material is determined with the help of the analytical hierarchy process. This chapter is designed to assist industry professionals and specialists in selecting the optimal group of materials for hydrogen production through water splitting.

Chapter 8 focuses on utilizing nanotechnology to advance the field of water splitting. This research area explores how nanoscale materials and structures can be engineered to enhance the efficiency and effectiveness of splitting water molecules into hydrogen and oxygen gases, a process crucial for renewable energy production and sustainable fuel generation. Key objectives include improving catalyst performance, optimizing light absorption for photoelectrochemical cells, and exploring new nanomaterials for electrolysis. Overall, nanotechnology in water splitting research aims to contribute to cleaner energy technologies and address global energy challenges.

Chapter 9 explores hydrogen utilization through fuel cells, emphasizing thermodynamics, kinetics, and various types of fuel cells (acidic, alkaline, phosphoric acid, and high-temperature). It covers electrocatalysts, proton exchange membranes, and porous electrodes. Challenges like hydrogen transport are discussed, highlighting the importance of fuel cells in achieving a carbon-free society.

Chapter 10 discusses catalysts for the anodic oxygen evolution reaction. It covers the advantages, disadvantages, and future perspectives of the catalyst types discussed in the literature. Additionally, the characterization of the catalyst in anodic oxygen evolution is included.
Chapter 11 delves into the characterization and analysis of electrolyzer performance, emphasizing mechanisms that enhance efficiency and inform material selection. This comprehensive study covers key aspects including current density, Faraday efficiency, energy efficiency, stability, as well as gas production rates and purity, providing crucial insights for improving system performance.

1

Thermodynamics of Electrochemical Water Splitting

Manash P. Nath[1,2], Manju Kumari Jaiswal[1,2], Suvankar Deka[1,2] and Biswajit Choudhury[1,2]*

[1]*Materials and Energy Laboratory, Physical Sciences Division, Institute of Advanced Study in Science and Technology (IASST), Assam, India*
[2]*Academy of Scientific and Innovative Research (AcSIR), Ghaziabad, India*

Abstract
The thermodynamics of water splitting elucidates the intricate balance of energy transformations fundamental to this crucial reaction, including photocatalytic and electrocatalytic water splitting. Electrocatalytic water splitting—triggered by the application of an external voltage—holds great potential for creating sustainable energy-producing hydrogen gas (cathode) and oxygen (anode). It requires overcoming the thermodynamic barriers, such as enthalpy change (ΔH) and Gibbs free energy change (ΔG), associated with the bond dissociation energies of water molecules and the formation of intermediates. Gibbs free energy barrier (ΔG) must be negative to proceed with the water splitting reaction spontaneously. At this hour, electrocatalysts play a pivotal role in decreasing the activation energy barrier enhancing the water splitting kinetics, and increasing the overall efficiency. Noble metal-based electrocatalysts such as platinum (Pt), ruthenium (Ru), and iridium (Ir) have been employed for their high catalytic activity with an extremely low free energy barrier. However, their expensiveness and scarcity have instigated researchers to search for low-cost earth-abundant electrocatalysts such as transition metal oxides, sulfides, phosphides, and high entropy alloys, among others. Therefore, the study provides an in-depth and comprehensive understanding of various thermodynamic parameters associated with the electrocatalytic water splitting process.

Keywords: Electrochemical, gibbs free energy, noble metal, layered materials

**Corresponding author*: biswajitchoudhury@iasst.gov.in; biswa.tezu@gmail.com

Inamuddin, Tariq Altalhi, Mohammad Luqman and Jorddy Neves Cruz (eds.) Water Splitting: Production of Hydrogen, (1–22) © 2025 Scrivener Publishing LLC

1.1 Introduction

The worldwide energy crisis and the related ecological problems have spurred scientists to seek the replacement of fossil fuels with renewable energy sources [1, 2]. Natural resources, such as energy generated by the sun or wind and water splitting technologies, offer an enticing means of producing hydrogen sustainably [3]. Water splitting involves the breaking down of water molecules into H_2 and O_2 under the effect of an external energy source such as a high temperature, light, or the application of an electric field [4]. From an environmental perspective, three approaches are mostly adopted for water splitting. These approaches are (a) photocatalytic, (b) electrocatalytic, and (c) photoelectrocatalytic. Photocatalytic water splitting requires a semiconductor with a high absorption cross section for incident solar photons to excite photocarriers and generate free electrons and holes [5]. For overall splitting, a semiconductor's conduction and valence band edge potential must be suitable enough to carry out the oxidation/reduction of water. In the ideal case, the band gap must exceed the minimum water splitting potential of 1.23 V [6]. Electrocatalytic water splitting involves H_2 generation (hydrogen evolution reaction or HER) and O_2 generation (oxygen evolution reaction or OER). The first electrolytic cell was proposed in 1789, and it comprises three parts: cathode, anode, and electrolyte [7]. Two moles of hydrogen and oxygen are produced from the splitting of 1 mol of water. It is an endothermic, non-spontaneous process with a standard Gibbs free energy change ($\Delta G = +237.3$ kJ/mol) [8]. Photoelectrocatalytic water splitting (PEC) is an electrocatalytic water splitting process with additional energy provided by light [9]. For photoelectrode materials to be commercially viable, they must have an STH, i.e., solar-to-hydrogen conversion efficiency of above 10% in addition to being sufficiently abundant on Earth and stable in water [10]. In 1972, Honda and Fujisima demonstrated the first PEC with TiO_2 photoanode [11].

The efficacy and feasibility of catalytic water splitting are enthalpy, entropy, and Gibbs free energy. Since the water splitting process is endothermic or uphill, its high ΔG of +237 kJ/mol is overcome by providing electrical energy equivalent to 1.23 V and increasing the temperature to accelerate the reaction [12]. To achieve it, the entire enthalpy requirement of $\Delta H = -283.85$ kJ/mol comes from the environment at temperature T as temperature T is necessary to achieve the required work $T\Delta S$, which is important for the electrolysis of water [13]. The first law of thermodynamics, "the law of energy conservation," describes how external energy (light, electricity, or heat) must be added to disintegrate the chemical bonds that bind water molecules, thereby converting them into H_2 and O_2 [14]. On the other hand, the second law of thermodynamics describes that the water splitting process is not 100% efficient with a positive entropy. A temperature equivalent to (T) >4,500 K can be used to split water, which is a maladroit process. Gibbs free energy change (ΔG), which describes the spontaneity of a water splitting reaction, becomes negative at T >4,500 K or E (V) >1.23 V, which indicates a favorable water splitting process [15]. Several attempts have been made to eliminate the internal energy losses that arise from overpotential and cell resistances to minimize the energy requirements of an electrolytic cell [16]. They include controlling the electrolyte's temperature, pressure, and chemical potential [16]. Thus, all three thermodynamic parameters, H, S, and G, are dependent on temperature (T) and pressure (P) and will describe the feasibility of an electrochemical water splitting system. In the case of HER, i.e., hydrogen evolution reaction, the change in the thermodynamic parameters (ΔH, ΔS, and ΔG) depends on the electrolytic conditions, i.e., acidic or alkaline. The abundance of protons (H^+), in the case of acidic electrolytes, makes the Volmer step rate determining [17]. In alkaline conditions, H_2O adsorption and dissociation occur before the Volmer step (adsorption of H). This is similar to OER, where the chemical environment (i.e., acidic or alkaline) determines the oxidation of water [18].

1.2 Thermodynamic Parameters

1.2.1 Enthalpy (H)

It is the system's overall heat content. Thus, the change in the enthalpy (ΔH) of a system consists of two parts, which are the requirements of thermal energy (ΔQ) and useful work done (ΔW). The overall equation can be written as follows:

$$\Delta H = \Delta Q \pm \Delta W \tag{1.1}$$

A positive change in enthalpy describes an endothermic reaction, whereas a negative change describes an exothermic reaction.

1.2.2 Entropy (S)

The entropy, a state function, measures the reaction system's overall degree of disorder or unpredictability. The relationship provides the shift in entropy (S).

$$\Delta S = \Delta Q/T \tag{1.2}$$

A more considerable change in entropy with input energy indicates that the system is away from the thermodynamic equilibrium, i.e., it is highly disordered, i.e., chaotic or spontaneous. It is directly dependent upon temperature. In the case of water splitting, H_2O (liquid) is broken down into gas H_2 and O_2, which increase the entropy of the system.

1.2.3 Gibbs Free Energy

Gibbs free energy describes the thermodynamic potential to determine the maximum work performed by a closed system in isothermal and isobaric conditions. Gibbs free energy change (ΔG) is the maximum amount of work that can be obtained from a closed system that can exchange heat and interact with its surroundings, i.e.,

$$\Delta G = -\Delta W \tag{1.3}$$

Therefore, upon switching Equations 1.2 and 1.3 in Equation 1.1, we obtain the following:

$$\Delta G = \Delta H - T\Delta S \tag{1.4}$$

The ΔG^{o} of water splitting is obtained under the standard temperature "T" = 398 K, pressure "P" = 1 bar, and concentration = 1 M. If ΔG is negative under these standard conditions, the water splitting process is both spontaneous and thermodynamically favorable. The equilibrium constant "K" of the system is relatable to the Gibbs free energy change (ΔG) by the following:

$$\Delta G = -RT\ln K \tag{1.5}$$

If ΔG decreases, the equilibrium constant "K" of the chemical reaction increases. If $\Delta G < 0$, the spontaneity of the chemical reaction increases.

1.3 Thermodynamics of Water Splitting

Water dissociation, which comprises liquid water and water vapor, decomposes into its elemental compositions (H_2 and O_2) according to the following:

Table 1.1 The thermodynamic parameters of liquid and gaseous H_2O.

Parameters	**Liquid (H_2O)**	**Vapor (H_2O)**
$\Delta H_{H2O}{}^{o}$ kJ mol^{-1}	+285.840	+241.80
$\Delta S_{H2O}{}^{o}$ J mol^{-1} K^{-1}	+163.15	+44.10
$\Delta G_{H2O}{}^{o}$ kJ mol^{-1}	+237.22	+228.66

$$H_2O \rightarrow H_2 + 1/2O_2 \tag{1.6}$$

Under the standard conditions of temperature (T^o = 298 K) and pressure (P^o = 1 bar) for water, the thermodynamic parameters of enthalpy ($\Delta H_{H2O}{}^{o}$), entropy ($\Delta S_{H2O}{}^{o}$), and Gibbs free energy ($\Delta G_{H2O}{}^{o}$) change are given in Table 1.1 [15].

The dissociation of water is highly endothermic and non-spontaneous, although the entropy conditions are favorable for water splitting. When there is a conversion in the state of matter from water (liquid) to gas (steam), the entropy increases from +44.1 to +228.66 J mol^{-1} K^{-1}, which decreases enthalpy and Gibbs free energy. All the three parameters ΔH^o, ΔS^o, and ΔG^o for water splitting depend on the temperature (T) and pressure (P) of the system through the following relation 1.7 [12]:

$$\Delta G\ (T, P) = \Delta H\ (T, P) - T\Delta S\ (T, P) \tag{1.7}$$

when water produces H_2 and O_2 due to water electrolysis, the electrical energy of $-\Delta G$ and a heat transfer of $-T\Delta S$ are required. The enthalpy voltage (V) and the water electrolysis voltage (E) connected with the enthalpy (ΔH) and free energy change (ΔG) are given by the following:

$$-\Delta H = nFV$$

$$-\Delta G = nFE \tag{1.8}$$

where "F" equals the Faraday constant of 96,485 C/mol, and n = 2 is the number of transferred electrons. Under ideal cases at STP conditions, "V" equals 1.23 V, and "E" equals 1.48 V [12]. But in non-ideal cases when an overpotential (η) is used, external heat is not required because of the participation of electrical energy in both $-\Delta G$ and $-T\Delta S$. Here, the voltage is called thermo-neutral voltage [12]. Thus, the enthalpy change (ΔH) can be expressed in terms of the enthalpies of the three phases involved as follows [16]:

$$\Delta H = H\,(\text{wet hydrogen}) + H\,(\text{wet oxygen}) - H\,(\text{reagent water}) \tag{1.9}$$

Here, the hydrogen and oxygen are assumed to be wet because water vapor is mixed with them. For the continuous production of H_2 and O_2, there is a continuous requirement for reagent water in the amount of $1 + 1.5P_w/(P - P_w)$, where the products of the process consist of 1 mol of hydrogen mixed with $P_w/(P - P_w)$ mol of water vapor at an overall pressure P, and 0.5 mol of oxygen mixed with $0.5P_w/(P - P_w)$ mol of water vapor, also at an overall pressure P [16].

1.4 Factor Dependence on the Thermodynamics of Water Splitting

1.4.1 Temperature

The thermodynamic parameters, consisting of H, S, and G, are all functions of the operating temperature (T). As seen in Figure 1.1, upon increasing the temperature in the system, the Gibbs free energy change (ΔG) decreases, although the enthalpy (ΔH) remains constant [15].

A decreasing value of ΔG indicates that water dissociation will consume less electrolysis voltage. The enthalpy and Gibbs free energy change in the water dissociation process, as given in relation 1.10 concerning temperature, are determined by the equations given below:

$$\Delta H\,(T) = H_{H2}\,(T) + 1/2H_{O2}\,(T) - H_{H2O}\,(T)$$

$$\Delta G\,(T) = G_{H2}\,(T) + 1/2G_{O2}\,(T) - G_{H2O}\,(T)$$

$$\Delta S\,(T) = S_{H2}\,(T) + 1/2S_{O2}\,(T) - S_{H2O}\,(T) \tag{1.10}$$

The ΔG, which is a function of the temperature ($\Delta G\,(T)$), can be directly used to determine the thermodynamic voltage of water splitting ($E\,(T)$) using the relation $E\,(T) = \Delta G\,(T)/nF$, where $F = 96{,}500\ C\ mol^{-1}$. When T

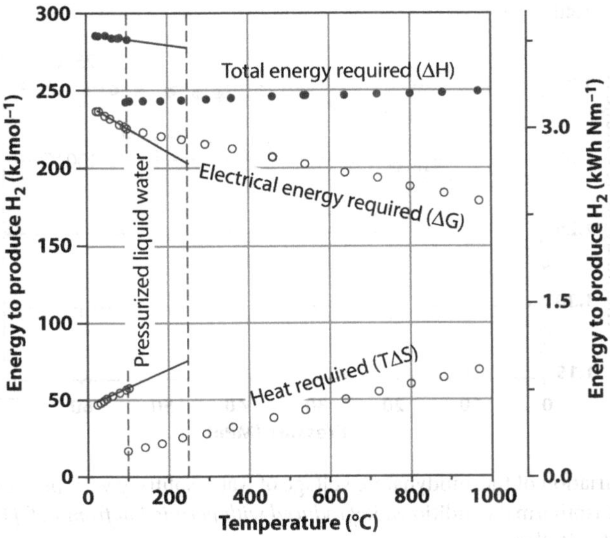

Figure 1.1 Production of H_2 under constant pressure of 1 bar and temperature ranging from 0°C to 1,000°C. *Reproduced with permission from Ref. [15]. Copyright (2015) Willey.*

> 3,000 K, the free energy electrolysis voltage E(T) drops from 1.23 V at 25°C to 1.18 V at 100°C and −0.9 V at 1,000°C. Electricity is not required to dissociate the water molecules [15].

1.4.2 Pressure

The thermodynamic parameters are quite dependent on the pressure of the electrolyzer. For commercial use, the electrolyzers are operated at pressures between 1 and 50 bars, or water splitting occurs at atmospheric pressures, and the output hydrogen gas is kept in compressed vessels for further use. Under isothermal conditions, the relationship between the thermodynamic voltage of water splitting (E) and the pressure over the electrolyzer is given by the following [15]:

$$(\delta E/\delta P)_T = 1/nF\ (\delta \Delta G/\delta P)_T \tag{1.11}$$

The variation of E with pressure (P) at a particular temperature T°C is shown in Figure 1.2 [14]. The effect of the pressure above the electrolyzer on the water splitting (E) thermodynamic voltage is significant (+200 mV at 25°C when the pressure is increased from 1 to 700 bar).

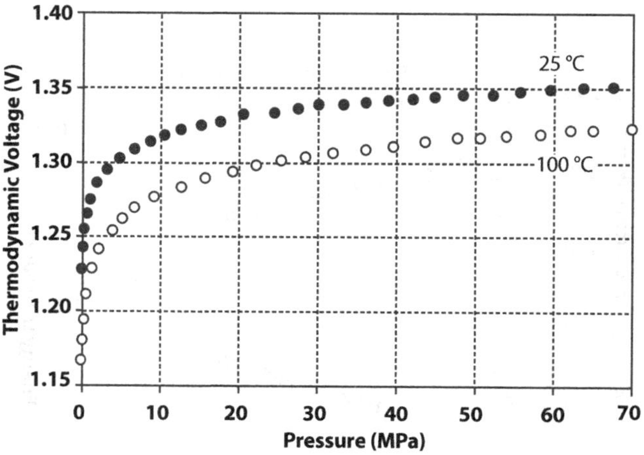

Figure 1.2 Variation of thermodynamic voltage of water splitting with pressure under three different isothermal conditions. *Reproduced with permission from Ref. [15]. Copyright (2015) Willey.*

1.5 Applications in Electrolysis

1.5.1 Hydrogen Evolution Reaction (HER)

The HER process occurs either through Volmer–Heyrovsky or Volmer-Tafel reaction on a HER active site. The Volmer or electrosorption step determines the adsorption of a nascent H in the HER active site (M) forming an adsorbed hydrogen (H_{ads}). It is followed by the release of the H_2 molecule either through Heyrovsky (electrochemical) or Tafel (chemical) desorption mechanisms [19]. The HER mechanism occurring in acidic and alkaline mediums is described below:

i) Electrochemical adsorption of hydrogen (Volmer reaction)

$$H^+ + M + e^- \rightarrow M\text{–}H_{ads} \text{ (acidic)}$$

Or

$$H_2O + M + e^- \rightarrow M\text{–}H_{ads} + OH^- \text{ (alkaline)}$$

ii) Electrochemical desorption of hydrogen (Heyrovsky reaction)

$$M\text{–}H_{ads} + H^+ + e^- \rightarrow M + H_2 \text{ (acidic)}$$

Or

$$M–H_{ads} + H^+ + e^- \rightarrow M + H_2 + OH^- \text{ (alkaline)}$$

iii) Chemical desorption of hydrogen (Tafel)

$$2M–H_{ads} \rightarrow M + H_2$$

In the first step, the change in enthalpy (ΔH) before and after H adsorption over an electrochemical active site is obtained, which is followed by the release of the H_2 molecule.

$$\Delta H_{H, ads} = H_{H, ads} - H_{site} - 1/2H_{H2} \quad (1.12)$$

where $\Delta H_{H, ads}$ is the change in the enthalpy of H adsorption over a site, $H_{H, ads}$ is the enthalpy of the site after H adsorption, H_{site} is the enthalpy of the active site before H adsorption, and H_{H2} is the enthalpy of the H_2 molecule. A more negative $\Delta H_{H, ads}$ indicates stronger binding strength between an intermediate and a catalyst's active site with better thermodynamic stability. The HER barrier depends on the intermediate state energy of an H adsorbed on the catalyst followed by the release of an H_2 molecule given by ref. [20]. Therefore, the HER performance can be assessed through ΔG_H, i.e., the Gibbs free energy change of the adsorbed hydrogen. The Gibbs free energy change of H adsorption is given by the following:

$$\Delta G_H = \Delta H_{H, ads} + \Delta H_{ZPE, ads} + T\Delta S_H \quad (1.13)$$

Here, $\Delta H_{H, ads}$ is the enthalpy change after H adsorption, $\Delta H_{ZPE, ads}$ is the zero point energy change after H adsorption, and ΔS_H is the entropy change of H adsorption given by $\Delta S_H = -0.5S_{H2}$, which is equivalent to 130 J mol^{-1} K^{-1}. "T" is the temperature under standard conditions, i.e., 298 K. ΔG_H is a crucial parameter to describe the HER activity of a catalyst in an acidic medium despite the existence of obstacles (Figure 1.3a) [21]. As shown in Figure 1.3b, alkaline HER requires the H_2O molecule to be adsorbed at the active site, dissociate into H and OH, adsorb H (Volmer step), and release H_2 [22]. It introduces an extra energy barrier (ΔG_B) that controls the overall reaction rate. Thus, its rate becomes considerably slower in alkaline than in acidic conditions. Hence, ΔG_H cannot directly determine a catalyst's HER activity in an alkaline medium. The Bronsted–Evans–Polanyi (BEP) relationship describes that the activation energy of a chemical reaction is directly proportional to its enthalpy change (ΔH). A low activation barrier indicates strong adsorption [23]. Therefore, in the case of alkaline HER, three factors are considered as follows: a) neither too

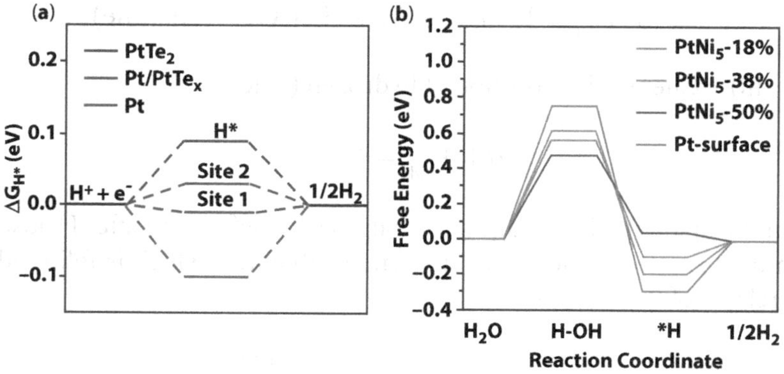

Figure 1.3 Gibbs free energy diagram of (a) acidic HER. *Reproduced with permission from Ref. [21]. Copyright (2021) Elsevier.* (b) alkaline HER. *Reproduced with permission from Ref. [22]. Copyright (2021) Wiley.*

high nor low H binding, b) low water dissociation barrier, and c) low OH binding energies. The ΔG_H for different materials is based on the Norskov reaction of H_2 formation, i.e., $H^+ + * \rightarrow 1/2H_2$. The resultant is a "Volcano" plot given in Figure 1.4 [24, 25]. It follows the Sabatier principle, which describes that for strong HER kinetics, the binding strength between the intermediates and the active site in the catalyst must "neither be too strong nor too weak." For the elements situated in the left, H adsorption becomes "strong" as ΔG_H values tend to be more negative, while for the elements located in the right, the adsorption becomes "weak" as ΔG_H values tend to be more positive. The ideal HER materials are those that are being placed at the peak of the volcano plot [25].

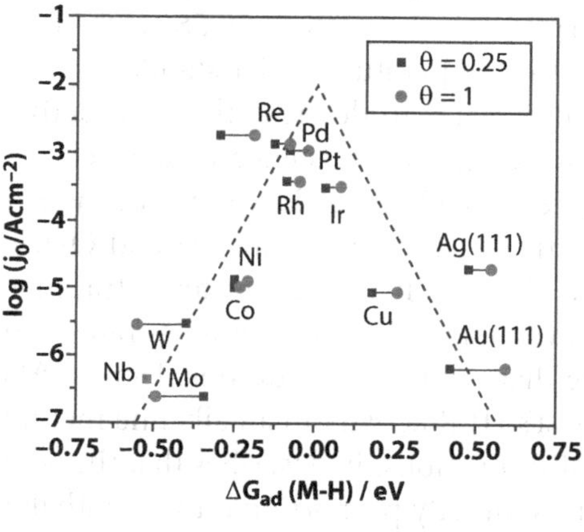

Figure 1.4 Volcano plots of materials in accordance with M–H bond strength. *Reproduced with permission from Ref. [25]. Copyright (2016) Wiley.*

The noble metal Pt is considered the most widely active catalyst toward acidic HER with ΔG_H nearly equal to zero as mentioned above. However, these noble metals are costly and rarely available, so many researchers are eager to investigate and develop a stable, effective electrocatalyst with lower noble metal concentrations for HER. In this regard, noble metals have been investigated for HER, including rhodium (Rh), ruthenium (Ru), and iridium (Ir) [26–34]. Chen *et al.* adopted a gas-controlled method using $H_2/O_2/Ar$ gases to synthesize free-standing Ru nanoparticles, nanowires, and nanosheets, respectively. He studied theoretically the adsorption of hydrogen atoms on Ru crystal facets viz. Ru (001), Ru (100), and Ru (101) by varying the hydrogen coverage over the Ru surface. He found that introduction of H_2 into the crystal planes of Ru makes the system more active as it accelerates the reduction rate of Ru and thus enhances the HER process. According to DFT calculations, the ΔG_H value obtained for Ru on its various crystal planes is in the following order: (101) > (100) > (001). This suggests that the surface effect of the crystals is mostly responsible for the HER performance of Ru NCs [26]. Zhang *et al.*, synthesized ultrafine Ru nanocrystals uniformly distributed on mesoporous Mo_2N nanorods (Ru–Mo_2N) and studied the HER activity of Ru–Mo_2N (i.e., Ru_3–Mo_2N (100)) and Ru_3–Mo_2N (111) in both acidic and alkaline medium. From DFT calculations, they found that at the interfaces between the Ru_3 cluster and Mo_2N, charge redistribution takes place, which results in slight increases in the oxidation state of Ru. Additionally, the PDOS plot revealed a downshift of the Ru nanoclusters' d-band center, which decreases the binding strength of the reactive intermediates on the catalyst surface and is useful for tuning ΔG_H values. The ΔG_H values calculated on Ru_3–Mo_2N (111) was found to be −0.07 eV, which is much lowered compared to the ΔG_H values of −0.29 eV (on Ru_3 cluster), −0.3 eV (Ru_3–Mo_2N (100)), −0.52 eV (Mo_2N (100)), and −0.79 eV (Mo_2N (111)). Thus, the Ru_3–Mo_2N (111) crystal facet rendered the most catalytically active HER performance [27]. A pH universal catalyst, $RuRh_2$ bimetallene with rich structural defects, was designed by the Mu *et al.* group to study HER in both acidic and alkaline conditions. As we know, the thermodynamics of HER in an acidic medium is governed by ΔG_H values; in contrast, the dissociation of water in alkaline media is dependent on low activation energies (i.e., ΔG_B). The DFT studies on the $RuRh_2$ bimetallene system were conducted to explore ΔG_H and ΔG_B values. It was found that among the seven models (viz. Ru, Rh, Pt, $RuRh_2$/Ru sites, $RuRh_2$/Rh sites, $RuRh_2$/Ru sites with strain effect, and $RuRh_2$/Rh sites with strain effect) adopted, the optimal ΔG_H value was found for the $RuRh_2$ bimetallene/Ru system with Rh sites, which helped in facilitating the Volmer step on the surface of the catalyst. On the other hand, a lower ΔG_B

value of 0.33 eV was obtained on the $RuRh_2$ bimetallene/Rh site (with strain-induced effect) system compared to the $RuRh_2$/Ru (with strain-induced effect) sites (0.35 eV), $RuRh_2$/Rh sites (0.34 eV), Ru (0.63 eV), and Rh (0.64 eV) indicating that the dissociation of H_2O on the $RuRh_2$ surface is much simpler than on the Ru or Rh surface, which could significantly increase HER activity [28]. Liu and co-workers constructed Ni@Ni_2P–Ru catalysts for HER catalysis. Initially, they utilized DFT computations to anticipate the advantageous outcome of adding Ru into the Ni_2P system. They simulated systems/models viz. Ni_2P ternary system, the outer Ru-plane sites (Ni_2P–Ru-plane), Ru–Ni_2P interface edge sites (Ni_2P–Ru-edge), and the Ru-cluster sites (Ni_2P–Ru-cluster) for studying the thermodynamics of HER in detail on the catalyst surface. They found that the simulated Ni_2P–Ru-cluster shows energetically favorable H adsorption with a ΔG_H value of 0.01 eV compared to Ru (−0.21 eV), Ni_2P (−0.19 eV), Ni_2P–Ru-plane (−0.16) and Ni_2P–Ru-edge (−0.09 eV), which is nearly equal to the optimum ΔG_H value and thus results in enhanced HER performance [29]. Li *et al.* performed the orbital modulation on Ir (111) crystal facet to study the behavior of H_2 adsorption/desorption on the surface of the catalyst for achieving enhanced HER performance. A theoretical study suggests that an electronegative carbon/nitrogen (C/N) matrix can be used to modify Ir NPs, which will further boost the HER processes on the surficial Ir sites. The PDOS distribution of Ir and IrNC with H adsorbates on the surface reveals that hydrogen bonding is greater at the surficial Ir sites on IrNC. The free energy of H adsorption is found to be 0.25 eV at Ir sites and 0.04 eV for IrNC by maintaining the balance of the electronegative environment with C/N sites. This implies that hydrogen adsorption/desorption would favor the interaction energy between H and the IrNC surface more than that of Ir [30]. Luo *et al.* synthesized iridium nanorods (Ir–NR/C) by annealing Ir nanoparticles at 300°C, which acts as a bifunctional electrocatalyst [31]. In acidic HER catalysis (Figure 1.5a), according to DFT investigations, the IrO_X layer formed on Ir nanorods helped lower the ΔG_H value from 0.26 to 0.02 eV indicating that the layer's presence aided in the chemisorption of hydrogen ions in acidic HER. In an alkaline medium, the HER was theoretically calculated based on the following pathway: $H_2O \rightarrow H{-}OH \rightarrow H^* \rightarrow {}^*H{-}OH \rightarrow H_2$ (Figure 1.5b). The ΔG_B values for Ir–NR/C and Ir/C was calculated to be 0.49 and 0.91 eV, respectively. Thus, greater energy was needed to split the water molecules for active sites containing absorbed hydrogen ions. These results suggest that the Ir–NR/C electrocatalyst indicates a higher intrinsic HER activity than the uncalcined Ir/C electrocatalyst [31]. A nano dendritic structure of Ir–tungsten (IrW ND) alloy was synthesized by Lv *et al.* as an effective catalyst for a

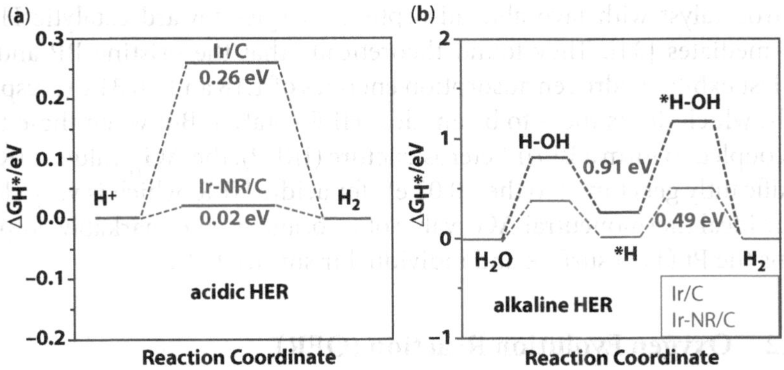

Figure 1.5 Computed free energy diagram of Ir/C and Ir-NR/C structures for HER in (a) acidic and (b) alkaline pH environment. *Reproduced with permission from Ref. [31]. Copyright (2020) Elsevier.*

pH-universal HER study [32]. Using first-principles calculations, he developed free energy diagrams to investigate the HER activity on IrW NDs. Both the binding energies of H (E_H) and OH (E_{OH}) adsorption on the catalyst surface were evaluated to get insights into the acidic and alkaline HER process. In acidic media, the most stable H adsorption was found on the top site of Ir for IrW NDs. The binding strength (E_H) for IrW NDs was found to be stronger than that for Ir but weaker than that for Pt indicating stronger adsorption of H on Ir active sites of IrW. In contrast, the E_{OH} value of IrW was calculated to be 0.61 eV, which is much lower than the E_{OH} values of Pt (0.88 eV) and Ir (1.13 eV). The increased affinity of the W active site for OH in an alkaline medium correlates with the lower barrier value of IrW. These characteristics allowed IrW to demonstrate superior HER activity in acidic and alkaline environments compared to Pt or Ir [32]. It has been reported that nitrogen-doped graphene sheets (Ir@N–G-750) coated with nanosized iridium nanoparticles exhibit strong HER activity in both acidic and alkaline environments. The first principle investigation of Wu *et al.* demonstrates how the pyridinic N synthesis might significantly lower the adhesion energy of carbon species doped with nitrogen [33]. The neighboring N atom's distance decreases and the Ir–N site begins to form as soon as the Ir cluster is introduced to the N–graphene system. The favorable active pyridinic N (−0.46 eV) adhered to the Ir surface modifies Gibbs free energy of hydrogen adsorption on Ir–N, in contrast to the pure Ir–graphene site (−2.72 eV) and the Ir–graphitic N (−0.77 eV). Consequently, N-doping helps to enhance the electrocatalytic performance of HER [33]. Mei *et al.* synthesized an efficient Ir-modified black phosphorous (BP) electrocatalyst with favorable adsorption energies toward catalytic HER intermediates [34]. They found theoretically that the pristine BP and Ir catalyst exhibit hydrogen adsorption energies of 1.18 and –0.31 eV, respectively, which limits them to be an ideal HER catalyst. But when these two are coupled to form a BP/Ir heterostructure (IRBP), the ΔG_H value of IRBP significantly gets tuned to the –0.02 eV for acidic HER, which is very close to the ideal thermoneutral ΔG_H value of zero, and even remarkably outperforms the Pt (111) surface and individual Ir surface [34].

1.5.2 Oxygen Evolution Reaction (OER)

The oxygen evolution reaction (OER) process is the other half-cell reaction of the water splitting process. It involves a four-electron transfer reaction with sluggish electrode kinetics. This results in an undesirable free energy landscape of intermediates, such as *OH and *OOH, accounting for a significant loss in energy and efficiency. Therefore, it is of utmost importance to determine the rate-determining step (RDS) (which is the slowest step in the reaction mechanism that determines the overall rate of the reaction) as well as to decode which mechanism governs the catalytic process. These include the following reactions [35]:

$$2H_2O \leftrightarrow OH^* + H^+ + e^- + H_2O$$

$$OH^* + H^+ + e^- + H_2O \leftrightarrow O^* + 2H^+ + 2e^- + H_2O$$

$$O^* + 2H^+ + 2e^- + H_2O \leftrightarrow OOH^* + 3H^+ + 3e^-$$

$$OOH^* + 3H^+ + 3e^- \leftrightarrow O_2 + 4H^+ + 4e^-$$

The first term is related to the OER thermodynamic barrier, and the other term is related to the OER kinetics. Gibbs free energy change is a quantitative indicator that will provide the required energy and direction to produce a product from a group of reactants. Therefore, Gibbs free energy change for the four OER reactions is calculated as follows [35]:

$$\Delta G_1 = \Delta G_{HO*} - \Delta G_{H2O} - eU - k_B T \ln (a_{H+}) + ZPE$$

$$\Delta G_2 = \Delta G_{O*} - \Delta G_{HO*} - eU - k_B T \ln (a_{H+}) + ZPE$$

$$\Delta G_3 = \Delta G_{HOO*} - \Delta G_{O*} - eU - k_B T \ln (a_{H+}) + ZPE$$

$$\Delta G_4 = \Delta G_{O2} - \Delta G_{HOO*} - eU - k_B T\ln(a_{H+}) + ZPE$$

In the above reactions the "ΔG" is the Gibbs free energy change between the reactants and products, "eU" is the applied overpotential, "$k_B T\ln(a_{H+})$" is related to the temperature-dependent entropy, and "ZPE" is the zero-point energy. Over the last few decades, ample attention has been projected to model the thermodynamics of electrochemical reactions. In this regard, the calculation of thermodynamic parameters proves to be a great boon. Apart from providing mechanistic insight, understanding the thermodynamics of a reaction can help in designing new materials that can lower the energy barriers and enhance the overall rate of OER. In a detailed analysis of the catalytic process, it is crucial to pinpoint the reaction energy associated with each elementary step. This determination is assessed from the deviation in adsorption energy between two intermediates [36]. To be precise, the OER performance is determined by Gibb's free energy descriptors, the M–OH to M–O transition ($\Delta G_{O*} - \Delta G_{HO*}$) and the M–O to M–OOH transition ($\Delta G_{HOO*} - \Delta G_{O*}$). With a constant energy difference of 3.2 eV between M–OOH and M–OH, and M–O serving as the intermediate, the correlation between $\Delta G_{O*} - \Delta G_{HO*}$ and $\Delta G_{HOO*} - \Delta G_{O*}$ acts as a "leverage." The OER performance is solely reflected by $\Delta G_{O*} - \Delta G_{HO*}$, and notably, optimal OER performance is observed only when the energy barriers are identical (both $\Delta G_{O*} - \Delta G_{HO*}$ and $\Delta G_{HOO*} - \Delta G_{O*}$ being 1.6 eV) [37, 38]. A very pivotal question was raised in front of Lee *et al.* between "*interfacing*" and "*doping*," which can be more advantageous for promoting high water oxidation in NiFe-layered double hydroxide [39]. Yes, indeed! The Gibbs free energy calculations assisted in coming to a solid conclusion

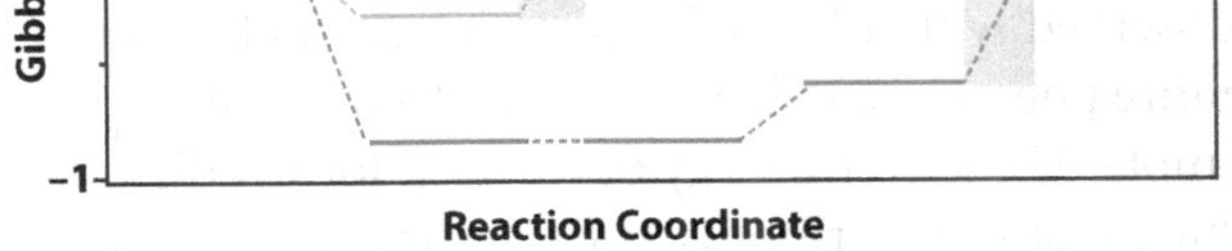

Figure 1.6 Gibb's free energy diagram obtained for OH*, O*, and OOH* adsorption over the active sites of NiFe–LDH, CeNiFe–LDH, and Ce@NiFe–LDH for OER performance at U = 1.23 V. *Reproduced with permission from Ref. [39]. Copyright (2020) Wiley.*

(Figure 1.6). After preferentially adsorbing OH*, O*, and OOH* on the Ni (110) surfaces sites of Ce@NiFe–LDH and CeNiFe–LDH, OH* to O* and OOH* to O_2 (g) were found to be the RDS in the two systems, respectively. The calculated values of ΔG for Ce@NiFe–LDH and CeNiFe–LDH were 0.56 and 0.61 Ev, respectively [39]. Eventually, the enhanced catalytic activity can be accounted for the dissimilar charge redistribution in NiFe–LDH upon Ce loading, which leads to optimal adsorption-free energy landscape of OER intermediates.

As a result, the calculated thermodynamic parameters, such as ΔS, ΔZPE, and ΔG, provide valuable insights into atomic-scale mechanisms and surface engineering for tuning transition metal oxyhydroxides to boost their catalytic activity. The two most prominent mechanisms that drive the OER reaction are (i) adsorbate evolution mechanism (AEM), which involves successive adsorption and desorption of oxygenated intermediates, and (ii) lattice oxygen mechanism (LOM) involving O—O coupling at the same metal center [40]. Though in most cases AEM is extensively used for describing the OER process, in the case of transitional metal oxides (TMO), a series of investigations have revealed that OER may proceed through LOM [18, 41, 42]. Therefore, unveiling the thermodynamics related to AEM and LOM, Alexandrov *et al.* performed an extensive computational study across several Ir-based oxides and compared them with the benchmark rutile structure IrO_2 polymorph [43]. They further considered two major reaction pathways for LOM—the one involving O–O coupling (LOM 1) and the other involving nucleophilic water attack (LOM 2). Intriguingly, they found that LOM 2 is thermodynamically more favorable and can even outperform the conventional AEM mechanism, which is supported by DFT calculations of reaction-free energies and *ab initio* molecular dynamic (AIMD) simulations of reaction activation barriers.

Coming to another fascinating aspect, thermodynamics has been used as a cornerstone by Thomas *et al.* to elucidate the global relationship between the corrosion of oxide catalysts and OER activity [44]. Most of the investigated metal oxide OER catalysts were found to undergo significant corrosion during the catalytic process that subsequently limits the anode lifetime. In addition to this, structural and compositional changes were also observed with a loss of long-range crystalline order. To understand the entire phenomena, they introduced a term $\Delta G^{R}_{Dissolution}$, which is the partial molar Gibbs free energy of reaction for the dissolution process. They conclude that the thermodynamic driving force per mole of oxygen atoms, ΔG^{R}_{LOER}, for the lattice oxygen evolution process is comparable to or even more negative than the thermodynamic driving force, ΔG^{R}_{OER}, under OER conditions. They proposed that the oxygen evolution activity of metal oxide catalysts was driven by the lattice OER process and the formation of a three-dimensional hydrous amorphous layer on the metal oxide surface. They further stated that instability of the oxygen anion is the primary reason for the observed structural changes and dissolution process. Further, if we dive into the energetic aspects of a reaction, enthalpy calculation can bestow some key findings. Wang *et al.* employed the notion of enthalpy change to comprehend how the incorporation of Ni into RuO_2 enhances OER stability by increasing the energy cost for Ru demetallation and subsurface oxygen loss (i.e., to understand the lattice oxygen stability). Initially, the researchers focused on assessing the enthalpy change during the surface Ru atom demetallation process on both RuO_2 and Ni–RuO_2 for the (110) plane. Strikingly, the findings reveal that the energy cost for Ru demetallation notably increased from 1.87 to 2.22 eV upon incorporation of Ni into the RuO_2 lattice. This observation indicates a heightened stability of the Ru surface in Ni–RuO_2 compared to that of pure RuO_2. Additionally, to evaluate the lattice oxygen stability, enthalpy changes were recorded for the subsurface oxygen loss process. The calculated enthalpy changes were 1.67 and 2.12 eV for RuO_2 and Ni–RuO_2 (110) surfaces, respectively. This suggests that the introduction of Ni dopants to the RuO_2 lattice also stabilized lattice oxygen during the OER process [45].

From the pool of thermodynamic parameters, entropy is another tool that entails considerable attention. It can provide a systematic guideline to understand the reaction kinetics and mechanism and presents a quantitative understanding of microscopic OER configurations. In this context, Nam *et al.* highlighted the entropic contribution to the water oxidation process particularly in the context of the biological system and the artificial heterogeneous systems. Their entropic-based approach explained how water transferred to the active site with detailed formation of the O–O bond. Their work further proposed that unfavorable activation entropy (ΔS) would be the crux for understanding the activity discrepancy between artificial heterogeneous catalysts and biological systems [46]. In recent years, high entropy alloys (HEA) and high entropy oxides (HEO) have garnered considerable interest as promising catalysts for OER. In this regard, conformational entropy is calculated for which HEA/HEO must possess a value greater than 1.5 R [47]. They found that HEO possessing a high conformational entropy value can help enhance the catalytic activity and stability through suitable binding energy of the oxidation intermediates and synergistic interactions among multiple metal elements. While electrocatalytic materials exhibit a high degree of complexity, and their catalytic activity is notably influenced by a myriad of parameters, it is noteworthy that a single microscopic parameter holds significant importance in providing valuable insights into the RDS of the OER and concurrently aids in the strategic design of materials. Despite the intricate nature of these materials, understanding and focusing on this microscopic parameter prove pivotal in unraveling the intricacies of OER mechanisms. To date, numerous descriptors have been employed to articulate and quantify the catalytic activities associated with OER. These descriptors encompass a wide range of factors, including the bonding energy of metal–oxygen (M–OH), the energy associated with the reaction step ($\Delta G_{O*} - \Delta G_{HO*}$), and various electronic structure parameters such as the occupation of d-orbitals, the filling of e_g-orbitals, and the center of the O p-band [37, 38]. The utilization of these descriptors serves as a comprehensive toolkit allowing researchers to discern and articulate the nuanced intricacies of OER catalysis. Aiding this exploration is the guidance provided by thermodynamic analysis, which serves as a compass for the further development of electrocatalysts. The overarching goal is to enhance the rate-determining step and minimize the energy barrier associated with the oxygen evolution process. This strategic development, guided by a nuanced understanding of microscopic parameters, represents a concerted effort to propel advancements in electrocatalytic materials contributing to the ongoing pursuit of sustainable and efficient energy technologies.

1.6 Conclusion

In conclusion, the study of the thermodynamics of water splitting holds profound implications for advancing sustainable energy technologies. Through the exploration of various thermodynamic principles, such as Gibbs free energy change (ΔG), enthalpy (ΔH), and entropy (ΔS) change, we can gain invaluable insights into the fundamental processes underlying water splitting reactions. These insights not only contribute to our understanding of the efficiency and feasibility of water splitting as a means of producing hydrogen fuel but also pave the way for the development of novel materials and catalysts to enhance the performance of water splitting devices. Moreover, by harnessing the power of thermodynamics, the challenges associated with scalability and cost effectiveness can be addressed.

AI disclosure statement: The author acknowledges the use of AI-based tools for minor language editing or grammatical correction. However, all intellectual contributions, analyses, and conclusions presented in this chapter are entirely the author's own.

References

1. Faber, M.S. and Jin, S., Earth-abundant inorganic electrocatalysts and their nanostructures for energy conversion applications. *Energy Environ. Sci.*, 7, 3519, 2014.
2. Jiao, Y., Zheng, Y., Jaroniec, M., Qiao, S.Z., Design of electrocatalysts for oxygen- and hydrogen-involving energy conversion reactions. *Chem. Soc. Rev.*, 44, 2060, 2015.
3. Walter, M.G., Warren, E.L., McKone, J.R., Boettcher, S.W., Mi, Q., Santori, E.A., Lewis, N.S., Solar water splitting cells. *Chem. Rev.*, 110, 6446, 2010.
4. Li, X., Zhao, L., Yu, J., Liu, X., Zhang, X., Liu, H., Zhou, W., Water splitting: From electrode to green energy system. *Nano Micro Lett.*, 12, 131, 2020.
5. Nishioka, S., Osterloh, F.E., Wang, X., Mallouk, T.E., Maeda, K., Photocatalytic water splitting. *Nat. Rev. Methods Primers*, 3, 42, 2023.
6. Wang, Q. and Domen, K., Particulate photocatalysts for light-driven water splitting: mechanisms, challenges, and design strategies. *Chem. Rev.*, 120, 919, 2020.
7. Yan, Y., Xia, B.Y., Zhao, B., Wang, X., A review on noble-metal-free bifunctional heterogeneous catalysts for overall electrochemical water splitting. *J. Mater. Chem. A*, 4, 17587, 2016.
8. Raveendran, A., Chandran, M., Dhanusuraman, R., A comprehensive review on the electrochemical parameters and recent material development of electrochemical water splitting electrocatalysts. *RSC Adv.*, 13, 3843, 2023.
9. Yu, J.M., Lee, J., Kim, Y.S., Song, J., Oh, J., Lee, S.M., Jeong, M., Kim, Y., Kwak, J.H., Cho, S., Yang, C., Jang, J.W., High-performance and stable photoelectrochemical water splitting cell with organic-photoactive-layer-based photoanode. *Nat. Commun.*, 11, 5509, 2020.
10. Bard, A.J. and Fox, M.A., Artificial Photosynthesis: Solar Splitting of water to hydrogen and oxygen. *Acc. Chem. Res.*, 28, 141, 1995.
11. Fujishima, A. and Honda, K., Electrochemical photolysis of water at a semiconductor electrode. *Nature*, 238, 37, 1972.
12. Onda, K., Kyakuno, T., Hattori, K., Ito, K., Prediction of production power for high-pressure hydrogen by high-pressure water electrolysis. *J. Power Sources*, 132, 64, 2004.
13. Rajan, A.G., Martirez, J.M.P., Carter, E.A., Coupled Effects of temperature, pressure, and pH on water oxidation thermodynamics and kinetics. *ACS Catal.*, 11, 11305, 2021.
14. Tsutsumi, A., Thermodynamics of water splitting: In energy carriers and conversion systems with emphasis on hydrogen, in: *Encyclopedia of life support systems*, vol. I, T. Ohta (Ed.), pp. 136–145, EOLSS Publications, France, 2009.
15. Millet, P., Fundamentals of water electrolysis, in: *Hydrogen production by electrolysis*, A.G. Jopek (Ed.), pp. 33–62, Wiley, Germany, 2015.
16. LeRoy, R.L., Bowen, C.T., Leroy, D.J., The thermodynamics of aqueous water electrolysis. *J. Electrochem. Soc.*, 127, 1954, 1980.

17. Mahmood, N., Yao, Y., Zhang, J.W., Pan, L., Zhang, X., Zou, J.J., Electrocatalysts for hydrogen evolution in alkaline electrolytes: Mechanisms, challenges, and prospective solutions. *Adv. Sci.*, 5, 1700464, 2018.
18. Fabbri, E. and Schmidt, T.J., Oxygen evolution reaction-The enigma in water electrolysis. *ACS Catal.*, 8, 9765, 2018.
19. Bockris, J.O'M. and Potter, E.C., The mechanism of the cathodic hydrogen evolution reaction. *J. Electrochem. Soc.*, 99, 169, 1952.
20. Skulason, E., Tripkovic, V., Bjorketun, M.E., Gudmundsdottir, S., Karlberg, G., Rossmeisl, J., Bligaard, T., Jonsson, H., Norskov, J.K., Modeling the electrochemical hydrogen oxidation and evolution reactions on the basis of density functional theory calculations. *J. Phys. Chem. C*, 114, 18182, 2010.
21. Chena, J., Qin, M., Ma, S., Fan, R., Zheng, X., Mao, S., Chen, C., Wang, Y., Rational construction of $Pt/PtTe_x$ interface with optimal intermediate adsorption energy for efficient hydrogen evolution reaction. *Appl. Catal., B*, 299, 120640, 2021.
22. Zhang, C. and Liang, X., H_2 *in situ* inducing strategy on Pt surface segregation over low Pt doped $PtNi_5$ nanoalloy with super high alkaline HER activity. *Adv. Funct. Mater.*, 31, 2008298, 2021.
23. Zheng, Y., Jiao, Y., Vasileff, A., Qiao, S.Z., The hydrogen evolution reaction in alkaline solution: from theory, single crystal models, to practical electrocatalysts. *Angew. Chem. Int. Ed.*, 57, 7568, 2018.
24. Norskov, J.K., Bligaard, T., Logadottir, A., Kitchin, J.R., Chen, J.G., Pandelov, S., Srimming, U., Trends in the exchange current for hydrogen evolution. *J. Electrochem. Soc.*, 152, J23, 2005.
25. Zeradjanin, A.R., Grote, J.P., Polymeros, G., Mayrhofer, K.J.J., A critical review on hydrogen evolution electrocatalysis: Re-exploring the volcano-relationship. *Electroanalysis*, 28, 2256, 2016.
26. Chen, L., Li, Y., Liang, X., $Ar/H_2/O_2$-controlled growth thermodynamics and kinetics to create zero-, one-, and two-dimensional ruthenium nanocrystals towards acidic overall water splitting. *Adv. Funct. Mater.*, 31, 2007344, 2021.
27. Zhang, X., Sa, R., Yang, S., Zhou, F., Jiang, Z., Wang, R., A non-carbon catalyst support upgrades the intrinsic activity of ruthenium for hydrogen evolution electrocatalysis via strong interfacial electronic effects. *Nano Energy*, 75, 104981, 2020.
28. Mu, X., Gu, J., Feng, F., Xiao, Z., Chen, C., Liu, S., RuRh bimetallene nanoring as high-efficiency pH-universal catalyst for hydrogen evolution reaction. *Adv. Sci.*, 8, 2002341, 2021.
29. Liu, Y., Liu, S., Wang, Y., Zhang, Q., Gu, L., Zhao, S., Xu, D., Li, Y., Bao, J., Dai, Z., Ru modulation effects in the synthesis of unique rod-like $Ni@Ni_2P$–Ru heterostructures and their remarkable electrocatalytic hydrogen evolution performance. *J. Am. Chem. Soc.*, 140, 2731, 2018.
30. Li, F., Han, G.F., Noh, H.J., Jeon, J.P., Ahmad, I., Chen, S., Yang, C., Bu, Y., Fu, Z., Lu, Y., Baek, J.B., Balancing hydrogen adsorption/desorption by orbital

modulation for efficient hydrogen evolution catalysis. *Nat. Commun.*, 10, 4060, 2019.

31. Luo, F., Guo, L., Xie, Y., Xu, J., Qu, K., Yang, Z., Iridium nanorods as a robust and stable bifunctional electrocatalyst for pH-universal water splitting. *Appl. Catal. B Environ.*, 279, 119394, 2020.
32. Lv, F., Feng, J., Wang, K., Dou, Z., Zhang, W., Zhou, J., Yang, C., Luo, M., Yang, Y., Li, Y., Gao, P., Guo, S., Iridium–tungsten alloy nanodendrites as pH-universal water-splitting electrocatalysts. *ACS Cent. Sci.*, 4, 1244, 2018.
33. Wu, X., Feng, B., Li, W., Niu, Y., Yu, Y., Lu, S., Zhong, C., Liu, P., Tian, Z., Cheng, L., Hu, W., Li, C.M., Metal-support interaction boosted electrocatalysis of ultrasmall iridium nanoparticles supported on nitrogen doped graphene for highly efficient water electrolysis in acidic and alkaline media. *Nano Energy*, 62, 117, 2019.
34. Mei, J., He, T., Bai, J., Qi, D., Du, A., Liao, T., Ayoko, G.A., Yamauchi, Y., Sun, L., Sun, Z., Surface-dependent intermediate adsorption modulation on iridium-modified black phosphorus electrocatalysts for efficient pH-universal water splitting. *Adv. Mater.*, 33, 2104638, 2021.
35. Musho, T., Li, J., Wu, N., Thermodynamics of the oxygen evolution electrocatalysis in a functionalized UiO-66 Metal-Organic Frameworks. *Int. J. Quantum Chem.*, 116, 1153, 2016.
36. Hu, C., Zhang, L., Gong, J., Recent progress made in the mechanism comprehension and design of electrocatalysts for alkaline water splitting. *Energy Environ. Sci.*, 12, 2620, 2019.
37. Chen, J., Yu, T., Li, R., Wang, Y., Shao, Z., Song, S., Recent advances in the understanding of the surface reconstruction of oxygen evolution electrocatalysts and materials development. *Electrochem. Energy Rev.*, 4, 566, 2021.
38. Man, I.C., Su, H.-Y., Vallejo, F.C., Hansen, H.A., Martinez, J., II, Inoglu, N.G., Kitchin, J., Jaramillo, T.F., Norskov, J.K., Rossmeisl, J., Universality in oxygen evolution electrocatalysis on oxide surfaces. *ChemCatChem*, 3, 1159, 2011.
39. Liu, M., Min, K.A., Han, B., Lee, L.Y.S., Interfacing or doping? Role of Ce in highly promoted water oxidation of NiFe-layered double hydroxide. *Adv. Energy Mater.*, 11, 2101281, 2021.
40. Zhou, D., Li, P., Lin, X., McKinley, A., Kuang, Y., Liu, W., Lin, W.-F., Sun, X., Duan, X., Layered double hydroxide-based electrocatalysts for the oxygen evolution reaction: identification and tailoring of active sites, and superaerophobic nanoarray electrode assembly. *Chem. Soc. Rev.*, 50, 8790, 2021.
41. Song, J., Wei, C., Huang, Z.-F., Liu, C., Zeng, L., Wang, X., Xu, Z.J., Review on fundamentals for designing oxygen evolution electrocatalysts. *Chem. Soc. Rev.*, 49, 2196, 2020.
42. Shan, J., Zheng, Y., Shi, B., Davey, K., Qiao, S.-Z., Regulating electrocatalysts via surface and interface engineering for acidic water electrooxidation. *ACS Energy Lett.*, 4, 2719, 2019.

43. Zagalskaya, A., Evazzade, I., Alexandrov, V., *Ab Initio* thermodynamics and kinetics of the lattice oxygen evolution reaction in Iridium Oxides. *ACS Energy Lett.*, 6, 1124, 2021.
44. Binninger, T., Mohamed, R., Waltar, K., Fabbri, E., Levecque, P., Kotz, R., Schmidt, T., Thermodynamic explanation of the universal correlation between oxygen evolution activity and corrosion of oxide catalysts. *Sci. Rep.*, 5, 12167, 2015.
45. Wu, Z.-Y., Chen, F.-Y., Li, B., Yu, S.-W., Finfrock, Y.Z., Meira, D.M., Yan, Q.-Q., Zhu, P., Chen, M.-X., Song, T.-W., Yin, Z., Liang, H.-W., Zhang, S., Wang, G., Wang, H., Non-iridium-based electrocatalyst for durable acidic oxygen evolution reaction in proton exchange membrane water electrolysis. *Nat. Mater.*, 22, 100, 2023.
46. Lee, K.-G., Balamurugan, M., Park, S., Ha, H., Jin, H., Nam, K.T., Importance of entropic contribution to electrochemical water oxidation catalysis. *ACS Energy Lett.*, 4, 1918, 2019.
47. Shen, Y. and Spearot, D.E., Mobility of dislocations in FeNiCrCoCu high entropy alloys. *Modell. Simul. Mater. Sci. Eng.*, 29, 085017, 2021.

2

Kinetics of Electrochemical Water Splitting

Suvankar Deka[1,2], Manju Kumari Jaiswal[1,2], Manash P. Nath[1,2] and Biswajit Choudhury[1,2*]

[1]*Materials and Energy Laboratory, Physical Sciences Division, Institute of Advanced Study in Science and Technology (IASST), Assam, India*
[2]*Academy of Scientific and Innovative Research (AcSIR), Ghaziabad, India*

Abstract
Escalating environmental concerns extends a clarion call to the scientific community to emphasize the development and advancement of renewable resources for green and sustainable energy generation. In this context, research on water splitting has garnered tremendous attention in recent years. OER and HER have their own distinct mechanisms that govern each process leading to varied kinetics even when assessed using the same material. Therefore, systematically investigating kinetic attributes associated with each reaction step is paramount in unveiling how a particular catalyst performs in OER and HER contexts. Eventually, this paves the way for strategic advancement in the design and optimization of catalysts to enhance the overall efficiency of water splitting. That being so, the basic understanding surrounding the kinetics of water splitting is crucial. Therefore, starting from the foundational principles of water splitting to exploring its intricate mechanisms in detail, this chapter imparts a comprehensive outlook to the readers regarding the multifaceted factors that govern the electrocatalytic water splitting process. Furthermore, the chapter is equipped with an in-depth discussion of various tools for studying different kinetic parameters associated with water splitting reactions for OER and HER, both in alkaline and acidic medium.

Keywords: Electrochemical, tafel slope, overpotential, HER, OER, layered materials

**Corresponding author*: biswajitchoudhury@iasst.gov.in; biswa.tezu@gmail.com

Inamuddin, Tariq Altalhi, Mohammad Luqman and Jorddy Neves Cruz (eds.) Water Splitting: Production of Hydrogen, (23–48) © 2025 Scrivener Publishing LLC

2.1 Introduction

This century has witnessed that the irrational utilization of fossil fuels for the generation of energy has brought menace to the global environment. With dwindling fuel stocks, an avalanche rise in energy demands has led to thinking of different energy conversion technologies to ameliorate the prevailing environmental issues and energy crisis [1, 2]. In the realm of energy research, water splitting sprouted out to be one of the most extensively explored areas. Water splitting can be reckoned as the only sustainable and eco-friendly way to generate high pure H_2 and O_2 on a larger scale without harming the flora and fauna. In recent years, three predominant approaches that have grabbed significant attention for water splitting are (i) Photocatalytic, (ii) Electrocatalytic, (iii) Photoelectrocatalytic, and (iv) Photobiological water splitting [3]. In this context, the two essential reactions that come into the picture are hydrogen evolution reaction (HER) and oxygen evolution reaction (OER). As we all know, hydrogen is often coined as the "fuel of the future" and stands at the forefront of the global transition toward a cleaner and more sustainable energy landscape. Being an incredibly versatile and abundant element, hydrogen promises to power diverse sectors, from transportation to industrial processes, with zero greenhouse gas emissions. Another intriguing facet that makes hydrogen so pivotal in the context of energy is because of its calorific value. Compared to most of renewable resource-derived fuels, such as petrol, diesel, methane, and liquified petroleum gas, hydrogen surpasses all by exhibiting an exceptional calorific value of approximately 150 MJ/kg.

In contrast, the former fuels possess a calorific value three times smaller than that of hydrogen, which typically falls within the range of 40–55 MJ/kg [4]. Unequivocally, this notable distinction underscores hydrogen as a suitable candidate in the energy generation sector compared to conventional renewable fuels. Now, diving into another byproduct of water splitting reaction—"oxygen" is notably generated with high purity often reaching levels well above 99.5% when extracted through electrochemical means [5]. In the realm of space exploration, the exceptional purity of oxygen holds significant importance, especially in rocket launch propulsion systems.

Nevertheless, the minimum prerequisite for driving electrochemical water splitting stands at a thermodynamic potential of 1.23 V, which corresponds to a minimum energy input of $\Delta G = 273.1$ kJ mol–^{1} under standard conditions, delineated by a temperature of 25°C and a pressure of 1 atm [6]. Despite this theoretical benchmark, the efficacy of energy conversion processes faces formidable impediments in the form of sluggish kinetics. An ideal electrocatalyst must exhibit the traits of cost effectiveness, high stability, and ability to operate at low overpotential and, crucially, must be derived from earth-abundant materials. Nevertheless, the challenge lies in achieving all these nine yards in a singular entity. Though various noble elements, such as Ir, Ru, and Pt, occupy the peak of the volcano plot exhibiting superior electrocatalytic activity, their high cost and scarcity limit their industrial utilization. That being so, extensive research has been emphasized on alternate classes of materials such as LDHs, sulfides, nitrites, and cobaltite. However, the exigent nature of both the OER and the HER introduces substantial overpotential requirements constituting a significant bottleneck in the pursuit of enhanced efficiency.

Furthermore, it is well known that OER and HER have their own distinct catalytic mechanisms; therefore, an active HER catalyst is recognized as a meager OER catalyst and vice versa [7]. The kinetic barriers encountered in these reactions reduce the overall efficiency and impede the timely and effective utilization of thermodynamic potential. This decisively suggests the significance and need for a fundamental understanding of water splitting kinetics as it forms the strategic basis for designing and tuning a catalyst that can effectively work for both OER and HER leading to a crucial enhancement in the overall efficiency process. Therefore, this chapter provides a comprehensive and subtle understanding of various kinetic parameters paving the way for developing more efficient and sustainable electrochemical water splitting systems to step into a carbon-neutral future.

2.2 Fundamentals of Water Splitting

The simplest equation to describe the splitting of water into its constituents H_2 and O_2 electrochemically is given by the following:

$$H_2O \rightarrow H_2 + \frac{1}{2}O_2, E^0 = 1.23V$$

While splitting water electrochemically, the reactions that are proposed to be happening at the cathode and anode for neutral, alkaline, as well as acidic medium are as follows [6]:

In neutral condition:

$$\text{At the anode, } H_2O \rightarrow 2H^+ + \frac{1}{2}O_2 + 2e^-$$

$$\text{At the cathode, } 2H_2O + 2e^- \rightarrow 2H_2 + 2OH^-$$

In alkaline conditions:

$$\text{At the anode, } 2OH^- \rightarrow H_2O + \frac{1}{2}O_2 + 2e^-$$

$$\text{At the cathode, } 2H_2O + 2e^- \rightarrow H_2 + 2OH^-$$

In acidic conditions:

$$\text{At the anode, } H_2O \rightarrow 2H^+ + \frac{1}{2}O_2 + 2e^-$$

$$\text{At the cathode, } 2H^+ + 2e^- \rightarrow H_2$$

The electrochemical reaction typically involves an interface charge transfer reaction and a mass transfer reaction. Generally, the evaluation of electrochemical performance, including activity and stability, is accomplished by employing a conventional three-electrode setup consisting of a working electrode, a reference electrode, and a counter electrode. However, to access the overall efficiency of water splitting, an electrode configuration is used where high-performance OER and HER electrocatalysts are affixed to the anode and cathode, respectively (i.e., a two-electrode setup) [8, 9]. Both electrode configurations are illustrated in Figure 2.1 [10]. Presently, tremendous efforts have been dedicated to tailoring dual-functional electrocatalysts for both OER and HER to expedite the overall water splitting process diminishing device complexity.

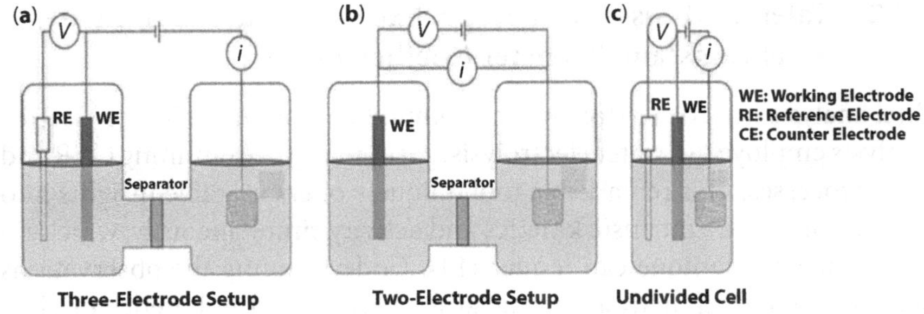

Figure 2.1 Different electrode configurations for accessing electrocatalytic water splitting. *Reproduced with permission from Ref. [10]. Copyright (2023) Wiley.*

2.3 Kinetic Parameters

2.3.1 Overpotential

Water splitting, a non-spontaneous process, requires an addition potential greater than the thermodynamic voltage of 1.23 V to overcome the inherent kinetic hindrance of the reaction. This extra potential is termed "overpotential" denoted by "η." The reversible thermodynamic potentials for HER and OER stand at 0 and 1.23 V, respectively. Therefore, employing the equations, $\eta_{HER} = E_{RHE} - 0$ V and $\eta_{OER} = E_{RHE} - 1.23$ V, one can calculate HER and OER overpotentials at the required current density without iR compensation [11]. The recognized standard current density in comparing electrocatalysts for the HER and OER across various pH conditions is set at 10 mA cm^{-2}.

For this reason, the figure of merit is approximately the current density anticipated for a solar-to-fuel conversion device with 10% efficiency under one sun illumination [12]. Overpotential reduction is crucial for enhancing the kinetics of OER an HER, and therefore, it is directly correlated to the catalyst's efficiency in improving the overall water splitting process. It has been witnessed that most of the research groups provide the iR-compensated overpotential at the benchmark current density. However, reports indicate numerous uncertainties in iR drop compensation methods and the extent to which iR drop compensations are considered [11, 13]. Moreover, since the electrolytic cell does not function according to iR-corrected overpotential, it is advisable to present iR drop uncompensated overpotential as it would be more appropriate than the overpotential compensated with iR drop.

2.3.2 Tafel Analysis—Tafel Slope, Exchange Current Density, Tafel Constant, Transfer Coefficient

Tafel analysis is a cornerstone in the comprehensive evaluation of electrocatalysts employed in water electrolysis, particularly in examining OER and HER processes. This remarkable tool uniquely offers valuable insights into an electrocatalyst's intrinsic kinetics and activity simultaneously, which no other known technique can render [14]. Understanding the observations derived from Tafel analysis is crucial for predicting the kinetics of electrocatalytic reactions at the catalytic interface. To comprehend the predictive capabilities of Tafel analysis, it is essential to revisit the foundational principles and delve into the derivation of the Tafel equation. The widely recognized Butler–Volmer equation encapsulates the relationship between current and overpotential at an electrical interface, which is given as follows [15–17]:

$$j = j_0\ [\exp\ (\alpha_A nF\eta)/RT) - \exp\ (-\alpha_C nF\eta)/RT] \tag{2.1}$$

where j stands for current density, j_0 is the exchange current density, α_A and α_C are the anodic and cathodic charge transfer coefficients, n is the number of electrons involved, F is the Faraday constant (96,485 C), η stands for overpotential, R is the ideal gas constant, and T is the absolute temperature in K.

When the overpotential is considerably higher, Equation 2.1 transforms into the following:

$$\eta = -RT/\alpha_C nF\ \ln\ (j) + RT/\alpha_C nF\ (\ln j_0) \tag{2.2}$$

$$\eta = RT/\alpha_A nF\ln\ (j) - RT/\alpha_C nF\ (\ln j_0) \tag{2.3}$$

These equations can be simplified into Equations 2.4 and 2.5, which resemble the equation of a straight line (y = mx+ c).

$$\eta = a - b\log j \tag{2.4}$$

$$\eta = a + b\ \log j \tag{2.5}$$

Therefore, plotting of log j vs. η will yield the well-known Tafel plot [17]. The slopes for cathodic and anodic polarization are given as follows:

$$\text{Slope for cathodic reaction} = 2.303\ RT/-\alpha_C nF \quad (2.6)$$

$$\text{Slope for anodic reaction} = 2.303\ RT/\alpha_C nF \quad (2.7)$$

Similarly, the exchange current density is determined by extending the linear fit from the Tafel plot until it intersects with the corresponding logarithmic current density at the reversible potential of the electrocatalytic investigation. Therefore, the exchange current densities for OER and HER were computed at the intercept potentials of 0 and 1.23 V vs. RHE, respectively. Usually, the exchange current density represents the reciprocal of the logarithmic current density of an electrocatalyst at the equilibrium potential of the studied electrocatalytic process. It is evident from Equations 2.6 and 3.7 that Tafel's slope has an inverse relationship with the charge transfer coefficient (α). This implies that a lower Tafel slope corresponds to a faster transfer of charges across the electrocatalytic interface. Recently, the scientific discourse has witnessed a heightened focus on extending the utility of the Tafel slope beyond its conventional role in anticipating the relative rates of electrochemical processes across electrocatalytic interfaces. Strikingly, within the context of HER, the Tafel slope is a key parameter intricately linked to the mechanism governing HER at a specific electrocatalytic interface. Generally, when the Tafel slope is very low (i.e., between 30 dec^{-1} and 40 mV dec^{-1}), the process is considered relatively facile following a faster Tafel pathway [18]. However, the prediction of the OER mechanism employing the Tafel slope is not as straightforward as it is for HER. It is primarily due to the relatively higher complexity arising from the numerous intermediate steps inherent in the anodic half-cell reaction of water electrolysis.

To sum up, Tafel's analysis provides four key indicators offering both kinetic and mechanistic insights. These markers include exchange current density, transfer coefficient, Tafel constant, and Tafel slope [17]. Among these, exchange current density and the Tafel constant serve as activity descriptors, while the Tafel slope and transfer coefficient are descriptors of kinetics. Tafel slopes are employed for predicting the rate-determining step (RDS), while the transfer coefficient indicates the ease with which the reaction occurs on the specific catalytic surface. The three adopted approaches for determining Tafel slope are voltammetry, chronoamperometry or chronopotentiometry, and electrochemical impedance spectroscopy [18, 19]. Among them, the most conventional one is through voltammetry techniques. However, Tafel plots deduced from voltammograms do not represent the true steady-state polarization curves. This may lead to huge inaccuracies in calculating the Tafel slope as well as exchange current densities because of the substantial current contribution from interface capacitance [11]. Therefore, it is advisable to achieve a steady-state condition employing chronoamperometry/chronopotentiometry after which the corresponding current density with *iR* correction potential should be picked to draw the Tafel curve [19].

2.3.3 Activation Energy

It serves as the energy threshold that molecules must overcome to undergo a reaction, and it plays a pivotal role in understanding the pace of water splitting. To illustrate, we can envision it as a mountain that a climber must ascend to reach the reaction's peak. Certainly, a gentle slope entails an easier approach to scaling the mountain. In a similar instance, a lower activation energy is favorable for accelerating the reaction rate. Hence, meticulous examination of activation energy can help us identify the bottlenecks associated with each step of water splitting. To calculate electrochemical activation energies (Ea), we have to extract the slope from the Arrhenius plot. For this purpose, LSV must be recorded at different temperatures, and subsequently, the Napierian logarithm of current density should be plotted against the reciprocal temperature. Consequently, fitting it with the Arrhenius equation, $\ln j = -(E_a/RT) + A$ [20], will yield the required activation energy. Therefore, understanding the activation energy reveals the intricate relationship between energy barriers and reaction rates paving the way to identify the slowest parts of a reaction.

2.3.4 Turnover Frequency

Turnover frequency (TOF) can be reckoned as a crucial and accurate metric for examining the intrinsic activity of an electrocatalyst. It illustrates how fast the catalyst is capable of catalyzing a desired electrochemical reaction [21]. Since HER and OER obey pseudo-first-order kinetics, the TOF is calculated per unit of time.

TOF is calculated using the following relation:

$$\mathrm{TOF} = (\mathrm{j} \times \mathrm{N_A})/(\mathrm{F} \times \mathrm{n} \times \Gamma) \tag{2.8}$$

where j denotes the current density at a particular overpotential (A cm^{-2}), N_A is the Avogadro number, F is the Faraday's constant, which is equal to 96,485 C mol^{-1}, n is the number of electrons, which is equal to 2 for HER and 4 for OER, and Γ is the surface concentration. The precise determination of TOF requires that the active sites be well defined. However, due to the presence of multiple metal sites in a composite system, the accurate determination of active sites becomes cumbersome due to the lack of reliable techniques. One widely accepted approach is to consider all metal sites as active sites, with the total moles of active materials deposited on the electrode calculated by Inductively Coupled Plasma-Optical Emission Spectroscopy (ICP-OES) [22]. This prevailing approach assumes 100% participation of all atoms in the catalyst, including those residing in the inner core of the catalytic material.

Nevertheless, a critical evaluation of this method reveals its inherent flaws, as it fundamentally misconstrues the intrinsic catalytic activity of the material. This erroneous assumption of universal atom participation can lead to a misrepresentation of the true catalytic potential of the material, particularly by overestimating the contribution of atoms located in the inner core that may not actively participate in the electrocatalytic process [23]. It has been reported that the aforementioned assumption often yields TOF values that significantly deviate from 10^2 to 10^3. Such a substantial and unacceptable difference in TOF values underscores the revisit of the methodology employed for determining surface active concentration. Therefore, it is imperative to move beyond the simplistic assumption of universal atom participation recognizing that not all atoms contribute equally to the catalytic process. In the case of OER, involving transitional metal-based electrocatalyst, the utmost accurate and reliable procedure of determining surface active concentration is to integrate the area under the redox peak associated with the redox reaction leading to the formation of oxyhydroxide from oxide/hydroxide [24]. In employing this method, the underlying rationale is that whichever metal site undergoes *in situ* oxidation and reduction, particularly just preceding the onset of OER, is exclusively deemed to be actively participating in the electrocatalytic OER process. This assumption is rooted in the logical inference that

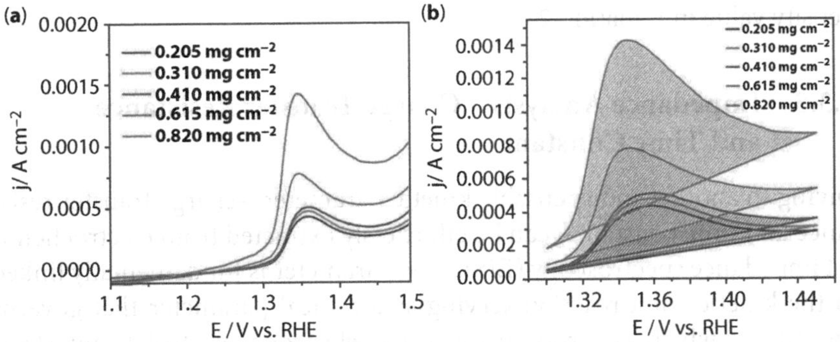

Figure 2.2 (a) LSV displaying NiOOH formation. (b) Integrated area for calculating surface active sites. *Reproduced with permission from Ref. [25]. Copyright (2019) American Chemical Society.*

all proposed OER mechanisms in the existing literature consistently emphasize the significance of the *in situ* oxidation and reduction events occurring at the active site within a single complete cycle. For instance, Figure 2.2a shows the formation of NiOOH peak in the LSV polarization curve [25]. This redox peak area (area shown with shaded portion in Figure 2.2b) is further integrated to calculate the surface-active metal sites, and subsequently, the associated TOF at different overpotentials were calculated [25]. To make it effortless for the readers, we will elucidate the steps in calculating TOF.

(i) Calculation of the surface concentration from redox features

Let us suppose that the integration of peak area associated with reduction of any metal site M^{3+} to M^{2+} at a scan rate of 0.05 V s^{-1} is 0.0035 VA.

Therefore, the associated charge is

$$= (\text{Calculated area/Scan rate in Vs}^{-1}) = 0.0035\ \text{VA}/0.05\ \text{Vs}^{-1} = 0.07\ \text{C}$$

Therefore, the number of electrons transferred is

$$= 0.07\ \text{C}/1.602 \times 10^{-19}\ \text{C} = 4.36 \times 10^{17}$$

Since the reduction of M^{3+} to M^{2+} is a single-electron transfer reaction, therefore, the number of electrons calculated above is the same as the number of surface-active sites. Therefore, the surface concentration of M^{2+} that participated in OER is 4.36×10^{17}.

(ii) TOF calculation

TOF at different overpotentials can be calculated by plugging the obtained value for surface-active species in conjunction with the current density value in Equation 2.8.

2.3.5 Impedance Analysis—Charge Transfer Resistance and Time Constant

Diving to another indispensable kinetic parameter—charge transfer resistance and time constant—can be effortlessly extracted from electrochemical impedance spectroscopy (EIS). This parameter is fundamentally linked to the kinetics of a reaction serving as a critical parameter that governs the rate at which electrons traverse the electrode-electrolyte interface. In contrast to other direct current (DC) electroanalytical methods, EIS remains unaffected by potential drops arising from series resistance [26].

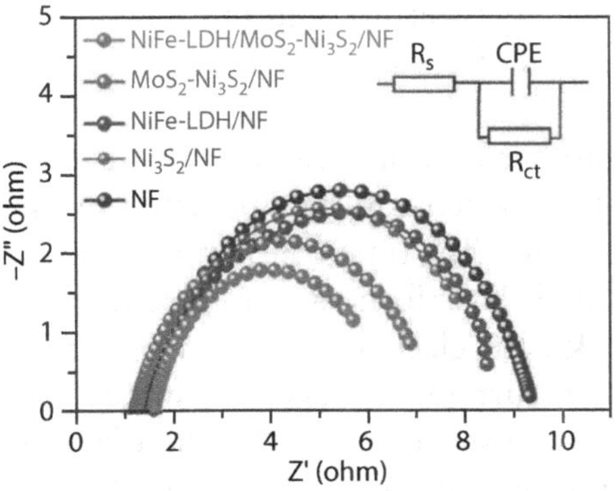

Figure 2.3 Nyquist plot of different catalysts obtained from EIS measurements. *Reproduced with permission from Ref. [27]. Copyright (2023) American Chemical Society.*

As a result, it unveils the inherent electrochemical characteristics of the interface being investigated without any distortion. However, a decisive question that creates ambiguity among the researchers is at what potential the EIS measurements should be carried out. Some researchers opt to conduct EIS measurements at the open circuit potential (OCP), while others choose the onset potential or a potential preceding the onset. Additionally, a subset of researchers investigates various potentials for distinct catalysts ensuring the delivery of the same current density. In contrast, a significant number of investigators prefer a uniform potential beyond the onset, where all the examined catalysts exhibit appreciable catalytic activity. Such divergence in these practices within the application of EIS to assess water splitting electrocatalysts often results in the potential misinterpretation of activity trends. In this regard, Anantharaj *et al.* has explicitly described a suitable panacea [26]. Their comparative EIS study reveals that EIS must be carried out at constant potentials in the catalytic turnover region to draw legitimate correlations.

Figure 2.3 displays the Nyquist plot of the EIS measurements in the form of depressed semi-circular shapes for various electrocatalysts [27]. The obtained data were then fitted with an equivalent circuit as shown in inset Figure 2.3. The most widely used circuit for fitting EIS data is Randels's circuit. After precisely fitting the obtained data, the value of charge transfer resistance (R_{ct}) in conjunction with solution resistance (R_s), pre-exponential factor (Q), and the exponent of the constant phase element (α) can be determined. Further, plugging these values into Brug's equation will yield the capacitance associated with the double layer. A lower

R_{ct} indicates faster charge transfer and more favorable reaction kinetics. Therefore, the diameter of the semi-circle decreases as the electrocatalytic activity increases.

Furthermore, another crucial parameter that can be derived from EIS is a time constant, τ, which can be calculated by multiplying R_{ct} with double-layer capacitance. The time constant is associated with the capacitive effects at the double layer formed at the electrode–electrolyte interface. It reflects how quickly the electrochemical system can respond to a slight change in potential or current. The lesser the value of τ, the faster the speed at which the charge transfer occurs [28]. Therefore, monitoring changes in charge transfer resistance over time or under different experimental conditions allows researchers to infer alterations in reaction kinetics. This information is instrumental in optimizing catalyst design, electrolyte composition, and operating conditions to enhance the overall efficiency of water splitting processes.

2.4 Unraveling Kinetics in the Realm of HER and OER

2.4.1 Hydrogen Evolution Reaction

First, let us delve into comprehending the HER process under acidic conditions. The HER process encompasses the reduction of protons within an acidic medium occurring on the electrode's surface with minimal external

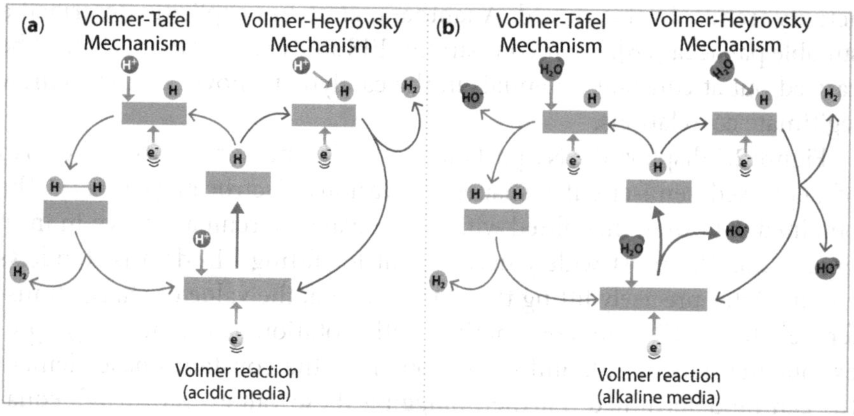

Figure 2.4 HER mechanism in (a) acidic and (b) alkaline medium. *Reproduced with permission from Ref. [29]. Copyright (2020) American Chemical Society.*

potential applied. The initial stage involves the Volmer reaction, where a proton combines with an electron to produce an adsorbed hydrogen atom (H*) on the surface of the electrode material (M), as depicted in Figure 2.4(a) [29].

(1) Electrochemical hydrogen adsorption (Volmer reaction)

$$H_3O^+ + M + e^- \rightleftharpoons M - H^* + H_2O$$

Second, the H_2 formation occurs via the Heyrovsky reaction or the Tafel reaction or both as shown below and presented in Figure 2.4(a).

(2) Electrochemical desorption (Heyrovsky reaction)

$$H^+ + e^- + M - H^* \rightleftharpoons H_2 + M$$

(3) Chemical desorption (Tafel reaction)

$$2M - H^* \rightleftharpoons H_2 + 2M$$

In the Tafel step, two H* in the vicinity combine on the surface of the electrode to evolve H_2.

In an alkaline medium, as shown in Figure 2.4(b), HER takes place via adsorption of water molecules on the surface of an electrode followed by dissociation and H adsorption (Volmer step) [29].

(1) Electrochemical hydrogen adsorption (Volmer reaction)

$$H_2O + M + e^- \rightleftharpoons M - H^* + OH^-$$

(2) Electrochemical desorption (Heyrovsky reaction)

$$H_2O + e^- + M - H^* \rightleftharpoons H_2 + OH^- + M$$

(3) Chemical desorption (Tafel reaction)

$$2M - H^* \rightleftharpoons H_2 + 2M$$

Now, considering a range of materials, let us try to decode how HER kinetics is entangled with inherent electronic, structural, and surface properties. An atomic-level configuration of a novel orthorhombic phase perovskite oxide electrocatalyst, $SrHf_{0.7}Ru_{0.3}O_{3-\delta}$ (SHRO) was designed and synthesized successfully via the sol–gel method by Bu *et al.* [30]. This SHRO perovskite oxide electrocatalyst shows an electron donation from Ru to Hf, which results in charge redistribution between $Hf^{(4-\delta)+}$ and $Ru^{(4+\delta)+}$ ions as $Hf^{4+} + Ru^{4+} \rightarrow Hf^{(4-\delta)+} + Ru^{(4+\delta)+}$, which is confirmed by XPS. Thus, this interaction between Ru–O–Hf results in a noteworthy enhancement in the conductivity of the system. SHRO exhibits an excellent HER activity in 0.5 M H_2SO_4 owing to an overpotential of 48 mV at 10 mA cm^{-2}. Also, SHRO exhibits a higher current density of 500 mA cm^{-2} at −0.18 V, which outperforms the benchmark Pt/C catalyst, which shows the same current density value but at a higher potential value (i.e., −0.4 V). The Tafel analysis shows a reduced slope value of approximately 26 mV dec^{-1} for SHRO, much smaller than that of Pt/C (39 mV dec^{-1}). The study also reveals that the presence of Hf in an SHRO system increases the hydrophilicity nature of the electrocatalyst and thus enhances the hydrogen adsorption/desorption kinetics for HER reaction. DFT calculation describes the mechanism of acidic HER on the SHRO surface as follows: Hf and Ru serve as bimetallic sites where H^+ adsorption takes place at the Hf site, and desorption occurs at the Ru sites. The O present in the system (Hf–O–Ru) acts as a bridge for the transfer of reaction intermediates formed from Hf to Ru [30]. Ha *et al.* synthesized alloys of transition metals viz. Co, Fe, Ni with iridium (Ir) and tungsten (W) to form an Ir-based ternary (Ir–M–W, M = Ni, Co, Fe) metal alloy nanoparticle (NPs) system and to study the variation in

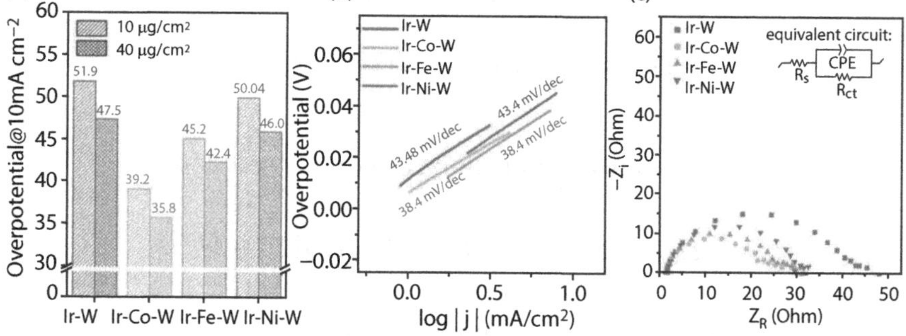

Figure 2.5 (a) Overpotential at a benchmark current density of 10 mA cm^{-2}. (b) Tafel slope. (c) Nyquist plot for Ir-based alloy nanoparticles. *Reproduced with permission from Ref. [31]. Copyright (2023) Elsevier.*

the surface electronic structure induced after the inclusion of a transition metal in the Ir–M–W NPs system and its catalytic activity toward acidic HER [31]. Among all the ternary alloys viz. Ir–Co–W, Ir–Ni–W, Ir–Fe–W, and binary alloy Ir–W, the lowest overpotential of 39.2 and 35.8 mV at 10 mA cm^{-2} with 10 μg cm^{-2} and 40 μg cm^{-2} of mass loading was exhibited by Ir–Co–W NPs, respectively, as shown in Figure 2.5a [31]. Also, the Tafel slope values were obtained in the range of 38.4 to 43.48 mV dec^{-1} for all Ir–M–W and Ir–W NPs, which indicated that the HER kinetics proceeds via the Volmer–Tafel reaction process (Figure 2.5b) [31]. Furthermore, XPS analysis suggests that Ir atoms of ternary Ir–M–W NPs have higher charge densities, are less oxidized, and have high electrochemically active Ir^{3+} species compared to the binary Ir–W NPs and, thus, results in superior catalytic activity. It is also observed from Nyquist measurements that Ir–Co–W NPs showed the smallest charge transfer resistance (R_{ct}) value of 21.76 Ω, which implies faster charge transfer rate and results in enhanced HER kinetics (Figure 2.5c) [31].

Recently, high-entropy alloys (HEA) with superior corrosion resistance have been used for electrolysis, particularly for studying acidic HER and OER processes. One such recent report by Song *et al.* shows the efficacy of quinary Ir-based HEA electrocatalyst viz. ZnNiCoIrX (X: Fe and Mn) toward HER in an acidic environment [32]. A facile sol–gel method followed by high-vacuum dealloying processes was employed to construct a nanoporous structure with high crystallinity using Zn as a sacrificial element. Different compositions of the HEA system showed different overpotential and Tafel slope values at −50 mA cm^{-2} viz. ZnNiIr (101 mV; 59.1 mV dec^{-1}) > ZnNiCoIr (72 mV; 49.5 mV dec^{-1}) > ZnNiCoIrFe (55 mV; 36.7 mV dec^{-1}) > ZnNiCoIrMn (50 mV; 30.6 mV dec^{-1}). Among them, the lowest Tafel slope value was exhibited by the ZnNiCoIrMn electrocatalyst, which was comparable to that of a commercial Pt/C (30.6 mV dec^{-1}) electrocatalyst. This further indicates the Tafel step as the rate-determining step for HER. The highest HER activity obtained for the ZnNiCoIrMn electrocatalyst was attributed to the change in the electronic structure of the Ir site with Mn incorporation, due to which a downshift of the d-band center took place. This downshift of the d-band center was proven beneficial as it resulted in weak adsorption energy of the reaction intermediates formed during the HER process at the catalyst's surface. The high-entropy effects of ZnNiCoIrMn limit the diffusion of elements and modify the system's cohesive and adsorption energies tailoring the electronic structure of active Ir sites [32].

Alkaline HER has been widely studied to achieve higher stability and durability of the catalyst. Some of the recently published reports of alkaline

HER electrocatalysts are described below. Lee *et al.* designed a robust electrocatalyst, Ru/RuO_2 aerogels, for catalyzing both HER and OER under alkaline conditions [33]. A simple sol–gel method using $NaBH_4$ as a reducing agent was used to produce Ru/RuO_2 aerogels. Ru was then converted to RuO_2 through calcination at 200°C. Both crystalline and amorphous nature with various surface defects, such as oxygen vacancies, kinked structures, dislocation, twin boundaries, and amorphous sites of Ru/RuO_2, were depicted by HR-TEM images. The optimal electrocatalytic HER activity was noticed for the Ru/RuO_2-2h (Ru/RuO_2 calcined for 2 h) sample, with an overpotential value of 34 mV at 10 mA cm^{-2} and the corresponding Tafel slope of 29.01 mV dec^{-1}. Also, the lowest charge transfer resistance obtained for the Ru/RuO_2-2h sample implies that the Ru/RuO_2 heterointerface shortens the distance of electron transfer and accelerates the rate of hydrogen release leading to faster electrocatalytic HER kinetics [33]. Similarly, studies were performed by Manna *et al.*, on Ru–Cu nano-heterostructures, which showed an efficient HER in alkaline water electrolyzers [34]. In this work, they engineered a hierarchical electrode compromising amorphous-TiO_2/Cu nanorods (NRs) decorated with Ru–Cu nano-heterostructures. A stable current density of −200 mA cm^{-2} up to 250 h, almost comparable to an industrial level, was observed for the Ru@Cu–TiO_2/Cu electrode with an overpotential of 74 mV in 1 M NaOH. According to experimental investigations, this enhanced activity of the Ru@Cu–TiO_2/Cu electrode toward the alkaline HER is attributed to the robust, porous nature of Cu NRs, which provides a very high electrical conductivity as well as a large electrochemically active surface area. The theoretical investigations depict that Ru nanocrystals act as effective water dissociation centers, and when it comes in contact with Cu clusters, it accelerates the Volmer step. Also, Cu clusters interacting with Ru sites show nearly $\Delta G_H \sim 0$, indicating faster Heyrovsky/Tafel processes [34]. Pan *et al.* studied synergistic effects between the nanocrystalline and amorphous cobalt oxides (CoO_x) for enhancing alkaline HER [35]. A facile electrodeposition technique was used to synthesize CoO_x over Ni foam (NF) with variable crystallinities. Three different electrocatalysts were prepared viz. (P) CoOx–NF, (CrP) CoO_x–NF, and (10CeCrP) CoO_x–NF to study HER activity. Among these, (10CeCrP) CoO_x–NF exhibited the highest catalytic activity achieving 200 mA cm^{-2} only at a potential of −0.354 V. Experimentally, the enhanced HER kinetics of CoO_x results from its amorphous nature, which helps in enhancing the pseudocapacitive K^+ adsorption and, thus, leads to high surface water affinity. Moreover, the mixed crystalline and amorphous structure of CoO_x in (10CeCrP) CoO_x-NF provides long-term catalytic stability which enables the catalyst with enhanced durability and sustained performance in alkaline HER. Also, high concentrations of hydrogen and water on the surface of CoO_x offer a large number of reactive species for HER and thus promote hydrogen transportation and conversion [36]. Yang *et al.* reported how hydrogen spillover can be beneficial in promoting alkaline HER kinetics by synthesizing amorphous molybdenum oxide (MoO_xRh–HNS) incorporated with rhodium hierarchical nanosheets (Rh–HNS). These MoO_xRh–HNS show a low Tafel slope of 22.5 mV dec^{-1} in alkaline electrolytes requiring only 102.5 mV overpotential to attain 200 mA cm^{-2}. This performance was decreased by 283.5 and 233.5 mV when compared to Pt/C and Rh-HNS [35]. Mu *et al.* developed Ru-incorporated nickel diselenide (Ru–$NiSe_2$) nanosheet arrays to perform overall water splitting in an alkaline medium. Ru–$NiSe_2$ was grown hydrothermally over a substrate (nickel foam). The Ru–$NiSe_2$ electrode shows 59 mV of overpotential with a Tafel slope of 72.2 mV dec^{-1} at 10 mA cm^{-1}. It also showed lower charge transfer resistance compared to the pristine counterparts and thus indicates faster HER kinetics. The theoretical calculation also suggests that the Ru–$NiSe_2$ system acts as a bifunctional electrocatalyst by tuning the Gibbs free energy of adsorption of intermediates during overall water splitting [37].

2.4.2 Oxygen Evolution Reaction

OER suffers from sluggish kinetics since it is linked to four elementary steps that involve multiple electron transfer processes. Under acidic conditions, water binds to the surface irreversibly removing one electron and one proton to create a surface hydroxide [38]. In alkaline conditions, the initiation of the water oxidation process involves the binding of an OH^- radical [39]. An activity descriptor for OER kinetics involves shifting of

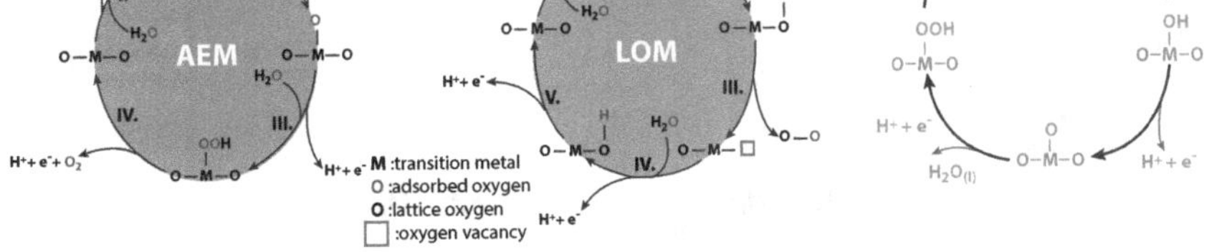

Figure 2.6 (a) Adsorbate Evolution Mechanism (AEM), (b) Lattice Oxygen Mechanism (LOM) for OER under alkaline conditions. *Reproduced with permission from Ref. [41]. Copyright (2020) Elsevier* (c) AEM under acidic conditions. *Reproduced with permission from Ref. [42]. Copyright (2021) Wiley.*

the cationic d-band center and O–p band center toward the Fermi level indicating strong OER activity; however, excessive proximity may decrease activity [40]. Since OER electrocatalysts' working conditions depend on pH, the OER mechanism varies in alkaline, acidic, and neutral conditions. A brief overview is illustrated in Figure 2.6 [41, 42]. As depicted in Figure 2.6(a), under alkaline conditions, the adsorbate evolution mechanism (AEM) involves the formation of M–OH* through the interaction of a hydroxyl radical (OH^-) with the active site "M". In the subsequent steps, four H^+ and e^- are released, leading to the production of O_2 [41, 43]. AEM's OER kinetics are determined by the difference in Gibbs free energy change ($\Delta G_{*OOH} - \Delta G_{*OH}$). In contrast, the Lattice Oxygen Mechanism (LOM), as depicted in Figure 2.6b, directly engages lattice oxygen in O–O coupling overcoming AEM's limitations without the adsorption energy scaling issue [41, 44, 45]. LOM's strong metal–oxygen covalency in metal oxides allows for electronic structure adjustments, with the deprotonation of *OH as the rate-determining step. Therefore, O_2 evolution involves oxygen from metal oxides' lattice O. Consequently, O_2 can directly evolve from the lattice oxygen atoms of a metal oxide catalyst.

Figure 2.6c displays the OER mechanism in the context of acidic conditions [42]. The H_2O molecule undergoes a reaction with the active site "M," facilitating the dissociation and adsorption of hydroxyl (OH*) species. As the anodic potential increases, OH* undergoes oxidation progressing to O* and eventually to OOH*. Concurrently, four protons and electrons are liberated during this process. Furthermore, to enhance our understanding of OER mechanisms, diving into the intricacies of OER kinetics across various

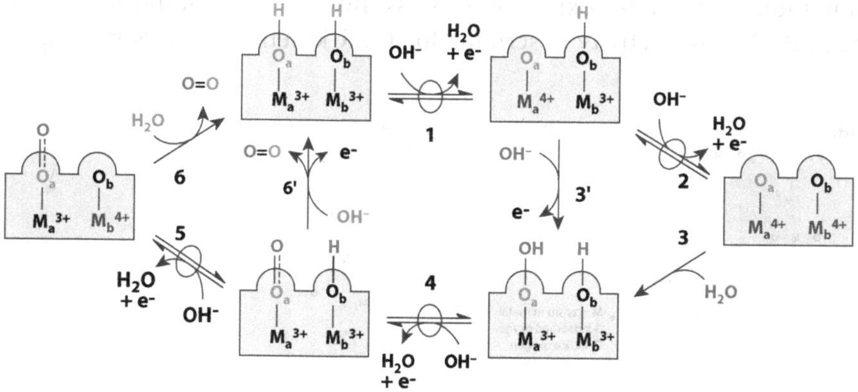

Figure 2.7 A generalized OER catalytic mechanism involving two redox centers in transition metal oxide systems. *Reproduced with permission from Ref. [46]. Copyright (2019) American Chemical Society.*

materials is essential. Let us begin with transition metal-based electrocatalysts, which are at the pinnacle due to their excellent OER activity. Figure 2.7 illustrates a generalized OER catalytic cycle happening on the surface of transition metal oxide structures [46]. To elucidate the process, let us consider a widely researched spinel-based electrocatalyst, Co_3O_4. OER on Co_3O_4 takes place on an active site with two redox centers, namely, $(M–O)_a$ and $(M–O)_b$ [46]. These centers, each composed of a 3d metal cation and a protonatable oxygen atom, are involved in oxidizing water. The catalytic cycle involves the oxidation of $(M–O)_a$ followed by $(M–O)_b$ leading to the formation of a reactive oxo species. Subsequent steps include the creation of an O–O bond between O_a and a water molecule forming a hydroperoxo intermediate. This intermediate undergoes electrochemical oxidation to a superoxo, and the $(M–O)_b$ center is again oxidized. The superoxo then transfers an electron to M_b eliminating O_2 in a non-electrochemical step. The catalytic cycle concludes with the dissociation of a second water molecule. The entire process can occur through either electrochemical or non-electrochemical pathways. Moreover, in such instances, the OER kinetics is primarily governed by steps 1 and 2 followed by the rate-limiting water-addition step. The enhanced OER activity in Co_3O_4 can also be attributed to the presence of mixed oxidation states of Co^{2+}/Co^{3+}.

To enhance the Oxygen Evolution Reaction (OER) kinetics in Co_3O_4, modifications, such as integration with N-doped carbonaceous materials [47–49], introduction of dopants like Li, Ni, Cu, Mg [50], and variations in morphology, have been explored. Doping with elements like Co^{2+} assists in forming CoOOH, an active site for OER kinetics. Different spinels, including $NiCo_2O_4$, have shown high OER catalytic activity. Heterostructures, like $NiCo_2O_4/ZnCo_2O_4$, and tuning the composition of Ni and Co contribute to increased electroactive sites and fast electrolyte/electron diffusion [51]. Generating oxygen defects (V_O) in $NiCo_2O_4$ and the presence of metal cations, especially Fe in $FexNi_{1-x}Co_2O_4$, display high OER activity. $MnxCo_{3-x}O_4$, a mixed-valent oxide, shows increased OER activity with tailored compositions, micro-nano-heterostructures, and conductive substrates [50].

Another interesting category of materials are perovskites, which showcases superior OER kinetics in acidic environments [52]. The enhancement of OER activity in acidic medium can be attributed to surface reconstruction and metal cation leaching. Jaramillo and colleagues identified that the OER activity of $SrIrO_3$ is attributed to the formation of IrO_X caused by Sr leaching on its surface [53]. For Ir-based perovskites, performed under acidic conditions, Sr leaching in $SrIrO_3$ thin films resulted in the creation of mosaic-shaped Sr-deficient IrO_X on the film's surface. The continuous generation of IrO_X active species proved crucial for sustaining electrocatalytic OER activity [54]. However, Ru-based perovskites, such as $SrRuO_3$, were found to exhibit a low OER activity in acidic solution because of their thermodynamic instability. Therefore, to mitigate this problem, Roja and his co-workers suggested that the incorporation of Na^+ into $SrRu(Na_xSr_{1-x}RuO_3)$ can enhance the activity as well as stability of OER in acidic conditions [55]. This substitution prevents structural distortion caused by the Jahn–Teller effect forming regular RuO_6 octahedra. Na substitution stabilizes Ru centers with higher oxidation states resulting in a lower overpotential within the range of 170 to 180 mV at benchmark 10 mA cm^{-2}. Another fascinating OER kinetic descriptor provided by Wang *et al.* for five new perovskites is the ratio of μ/t, where "μ" and "t" represent octahedral and tolerance factors, respectively [56]. Anionic vacancies engineered in perovskite oxides were also found to enhance OER kinetic activity by facilitating oxygen incorporation and transport [57]. Changes in the electronic structure alter the relative position of the metal d-band in relation to oxygen p-band centers. The presence of oxygen vacancies, as demonstrated in $SrMnO_{3-\delta}$, particularly $SrMnO_{2.5}$ (where $\delta = 0.5, 0.4$), is found to exhibit enhanced OER performance [58]. It is because oxygen vacancies cause the d-band center (M_d) to move closer to the Fermi level (E_f) reducing the energy gap between the metal d and oxygen p band centers and boosting electrocatalytic OER activity, which has been confirmed through a DFT study [57].

Layered double hydroxides (LDH) are another class of widely studied materials for OER. In monometallic LDHs, the OER activity is directly linked to the abundance of d states. By increasing the density of d-level states in the sequence $Mn^{2+} > Fe^{2+} > Co^{2+} > Ni^{2+}$, the OER activity can be enhanced [59]. In various monometallic LDHs, such as $Ni(OH)_2$ and $Co(OH)_2$, Fe emerges as a pivotal factor in augmenting OER performance. A noteworthy finding reported by Młynarek *et al.* indicates a substantial enhancement in OER activity with the introduction of less than 1% Fe impurities into $Ni(OH)_2$ [60]. Incorporation of Fe into the $Ni(OH)_2$ interface is shown to elevate the electrical conductivity of its catalytic interface in alkaline media, and this effect intensifies proportionally with the percentage of Fe doping. Additionally, the $Ni(OH)_2$ lattice undergoes surface reconstruction due to the incorporation of Fe into its lattice further elucidating the intricate mechanisms contributing to the observed improvements in electrochemical performance [61, 62]. Apart from NiFe LDH systems, the integration of iron into the hydroxide and oxyhydroxide films of cobalt demonstrates a remarkable 100-fold improvement in oxygen evolution reaction (OER) performance. As the concentration of Fe increases, there is an anodic shift in the Co^{2+} to Co^{3+} transition accompanied by notable surface reconstruction at the catalytic sites. This phenomenon contributes significantly to the enhancement of OER performance in cobalt-based systems [63]. The most effective catalysts for the Oxygen Evolution Reaction (OER) are NiFe-based Layered Double Hydroxides (LDH) and layered oxides, outperforming other metal combinations such as Ni–Co, Ni–Ni, Co–Co, Co–Fe, Co–Mn, Ni–Mn, Ni–V, and Co–Cr. Considering the role of morphologies and heterojunction in enhancing the OER activity of NiFe LDH, it has been found that the 3d NiFe hydrous electrode has notably demonstrated remarkable performance achieving a Tafel slope of 28 mV dec^{-1} [64]. In the context of heterojunctions, NiFe LDH/carbon dots [65] and NiFe LDH/CNT [66] have displayed noteworthy OER activity in their respective studies.

2.5 Conclusion

The imperative quest for sustainable and green energy sources in the face of escalating environmental concerns has fueled significant interest in water splitting research. Since OER and HER are governed by their distinct mechanisms, optimizing a single material for both OER and HER is utterly cumbersome and has become the question of the hour. By recognizing the unique attributes of these mechanisms, this chapter systematically investigates various kinetics parameters related to OER and HER, and a comprehensive study of a range of materials has been provided. Furthermore, the chapter emphasizes the critical role of understanding water splitting kinetics as a fundamental step toward advancing the overall efficiency of the water splitting process.

AI disclosure statement: The author acknowledges the use of AI-based tools for minor language editing or grammatical correction. However, all intellectual contributions, analyses, and conclusions presented in this chapter are entirely the author's own.

References

1. Xu, X., Zhou, Q., Yu, D., The future of hydrogen energy: bio-hydrogen production technology. *Int. J. Hydrog. Energy*, 47, 33677, 2022.
2. Ng, B.J., Putri, L.K., Kong, X.Y., Teh, Y.W., Pasbakhsh, P., Chai, S.P., Z-scheme photocatalytic systems for solar water splitting. *Adv. Sci.*, 7, 1903171, 2020.

3. Yin Tee, S., Yin Win, K., Siang Teo, W., Koh, L.-D., Liu, S., Peng Teng, C., Han, M.-Y., Tee, S.Y., Win, K.Y., Koh, L., Liu, S., Teng, C.P., Han, M., Teo, W.S., Recent progress in energy-driven water splitting. *Adv. Sci.*, 4, 1600337, 2017.
4. Heat values of various fuels, *World Nuclear Association*, https://world-nuclear.org/information-library/facts-and-figures/heat-values-of-various-fuels.aspx (accessed 11 February 2024).
5. Zhang, Y., Xie, K., Zhou, F., Wang, F., Xu, Q., Hu, J., Ding, H., Li, P., Tan, Y., Li, D., Zhu, J., Zhou, H., Zhao, C., Lin, S., Wu, Y., Electrochemical oxygen generator with 99.9% oxygen purity and high energy efficiency. *Adv. Energy Mater.*, 12, 2201027, 2022.
6. Bodhankar, P.M., Sarawade, P.B., Singh, G., Vinu, A., Dhawale, D.S., Recent advances in highly active nanostructured NiFe LDH catalyst for electrochemical water splitting. *J. Mater. Chem. A Mater.*, 9, 3180, 2021.
7. Wang, H., Lee, H.W., Deng, Y., Lu, Z., Hsu, P.C., Liu, Y., Lin, D., Cui, Y., Bifunctional non-noble metal oxide nanoparticle electrocatalysts through lithium-induced conversion for overall water splitting. *Nat. Commun.*, 6, 1, 2015.
8. Sun, H., Xu, X., Kim, H., Jung, W.C., Zhou, W., Shao, Z., Electrochemical water splitting: bridging the gaps between fundamental research and industrial applications. *Energy Environ. Mater.*, 6, e12441, 2023.
9. Wei, C., Rao, R.R., Peng, J., Huang, B., Stephens, I.E.L., Risch, M., Xu, Z.J., Shao-Horn, Y., Recommended practices and benchmark activity for hydrogen and oxygen electrocatalysis in water splitting and fuel cells. *Adv. Mater.*, 31, e1806296, 2019.
10. Boucher, D.G., Carroll, E., Nguyen, Z.A., Jadhav, R.G., Simoska, O., Beaver, K., Minteer, S.D., Bioelectrocatalytic synthesis: concepts and applications. *Angew. Chem.*, 135, 202307780, 2023.
11. Anantharaj, S., Ede, S.R., Karthick, K., Sam Sankar, S., Sangeetha, K., Karthik, P.E., Kundu, S., Precision and correctness in the evaluation of electrocatalytic water splitting: revisiting activity parameters with a critical assessment. *Energy Environ. Sci.*, 11, 744, 2018.
12. McCrory, C.C.L., Jung, S., Peters, J.C., Jaramillo, T.F., Benchmarking heterogeneous electrocatalysts for the oxygen evolution reaction. *J. Am. Chem. Soc.*, 135, 16977, 2013.
13. Anantharaj, S. and Noda, S., iR drop correction in electrocatalysis: everything one needs to know! *J. Mater. Chem. A*, 10, 9348, 2022.
14. Li, D., Lin, C., Batchelor-McAuley, C., Chen, L., Compton, R.G., Tafel analysis in practice. *J. Electroanal. Chem.*, 826, 117, 2018.
15. Bard, A.J., Faulkner, L.R., White, H.S., *Electrochemical methods: fundamentals and applications*, UK, John Willey & Sons Ltd, 2022.
16. Bockris, J.O., Reddy, A.K.N., Gamboa-Aldeco, M., *Modern Electrochemistry 2A: Fundamentals of Electrodics*, New York, Springer, 2000.

17. Anantharaj, S. and Noda, S., How properly are we interpreting the Tafel lines in energy conversion electrocatalysis? *Mater. Today Energy*, 29, 101123, 2022.
18. Shi, Y. and Zhang, B., Recent advances in transition metal phosphide nanomaterials: synthesis and applications in hydrogen evolution reaction. *Chem. Soc. Rev.*, 45, 1529, 2016.
19. Anantharaj, S., Ede, S.R., Sakthikumar, K., Karthick, K., Mishra, S., Kundu, S., Recent trends and perspectives in electrochemical water splitting with an emphasis on sulfide, selenide, and phosphide catalysts of Fe, Co, and Ni: a review. *ACS Catal.*, 6, 8069, 2016.
20. Liu, M., Min, K.A., Han, B., Lee, L.Y.S., Interfacing or doping? role of Ce in highly promoted water oxidation of NiFe-layered double hydroxide. *Adv. Energy Mater.*, 11, 2101281, 2021.
21. Anantharaj, S., Karthik, P.E., Kundu, S., Self-assembled IrO_2 nanoparticles on a DNA scaffold with enhanced catalytic and oxygen evolution reaction (OER) activities. *J. Mater. Chem. A*, 3, 24463, 2015.
22. Qian, L., Lu, Z., Xu, T., Wu, X., Tian, Y., Li, Y., Huo, Z., Sun, X., Duan Qian, X.L., Lu, Z., Xu, T., Wu, X., Tian, Y., Li, Y., Sun, X., Duan, X., Huo, Z., Trinary layered double hydroxides as high-performance bifunctional materials for oxygen electrocatalysis. *Adv. Energy Mater.*, 5, 1500245, 2015.
23. Dionigi, F., Strasser, P., Dionigi, F., Strasser, P., NiFe-based (Oxy)hydroxide catalysts for oxygen evolution reaction in non-acidic electrolytes. *Adv. Energy Mater.*, 6, 1600621, 2016.
24. Anantharaj, S., Reddy, P.N., Kundu, S., Core-oxidized amorphous cobalt phosphide nanostructures: an advanced and highly efficient oxygen evolution catalyst. *Inorg. Chem.*, 56, 1742–1756, 2016.
25. Anantharaj, S. and Kundu, S., Do the evaluation parameters reflect intrinsic activity of electrocatalysts in electrochemical water splitting? *ACS Energy Lett.*, 4, 1260, 2019.
26. Anantharaj, S. and Noda, S., Appropriate use of electrochemical impedance spectroscopy in water splitting electrocatalysis. *Chem. Electro. Chem.*, 7, 2297, 2020.
27. Wang, S., Ning, X., Cao, Y., Chen, R., Lu, Z., Hu, J., Xie, J., Hao, A., Construction of an advanced NiFe-LDH/MoS_2-Ni_3S_2/NF heterostructure catalyst toward efficient electrocatalytic overall water splitting. *Inorg. Chem.*, 62, 6428, 2023.
28. Bao, F., Kemppainen, E., Dorbandt, I., Bors, R., Xi, F., Schlatmann, R., van de Krol, R., Calnan, S., Understanding the hydrogen evolution reaction kinetics of electrodeposited nickel-molybdenum in acidic, near-neutral, and alkaline conditions. *Chem. Electro. Chem.*, 8, 195, 2021.
29. Zhu, J., Hu, L., Zhao, P., Lee, L.Y.S., Wong, K.Y., Recent Advances in electrocatalytic hydrogen evolution using nanoparticles. *Chem. Rev.*, 120, 851, 2020.
30. Wang, Y., Lu, Q., Li, F., Guan, D., Bu, Y., Atomic-scale configuration enables fast hydrogen migration for electrocatalysis of acidic hydrogen evolution. *Adv. Funct. Mater.*, 33, 2213523, 2023.

31. Kim, M., Kang, H., Hwang, E., Park, Y., Jeong, W., Hwang, Y.J., Ha, D.H., Facile colloidal synthesis of transition metal (Co, Fe, and Ni)-added Ir-W NPs for HER in acidic electrolyte. *Appl. Surf. Sci.*, 612, 155862, 2023.
32. Kwon, J., Sun, S., Choi, S., Lee, K., Jo, S., Park, K., Kim, Y.K., Park, H.B., Park, H.Y., Jang, J.H., Han, H., Paik, U., Song, T., Tailored electronic structure of Ir in high entropy alloy for highly active and durable bifunctional electrocatalyst for water splitting under an acidic environment. *Adv. Mater.*, 35, 2300091, 2023.
33. Nicole, S.L.D., Li, Y., Xie, W., Wang, G., Lee, J.M., Heterointerface and tensile strain effects synergistically enhances overall water-splitting in Ru/RuO_2 Aerogels. *Small*, 19, 2206844, 2023.
34. Zuo, Y., Bellani, S., Saleh, G., Ferri, M., Shinde, D.V., Zappia, M.I., Buha, J., Brescia, R., Prato, M., Pascazio, R., Annamalai, A., de Souza, D.O., De Trizio, L., Infante, I., Bonaccorso, F., Manna, L., Ru-Cu nanoheterostructures for efficient hydrogen evolution reaction in alkaline water electrolyzers. *J. Am. Chem. Soc.*, 145, 21419, 2023.
35. Pan, S., Li, C., Xiong, T., Xie, Y., Luo, F., Yang, Z., Hydrogen spillover in MoO_xRh hierarchical nanosheets boosts alkaline HER catalytic activity. *Appl. Catal., B*, 341, 123275, 2024.
36. Feng, J., Qiao, L., Zhou, P., Bai, H., Liu, C., Leong, C.C., Chen, Y.Y., Ip, W.F., Ni, J., Pan, H., Nanocrystalline CoO_x glass for highly-efficient alkaline hydrogen evolution reaction. *J. Mater. Chem. A*, 11, 316, 2022.
37. Qin, R., Wang, P., Li, Z., Zhu, J., Cao, F., Xu, H., Ma, Q., Zhang, J., Yu, J., Mu, S., Qin, R., Wang, P., Li, Z., Zhu, J., Cao, F., Xu, H., Ma, Q., Zhang, J., Yu, J., Mu, S., Foshan, S.M., Ru-incorporated nickel diselenide nanosheet arrays with accelerated adsorption kinetics toward overall water splitting. *Small*, 18, 2105305, 2022.
38. Birss, V.I., Damjanovic, A., Hudson, P.G., Oxygen evolution at platinum electrodes in alkaline solutions:II . mechanism of the reaction. *J. Electrochem. Soc.*, 133, 1621, 1986.
39. Zeng, K. and Zhang, D., Recent progress in alkaline water electrolysis for hydrogen production and applications. *Prog. Energy Combust. Sci.*, 36, 307, 2010.
40. Xu, W., Apodaca, N., Wang, H., Yan, L., Chen, G., Zhou, M., Ding, D., Choudhury, P., Luo, H., A-site excessive $(La_{0.8}Sr_{0.2})_{1+x}MnO_3$ perovskite oxides for bifunctional oxygen catalyst in alkaline media. *ACS Catal.*, 9, 5074, 2019.
41. Gu, X.K., Camayang, J.C.A., Samira, S., Nikolla, E., Oxygen evolution electrocatalysis using mixed metal oxides under acidic conditions: challenges and opportunities. *J. Catal.*, 388, 130, 2020.
42. Pu, Z., Liu, T., Zhang, G., Ranganathan, H., Chen, Z., Sun, S., Electrocatalytic oxygen evolution reaction in acidic conditions: recent progress and perspectives. *Chem. Sus. Chem.*, 14, 4636–4657, 2021.

43. Song, J., Wei, C., Huang, Z.F., Liu, C., Zeng, L., Wang, X., Xu, Z.J., A review on fundamentals for designing oxygen evolution electrocatalysts. *Chem. Soc. Rev.*, 49, 2196, 2020.
44. Adamu, H., Abba, S.I., Anyin, P.B., Sani, Y., Yamani, Z.H., Qamar, M., Tuning OER Electrocatalysts toward LOM Pathway through the lens of multi-descriptor feature selection by artificial intelligence-based approach. *ACS Mater. Lett.*, 5, 299, 2023.
45. Rong, X., Parolin, J., Kolpak, A.M., A fundamental relationship between reaction mechanism and stability in metal oxide catalysts for oxygen evolution. *ACS Catal.*, 6, 1153, 2016.
46. Plaisance, C.P., Beinlich, S.D., Reuter, K., Kinetics-based computational catalyst design strategy for the oxygen evolution reaction on transition-metal oxide surfaces. *J. Phy. Chem. C*, 123, 8287, 2019.
47. Li, X., Fang, Y., Lin, X., Tian, M., An, X., Fu, Y., Li, R., Jin, J., Ma, J., MOF derived Co_3O_4 nanoparticles embedded in N-doped mesoporous carbon layer/MWCNT hybrids: extraordinary bi-functional electrocatalysts for OER and ORR. *J. Mater. Chem. A*, 3, 17392, 2015.
48. Kumar, K., Canaff, C., Rousseau, J., Arrii-Clacens, S., Napporn, T.W., Habrioux, A., Kokoh, K.B., Effect of the oxide-carbon heterointerface on the activity of Co_3O_4/NRGO nanocomposites toward ORR and OER. *J. Phys. Chem. C*, 120, 7949, 2016.
49. Liang, Y., Li, Y., Wang, H., Zhou, J., Wang, J., Regier, T., Dai, H., Co_3O_4 nanocrystals on graphene as a synergistic catalyst for oxygen reduction reaction. *Nat. Mater.*, 10, 780, 2011.
50. Zhao, Q., Yan, Z., Chen, C., Chen, J., Spinels: controlled preparation, oxygen reduction/evolution reaction application, and beyond. *Chem. Rev.*, 117, 10121, 2017.
51. Huang, Y., Miao, Y.E., Lu, H., Liu, T., Hierarchical $ZnCo_2O_4$@$NiCo_2O_4$ core–sheath nanowires: bifunctionality towards high-performance supercapacitors and the oxygen-reduction reaction. *Chem. Eur. J.*, 21, 10100, 2015.
52. Song, H.J., Yoon, H., Ju, B., Kim, D.-W., Song, H.J., Yoon, H., Ju, B.D., Kim, W., Highly efficient perovskite-based electrocatalysts for water oxidation in acidic environments: a mini review. *Adv. Energy Mater.*, 11, 2002428, 2021.
53. Seitz, L.C., Dickens, C.F., Nishio, K., Hikita, Y., Montoya, J., Doyle, A., Kirk, C., Vojvodic, A., Hwang, H.Y., Norskov, J.K., Jaramillo, T.F., A highly active and stable $IrO_x/SrIrO_3$ catalyst for the oxygen evolution reaction. *Science*, 353, 1011, 2016.
54. Lee, K., Osada, M., Hwang, H.Y., Hikita, Y., Oxygen Evolution Reaction Activity in IrO_x /$SrIrO_3$ Catalysts: correlations between structural parameters and the catalytic activity. *J. Phys. Chem. Lett.*, 10, 1516, 2019.
55. Retuerto, M., Pascual, L., Calle-Vallejo, F., Ferrer, P., Gianolio, D., Pereira, A.G., García, A., Torrero, J., Fernández-Díaz, M.T., Bencok, P., Peña, M.A., Fierro, J.L.G., Rojas, S., Na-doped ruthenium perovskite electrocatalysts

with improved oxygen evolution activity and durability in acidic media. *Nat. Commun.*, 10, 1, 2019.
56. Weng, B., Song, Z., Zhu, R., Yan, Q., Sun, Q., Grice, C.G., Yan, Y., Yin, W.J., Simple descriptor derived from symbolic regression accelerating the discovery of new perovskite catalysts. *Nat. Commun.*, 11, 1, 2020.
57. Lee, H., Gwon, O., Choi, K., Zhang, L., Zhou, J., Park, J., Yoo, J.W., Wang, J.Q., Lee, J.H., Kim, G., Enhancing bifunctional electrocatalytic activities via metal d-band center lift induced by oxygen vacancy on the subsurface of perovskites. *ACS Catal.*, 10, 4664–4670, 2020.
58. Hona, R.K. and Ramezanipour, F., Effect of the oxygen vacancies and structural order on the oxygen evolution activity: a case study of $SrMnO_{3-\delta}$ featuring four different structure types. *Inorg. Chem.*, 59, 4685, 2020.
59. Subbaraman, R., Tripkovic, D., Chang, K.C., Strmcnik, D., Paulikas, A.P., Hirunsit, P., Chan, M., Greeley, J., Stamenkovic, V., Markovic, N.M., Trends in activity for the water electrolyser reactions on 3d M(Ni,Co,Fe,Mn) hydr(oxy)oxide catalysts. *Nat. Mater.*, 11, 550, 2012.
60. Młynarek, G., Paszkiewicz, M., Radniecka, A., The effect of ferric ions on the behaviour of a nickelous hydroxide electrode. *J. Appl. Electrochem.*, 14, 145, 1984.
61. Trotochaud, L., Ranney, J.K., Williams, K.N., Boettcher, S.W., Solution-cast metal oxide thin film electrocatalysts for oxygen evolution. *J. Am. Chem. Soc.*, 134, 17253, 2012.
62. Trotochaud, L., Young, S.L., Ranney, J.K., Boettcher, S.W., Nickel-Iron oxyhydroxide oxygen-evolution electrocatalysts: The role of intentional and incidental iron incorporation. *J. Am. Chem. Soc.*, 136, 6744, 2014.
63. Burke, M.S., Kast, M.G., Trotochaud, L., Smith, A.M., Boettcher, S.W., Cobalt-Iron (oxy)hydroxide oxygen evolution electrocatalysts: the role of structure and composition on activity, stability, and mechanism. *J. Am. Chem. Soc.*, 137, 3638, 2015.
64. Lu, X. and Zhao, C., Electrodeposition of hierarchically structured three-dimensional nickel–iron electrodes for efficient oxygen evolution at high current densities. *Nat. Commun.*, 6, 1, 2015.
65. Tang, D., Liu, J., Wu, X., Liu, R., Han, X., Han, Y., Huang, H., Liu, Y., Kang, Z., Carbon quantum dot/NiFe layered double-hydroxide composite as a highly efficient electrocatalyst for water oxidation. *ACS Appl. Mater. Interfaces*, 6, 7918, 2014.
66. Gong, M., Li, Y., Wang, H., Liang, Y., Wu, J.Z., Zhou, J., Wang, J., Regier, T., Wei, F., Dai, H., An advanced Ni-Fe layered double hydroxide electrocatalyst for water oxidation. *J. Am. Chem. Soc.*, 135, 8452, 2013.

3

Perovskite Electrocatalyst for Water Splitting

Nawishta Jabeen[1], Imtiaz Ahmad Khan[2], Adeela Naz[2] and Ahmad Hussain[3*]

[1]Department of Physics, Fatima Jinnah Women University, Rawalpindi, Pakistan
[2]School of Materials Science and Engineering, Harbin Institute of Technology, Shenzhen City, China
[3]Department of Physics, The University of Lahore, Sargodha Campus, Sargodha, Pakistan

Abstract

Future energy and transportation applications may find hydrogen fuel from electrochemical water splitting, which uses renewable electricity as a clean, significant, and sustainable source. To provide increased activity for the water splitting reaction and stability for long-term operation, a lot of study must be done on the creation of an effective and stable electrocatalyst because of the sluggish reaction rates occurring at both the anode and cathode. By supplying alternative perovskite-derived replacements for valuable metals while maintaining essential electrocatalyst traits, this chapter provides a comprehensive review of the field's developments. Herein, a comprehensive analysis of the most current studies on perovskite oxides for cathodic, or anodic, and oxygen-reduction processes is presented. This includes an examination of the mechanism underlying the catalytic reactions of perovskite oxides, their controlled composition, and the function of different design strategies in achieving high catalytic performance. Additionally, the chapter offers a glimpse into the related density functional theory, which can offer a deep comprehension of the mechanics underlying these reactions and the necessity of computational research to maximize catalysts' active area. Moreover, this chapter also provides guidance to help scholars investigate important research topics in the most recent generation of perovskites, which have demonstrated improved catalytic performance and can address unforeseen difficulties.

Keywords: Perovskite, water splitting, electrocatalyst, renewable, electrode

**Corresponding author*: ahmad.hussain@phys.uol.edu.pk; ahmed_00277@yahoo.com

Inamuddin, Tariq Altalhi, Mohammad Luqman and Jorddy Neves Cruz (eds.) Water Splitting: Production of Hydrogen, (49–74) © 2025 Scrivener Publishing LLC

3.1 Introduction

The exploration of renewable energy has garnered substantial interest due to the swiftly escalating global energy demand and the consequential apprehensions regarding climate change. Technologies related to the conversion and storage of energy, with a specific emphasis on electrochemical devices, have the potential to facilitate the widespread utilization of sustainable energy by mitigating the unpredictability and intermittent nature inherent in renewable energy production [1, 2]. These progressive electrochemical advancements have garnered considerable attention, particularly in the realm of fuel cells and metal–air batteries. The sluggish kinetics and noteworthy overpotentials associated with oxygen evolution reaction (OER), oxygen reduction reaction (ORR), and hydrogen evolution reaction (HER) contribute to the elevated energy densities achievable in metal–air batteries and regenerative fuel cells [3, 4]. As the most advanced electrocatalysts available, exclusive use has been made of catalysts derived from noble metals. RuO_2 and IrO_2 have shown significant activity toward the OER, whereas Pt has been studied as a potential efficient electrocatalyst towards the HER and ORR. Precious metals are capable of significant catalytic activity, but their high price and scarcity have unavoidably prevented their widespread use in a variety of electrochemical systems.

Consequently, it is essential to produce economically viable and catalytically active electrocatalysts for the HER and oxygen electrocatalysis (ORR/OER) [5]. Over the past few decades, there has been significant research dedicated to the exploration of carbonaceous materials, as well as transition metal oxides, like perovskite and spinel oxides, alongside transition metal phosphides, selenides, sulfides, and carbides [6]. These materials have been extensively investigated for their potential applications in the ORR, OER, and HER. The focus of these studies has encompassed considerations of cost, intrinsic reactivity, and stability when exposed to alkaline solutions, and natural reserves, perovskite oxides (ABO_3), outperform the other noble-metal-free substitutes listed above [7, 8]. Their exclusive abilities of chemical defects and electrical structure mark them striking ORR/OER candidates for hydrogen storage applications. Numerous researchers on perovskites have demonstrated a direct relationship among intrinsic catalytic ability of the ORR and OER, accompanied by the covalent bonding among boron (B) atoms and the possession of electrons in the antibonding orbitals at B sites [9, 10]. Optimum ORR and OER efficiency can be attained by perovskite oxides displaying almost ample electron occupancy and considerable covalency among B and O atoms. The catalytic activity for both ORR and OER can be particularly sensitive by deliberately engaging partial cation replacement and inducing oxygen deficiency throughout the manufacturing process. This outcomes result in the expansion of bifunctional electrocatalysts showcasing higher performance in both the ORR and OER domains, such as $PrBa_{0.5}Sr_{0.5}Co_{1.5}Fe_{0.5}O_{5+\delta}$ (PBSCF) and $La_{0.3}(Ba_{0.5}Sr_{0.5})_{0.7}$–$Co_{0.8}Fe_{0.2}O_{3-\delta}$ (LBSCF), which have been developed [11, 12]. Additionally, the most recent focus has been on the utilization of perovskites to speed up the HER in alkaline medium. HER electrocatalysts in alkaline settings have been studied for materials based on $NdBa_2$–$Mn_2O_{5.5}$ (NBM) and $Ba_{0.5}Sr_{0.5}Co_{0.8}Fe_{0.2}O_3$ (BSCF), among others [13]. This opens the possibility of using perovskites in overall water splitting systems. Even yet, they still need to put in a lot of work to boost their catalytic activity for the HER because it is still lower than that of several electrocatalysts, including precious metals and transition metal sulfides and phosphides.

Creating a solitary component electrocatalyst with advantageous adsorption/desorption energies for various species proves challenging in the realm of versatile electrocatalysis. This complexity arises due to the disparate potentials at which the hydrogen catalytic reaction and the oxygen catalytic reaction take place leading to the formation of distinct intermediate species (such as OH, O_2, OOH, H_2O, H*, and H_2). Developing such a catalyst demands a meticulous approach to accommodate the diverse requirements of these reactions. Anticipated is the catalytic efficacy within pliant electrochemical apparatus, which is expected to increase with a sensible electrocatalyst design that includes plenty of catalytic sites and several active components. One important technique for creating a robust and stable interfacial contact between the components is heterostructure engineering. In particular, our recent study has demonstrated that the creation of a strong linked heterostructure is made possible. Through the

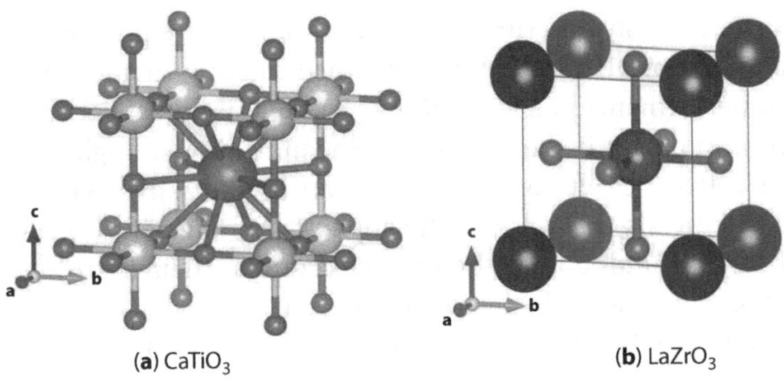

Figure 3.1 Different $CaTiO_3$, $LaZrO_3$ perovskite structures.

incorporation of a secondary stage by means of *in situ* exsolution, there is also a generation of a noteworthy quantity of surface oxygen gaps. Basic crystallographic structures of proveskites are demonstrated in Figure 3.1.

Shao *et al.* have demonstrated that active sites that catalyze synergistic reactions may emerge at the interface between perovskite and metal nanoparticles (NPs). Here, the perovskite oxide aids in the dissociation of water, while the Ni NPs exsolved *in situ* assume a crucial function in adsorbing hydrogen species [14]. Leveraging the *in situ* exsolution technique enables perovskites to serve as versatile electrocatalysts, suitable for both overall water splitting and rechargeable fuel cells.

Bulk-phase perovskite oxides possess high sintering temperature and time owing to their high calcination temperature requirements during phase development. Consequently, perovskite oxides typically possess a lower surface area, which causes them to have deprived mass activity for oxygen electrocatalysis. Additionally, the growth of the perovskites on carbon-based substrates helps prolong the surface application of the perovskite-phase, as well as increase the conductivity. Until now, current researchers on the topics ORR/OER activity of perovskites (or conducting metal oxides) in alkaline media have engaged carbon as a secondary electrode constituent to remove any conductive limits of the perovskite phase.

Generally speaking, the feedback-loop approach can be used to rationally design electrocatalysts. This involves the following steps: (1) carrying out generating numerous designs of experiments (DOEs) and constructing a material repository containing catalytically active elements. (2) The subsequent phase involves harnessing the informational assets within the repository to produce diverse catalysts possessing targeted characteristics through the implementation of varied synergistic approaches and (3) methodically evaluating their electrochemical parameters and correlations and outcomes. The material library receives valuable feedback from these results enabling the completion of the cycle.

Even if previously described approaches have made progress, there are still a lot of serious issues with the obtained catalysts' inadequate electrochemical performance. Electrocatalysts' large-scale practical applicability would be limited, for instance, through the utilization of highly acidic and alkaline electrolytic solutions, as an illustration, and active materials based on precious metals (Ir and Pt). The application range of non-noble electrocatalysts will undoubtedly be restricted due to their narrow pH operating range, high energy loss, substantial overpotential, and poor stability. Figure 3.2 is depicting the design strategy for the electrocatalysts.

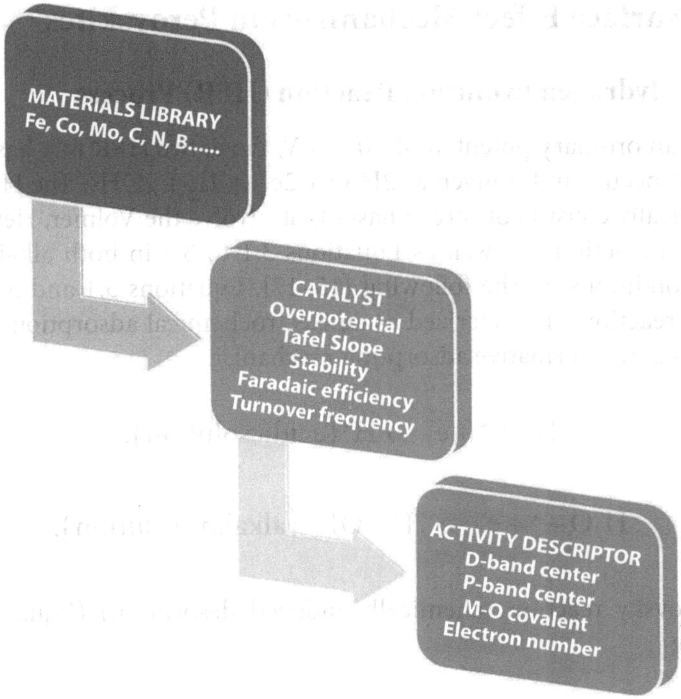

Figure 3.2 Illustration depicting the design strategy for the electrocatalyst. A systematic loop is instituted encompassing the catalyst's creation from the material repository, assessment of electrochemical capabilities, and validation of the activity descriptor. Please take note of the following: M denotes metal, and eg signifies the orbital formed during the cleavage of the d orbital in the process of complex formation.

3.2 Efficient and Stable Electrocatalysts for Water Splitting

By harnessing a sustainable energy reservoir, electrolysis through water splitting has attracted considerable attention as a potentially eco-friendly and energy-conserving method. Enhancing the energy conversion efficiency of hydrogen and oxygen evolution on electrodes, while lowering the requisite energy barrier, demands resilient and enduring electrocatalysts. In recent times, perovskites have emerged as a novel group of effectual electrocatalysts for ORR and OER in alkaline solutions for utilization in metallic batteries and fuel cells. Nevertheless, their HER performance in alkaline solutions were infrequently reported before.

3.3 Surface Effect Mechanisms in Perovskites

3.3.1 Hydrogen Evolution Reaction (HER) Process

Having an ordinary potential of –0.829 V, the entire HER in a basic environment occurs and is given as $2H_2O + 2e^- \rightarrow H_2 + 2OH^-$. The HER process typically consists of three phases that involve the Volmer, Heyrovsky, and Tafel reactions, as well as Equations 3.1 to 3.5 in both alkaline and acidic conditions. In the following [15–17], Equations 3.1 and 3.2 of the Volmer reaction, characterized as an electrochemical adsorption process, undergo a transformative adsorption mechanism.

$$H^+ + * + e^- \rightarrow H^*(\text{acidicsolution}). \tag{3.1}$$

$$H_2O + * + e^- \rightarrow H^* + OH^-(\text{alkaline solution}). \tag{3.2}$$

Heyrovsky reaction (chemically induced desorption) (Equations 3.3 and 3.4):

$$H^* + H^+ + e^- \rightarrow * + H_2(acidic\ solution), \tag{3.3}$$

$$H^* + H_2O + e^- \rightarrow * + H_2 + OH^-(alkaline\ solution), \tag{3.4}$$

Tafel reaction, which is chemically desorbed, is represented by the equation below.

$$2H^* \rightarrow 2 * H_2. \tag{3.5}$$

Heyrovsky and Tafel both mention the Volmer reaction's reaction pathways, which include the desorption of OH^-, the development of the hydrogen molecule following adsorption of water molecules at vacant active sites on a molecular level, coupled with the electrochemical reduction of the adsorbing water molecule to yield H^* and OH^-. As per the perovskite oxide catalyst mechanism, the negatively charged surface oxygen serves as the catalyst's reactive site rather than the metal. Recently, Xu *et al.* [18] showed the effectiveness of the perovskite $Pr_{0.5}(Ba_{0.5}Sr_{0.5})_{0.5}Co_{0.8}Fe_{0.2}O_{3-\delta}$ in the catalysis of HER proposing a mechanism for alkaline HER. According to this mechanism, an adjacent metal site becomes reduced or oxidized (with

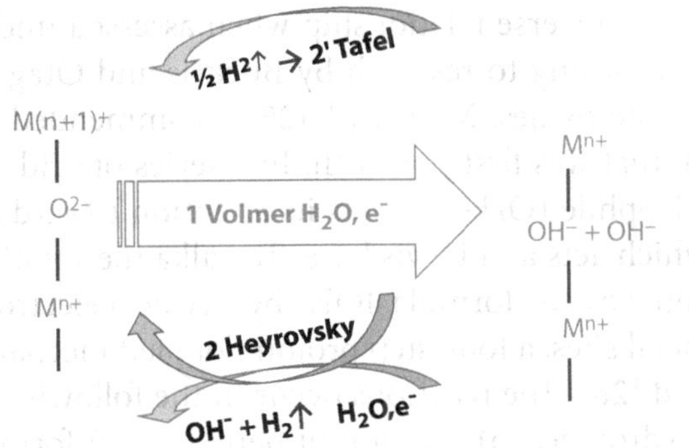

Figure 3.3 The suggested mechanism for the HER occurring at the perovskite oxide surface. The cathodic response associated with HER is depicted in a visual representation.

an oxidation state change) when the electrolyte's H_2O molecule engages with an electron. A chemical interplay ensues at a linked oxygen site giving rise to the formation of an adsorbed H* intermediate. While it is difficult to determine whether the Heyrovsky reaction or the Tafel reaction is used in the second phase, OER/ORR exhibits several reaction intermediates. However, for advanced catalysts (like Pt), zero overpotential results from the single intermediate driving the two-electron transfer HER [19] indicating a goal to improve perovskite catalysts. The binding potential of the H* intermediate, also known as the hydrogen adsorption free energy (ΔGH), strongly influences the rate of HER [20]. An optimal hydrogen-binding HER catalyst should have a ΔGH value that is almost zero, without binding hydrogen too strongly or too weakly [21]. Figure 3.3 is presenting the mechanism for the HER occurring at the perovskite oxide surface.

3.3.2 OER/ORR Reaction Mechanism

In 1977, T. Seung and Jasem [17] proposed the idea that linking transition metals to the redox electrode at potentials below those of the oxygen electrode can lead to the generation of more active sites exhibiting heightened levels of oxidation. Consequently, this approach gave rise to the development of an active catalyst for OER and ORR, along with an exceptional catalyst featuring enhanced electrical conductivity. During his investigation of perovskite-type oxides from 1980 to 1984, Trasatti [22, 23] unearthed a correlation resembling a volcanic pattern between the OER overpotential and the enthalpy associated with the transition from a higher oxidation state to a lower oxidation state. During the interim, the correlation between the enthalpy of $M(OH)_3$ hydroxide production and current density exhibits an inverse relationship when assessed under a consistent overpotential, according to research by Bockris and Otagawa [24] on 18 different perovskite oxides. Man *et al.* [25] recommended a reworking of the mechanism that was first put forth. In a series of acid–base reactions, an oxygen nucleophile (OH–) directs its attention toward a metal-bound electrophile, which acts as a Lewis base. The alkaline solution's ORR reaction is represented by the formula $4OH^-$ by a typical electrode potential of ~0.401 V. At metal sites, a four-step proton-coupled mechanism drives the reaction forward [26]. The reactions occur in the following order: (a) displacement of hydroxide, (b) creation of peroxide, (c) formation of oxide, and (d) regeneration of hydroxide. H_2O, or molecular O_2, and electrons are involved in the first stage. When they react with metal surface sites, oxygenated compounds are indicated by the symbols O*, OH*, OOH*, OO*, and, where * signifies a reactive site on the catalytic surface , they play crucial roles in catalytic reactions such as oxidation, reduction, and water-splitting processes, influencing reaction pathways, intermediates, and overall efficiency. In conjunction with these oxygenated adsorbates, characterized by oxidation states spanning from n+ to n+ 1, the surface metal sites experience redox reactions involving oxidation and reduction. The kinetics of the ORR exhibit notable sensitivity to both binding energy of oxygen and surface metal sites. The shift from OH* to OO* is denoted as the rate-determine step, when influenced by a weak oxygen-binding energy. Conversely, the transformation from O* to OH* in the final step is identified as the rate-limiting step (RLS) in the presence of an exceptionally robust oxygen-binding energy. The Sabatier principle highlights the moderate oxygen-binding energy inherent in an ideal perovskite catalyst [26, 27]. According to Suntivich *et al.* [28], the rate-determining stages are significantly impacted by the eg orbital filling in metal cations. They contended that, should the eg filling be more than 1, the rate would be determined by the creation of an O–O bond in the OOH* adsorbate. On the other hand, RDS can occur when the oxyhydroxide forms a peroxide ion in an amount smaller than 1. The OH^- decay and regeneration process for OER can be compared to a comparable mechanism. OER, or the oxygen evolution reaction, can be expressed as $4OH^- \rightarrow O_2 + 4e^- + 2H_2O$ on a perovskite surface in an alkaline medium. Hydroxide deprotonation (a), peroxide generation (b), peroxide deprotonation (c), and hydroxide creation (d) are the four separate phases that this reaction goes through. The interaction between the electrolyte's OH^- and metal sites initiates a redox reaction on metal surfaces leading to a variation in oxidation condition from n+ to n+1. The sequence of reactive intermediates includes O*, OH*, OO*, and OOH*. Additionally, the kinetics of the OER are influenced by the extent to which oxygen binds to metal sites. The RLS, where binds are very weakly, is represented by the peroxide production (O* to OOH*). On the other hand, oxygen acts as an RDS during the deprotonation stage (OOH* to OO*), when it bonds too firmly [28]. Many of the found perovskites have non-zero overpotentials, and this is because of scaling relations. This results in the surface oxygen adsorbate in OER and ORR having a substantially linked binding strength. In the previously described ORR/OER mechanisms, DFT measurements reveal various forms of reaction intermediates present on the surfaces of perovskite oxides. In the case of ORR, the sequence comprises O2*, O*, OH*, or OOH*, whereas during OER, the sequence is inverted on metal sites (O2*, OOH*, O*, and OH*). Despite these variations, the computed fundamental steps exhibit similarities. This disparity can be ascribed to the distinct surface properties inherent in perovskite oxides versus metal surfaces. The correlation between the adsorption energies of oxygen-binding species participating in the reaction mechanism can be leveraged, as demonstrated by Equations 3.6 to 3.9 [25]. A single descriptor describes the divergence in theoretical overpotential among dual oxides at their surfaces; this descriptor may manifest as either the oxygen adsorption energy (ΔGo) or the relative stability of OH* compared to OOH* (ΔGOH – ΔGOOH) [25] Thus, using Sabatier's principle, a volcano-shaped connection was found, with $LaNiO_3$ and $SrCoO_3$ at the top due to their lowest overpotentials, which are obtained as a result of excellent oxygen binding. Because of their strong oxygen bonds, $LaCoO_3$, $LaFeO_3$, $LaMnO_3$, $LaCrO_3$, and $LaVO_3$ are positioned on the left side of the volcano, whereas $LaCuO_3$ is positioned on the right side because of their weak oxygen bonds.

$$* + \mathrm{H_2O} \rightarrow \mathrm{OH}^* + e^- + \mathrm{H}^+, \tag{3.6}$$

$$\mathrm{OH}^* \rightarrow \mathrm{O}^* + e^- + \mathrm{H}^+, \tag{3.7}$$

$$O^* + H_2O \rightarrow OOH^* + e^- + H^+, \tag{3.8}$$

$$OOH^* \rightarrow O_2 + e^- + H^+. \tag{3.9}$$

The notable catalytic efficacy of $La_{1-x}Sr_xO_3$ and analogous compounds can be ascribed to the existence of lattice oxygen in the OER mechanism. This is in contrast to the previously proposed protocol, as subsequent studies deviate from it. Furthermore, a distinctive OER process (expressed in

Equations 3.10–3.11) was unearthed through a DFT-based investigation. In this process, lattice oxygen plays a crucial role by inducing the formation of surface oxygen vacancies (Vo). A promising avenue involves displacing surface lattice oxygen from the plane presenting an effective means to engage with OH* on $LaNiO_3$ and other precarious perovskites for the generation of Vo and OO*.

$$OH^* \rightarrow (V_\circ + OO^*) + H^+ + e^-, \tag{3.10}$$

$$\left(V_\circ + OO^*\right) + H_2O \rightarrow O_2 + V_\circ + OH^* + H^+ + e^-., \tag{3.11}$$

Henceforth, the categorization of highly active perovskites hinges on the delineation of two distinct OER mechanisms. The lattice oxygen contribution mechanism (LOM) is expounded through Equations 3.10 to 3.13, while Equations 3.6 to 3.9 elucidate the adsorbate evolution mechanism (AEM) [25]. The potential preference between AEM and LOM emerges as a pivotal consideration for specific perovskites. Externally elevating the quantity of reactive sites and enhancing activity is contingent upon the geometric effect. To attain this, innovating novel approaches for synthesizing perovskite nanostructures and their composites proves instrumental. Given the interconnection of these variables, adjusting the composition can directly influence both oxygen vacancy and crystal structure. Notably, instances may arise wherein both extrinsic and intrinsic qualities exhibit simultaneous augmentation [21].

3.4 ORR, OER, and HERS in Perovskite Catalysts

Perovskites gained recognition as a promising material for further exploration in the late 1990s, as researchers redirected their attention toward more cost-effective catalysts to drive reactions. Enhancing the activity of a perovskite electrocatalyst can be achieved through two distinct methods. One approach involves targeting the inherent boost in activity at each reactive site by delving into the physic–chemical attributes of the electronic structures of perovskites. Such behavior qualities encompass diverse crystal morphologies, oxygen vacancies, and compositions resulting from substitutions. Contrary to the presence of a phase transition between Sr and La in LSF-0.2, SF, and LF, the electrochemical characteristics of LSF exhibit noteworthy features. Jin and Gao innovatively crafted an urchin-like

$La_{0.8}Sr_{0.2}MnO_3$ (LSM) perovskite using a co-precipitation technique with a urea precipitator for both OERs and ORR. This method yields a distinctive strontium-doped perovskite structure, where strontium doping significantly enhances its activity. The resulting cathodic current density, specifically 3.7 mA/cm^2 at 0.9 V (against Ag/AgCl), the expansive specific surface area of LSM, and the linear sweep voltammetry (LSV) values for OER collectively exert a considerable influence on the bi-functional catalytic activity. Consequently, the findings indicate its efficacy as a beneficial catalyst for high energy density lithium–air batteries. The "Nucleation Growth Assembly" process produces a structure resembling an urchin. When compared to optimal Pt/C catalyst, urchin-like LSM exhibits a lower initial potential and restrictive current density. However, it boasts twice the diffusion-limiting current density of LSM. The accomplishment of BET surface area explains the enhanced performance compared to LSM. The spherical particles that comprise the LSM powder have thorns arranged perpendicularly on their surface. The diameter of the sphere varied from 3 to 5 m, and the length of the thorns varied from 0.2 to 0.6 m. The linear and parallel graphs for applied potentials of the urchin-like LSM represented the first-order O_2 kinetics and dependence on the subsequent electrocatalyst. By analyzing perovskites of the $La_{1-x}Sr_xMnO_3$ type for even values of X (0–1), Tulloch *et al.* [29] made a similar contribution and found that $La_{0.4}Sr_{0.6}MnO_3$ was the most active toward ORR. It was found that the catalyst was extremely crystalline, highly porous, and had a surface area of just 1 m^2 g^{-1}. It was shown that when manganese gets closer to its valence state, Mn^{4+},Sr^{2+} doping increases. However, there was an inverse relationship between the conductivity and BET surface area at the same time. Because of the direct four-electron transfer, $La_{0.4}Sr_{0.6}MnO_3$ demonstrates the best ORR performance; nevertheless, for water splitting, it does not have dual functionality.

The alteration of components at the B-site/A-site has a profound impact on the electronic configuration of the catalyst, influencing its operational efficiency. The amalgamation of metals from the transition and rare earth (RE) categories alters the overall positive charge influence of the metallic cation situated at the A-sites. Such behavior not only adjusts the oxidation state of transition metal cations but also has the potential to induce an oxygen vacancy, thereby altering their spin state [30]. Numerous investigations have examined the characteristics of transition metals introduced at the B-site in comparison to RE metals introduced at the A-site.

3.4.1 Oxygen-Deficient Perovskites

Their intrinsic HER and OER/ORR activities are largely dependent on the prompt filling or generation of lattice oxygen vacancies that arise from the production of oxygen-deficient oxides. Oxygen shortage concentration strongly influences catalytic activity leading to a non-cubic structural distortion. The $La_{1-x}Sr_xCoO_{3-\delta}$ series' OER activity is strongly correlated with the number of oxygen vacancies; $SrCoO_{2.7}$ exhibits the highest activity and highest vacancy content. To develop the special qualities for ORR, Yuasa *et al.* [31] published work LSFM using the reverse miscella approach to prepare (x = 0–0.4) and (y = 0–0.8). At current densities above 80 mA cm^{-2}, the $La_{0.6}Sr_{0.4}Mn_{0.8}Fe_{0.2}O_{3+\delta}$-loaded electrode showed higher ORR activity than Pt/C; however, the activity dropped as current densities declined. Although performance increased as x increased or decreased as y increased, oxides were not sufficiently prepared for a higher x. In the durability test, the perovskite-type oxide with x = y = 0.2 showed noticeably better stability. Singh *et al.* [32] used the general formula $MMoO_4$ to study this type of perovskite (where M = Ni, Co, or Fe). Fe doping of $MMoO_4$ on-base oxides, with M as Ni and Co, increases the current density for $CoMoO_4$ and 15 times $NiMoO_4$ leading to increased catalytic activity. Furthermore, the orthorhombic crystal structure was found by employing the co-precipitation method to boost the electrocatalytic activity of $Cr_2(MoO_4)_3$ and its Fe-substituted derivatives. Fe doping in $Cr_2(MoO_4)_3$ increases the specific activity by an order of magnitude and decreases the Tafel slope improving the catalytic performance. Creating oxygen flaws results in the formation of several kinds of structures. According to Zhou *et al.* [33] ORR and OER activities are directly impacted by the crystal structure. The investigations were conducted on several $LaNiO_3$ crystalline structures (δ) that were produced through heating at 400°C, 600°C, and 800°C temperatures. Compared to the rhombohedral ones, the cubic structure LN showed a lower Tafel slope and higher mass activity, which may have been caused by the unique Fermi surface of the cubic structure. Numerous electrochemical characteristics pertaining to developed catalysts, such as their charge transfer resistance, current density, and overpotential, influence their overall catalytic activity and efficiency. These properties are critical in optimizing catalyst design for applications in fuel cells, water splitting, and other electrochemical processes. When the rhombohedral to cubic phase changed, the ORR/OER performance improved. Figure 3.4 illustrates how a cubic structure's Ni–O bond length elongation promotes strong OER/ORR performance. The case results showed that the cubic structure with a perfect stoichiometry produced superior outcomes; therefore, further work is needed to enhance the characteristics of oxygen-deficient perovskite examined in conjunction with perovskite that lacks oxygen. The perovskite-type oxides' Mn oxidation state plays a crucial role in water splitting, as it influences the material's catalytic activity, stability, and electron transfer properties. In particular, Mn in perovskites can exhibit mixed valence states (e.g., Mn^{3+}/Mn^{4+}), which facilitate oxygen evolution reaction (OER) kinetics by enabling efficient charge transfer and surface redox processes. Creating oxygen defects is a simple process that can be used to increase the electrocatalytic activity of non-stoichiometric $CaMnO_{3-\delta}$ perovskite. Electrochemical investigations demonstrated that the catalyst's catalytic activity is highly dependent on oxygen shortage. The electrochemical characteristics for OER and ORR determine efficiency in water splitting. Although these oxygen-deficient compounds' structural morphologies resemble those of pure oxides, they outperform pristine oxides in OER and ORR. The highest ORR catalytic activity for the n valency is approximately 3.5. Because of its enhanced EC and oxygen activation, this non-precious catalyst is a viable option for ORRs and OER. The comparison of the produced catalysts' polarization curve findings is shown in Figure 3.4. For the ORR process, the curve lies below 1.1 V, whereas for the OER process, it lies over 1.4 V. Comparing the oxygen-deficient perovskite samples to pure oxides, they demonstrated notable activity. $Ca_2Mn_2O_5$ was found to be an electrocatalyst for OER in an alkaline medium, and the inquiry proceeded with the result that it showed increased catalytic activity, which was in contrast to the previous findings by Zhang *et al.* [34]. OER at 1.70 V (versus RHE) displayed a mass activity of 30.1 A g^{-1} for $Ca_2Mn_2O_5$. The catalyst's heightened efficacy stems from manganese's elevated spin electronic configuration in a crystalline form and the augmented porosity induced by oxygen vacancies. This disparity emerged during catalyst development using the innovative reductive annealing technique. Distinctive attributes of this catalyst, rendering it comparable to precious metals in catalytic performance, arise from the crystal structure facilitating smoother OHM ion movement. By adeptly synthesizing diminutive nanoparticles as an electrocatalyst, the characteristics of oxygen-deficient perovskite will vary further. A barium-based bifunctional electrocatalyst, deficient in oxygen, was investigated. $BaTiO_{3-x}$ nanoparticles, produced through the sol–gel approach, exhibited superior OER performance compared to precious metals manufactured similarly. Owing to more oxygen vacancies in ORR, it displayed larger current densities at low potentials (<1.6 V) and more negative onset potentials than IrO_2 resulting in accelerated adsorption and charge transfer phenomena. Figure 3.4 illustrates the ORR/OER for BTO-950AIR, BTO-1300AR, and BTO-1300VAC. Notably, BTO-1300VAC demonstrated robust activity and stability, particularly in ORR at higher potentials, albeit

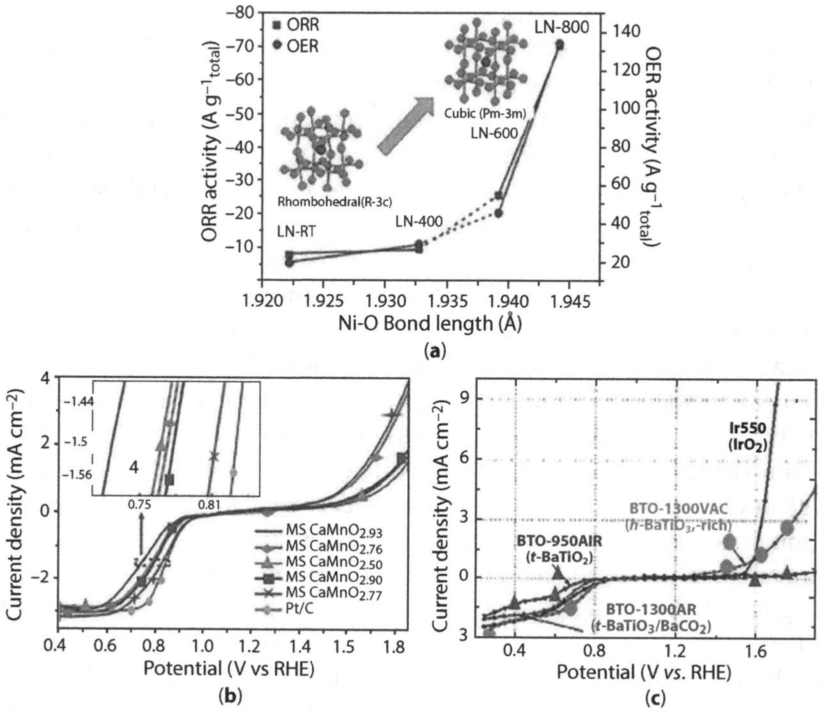

Figure 3.4 (a) The correlation between crystallographic structure and ORR/OER is depicted (adapted with authorization from Reference [33], Copyright © 2013, American Chemical Society). (b) An ORR/OER bi-functional catalyst showcasing its activities (used with permission from Reference [34], Copyright © 2014, American Chemical Society). (c) Steady-state polarization graphs for both OER and ORR are presented, with OER representing the anodic reaction and ORR denoting the oxygen-reduction reaction.

experiencing an early decline in current density during OER. Significant contributions toward enhancing energy conversion and storage capabilities for metal–air and fuel cell batteries stem from the catalyst's ORR activity. Transition metal substitution at the cation site, particularly when coupled with oxygen defects, is known to amplify catalytic activity. Wang *et al.* explored Ni doping in manganite-type perovskites $La_{0.8}Sr_{0.2}Mn_{1-x}Ni_xO_3$ demonstrating increased catalytic activity due to detectable Ni^{3+} and enhanced BET surface area [35]. Utilizing the solid state reaction, aerogel technique, and sol–gel, Christ *et al.* focused on creating Al-doped $Ca_xLa_{1-x}Al_{1-x}Mn_xO_{3-\delta}$ perovskite oxides for increased surface area in ORR. Among the specimens, $Ca_{0.9}La_{0.1}Al_{0.1}Mn_{0.9}O_{3-\delta}$ exhibited optimal electrochemical parameters, with the aerogel method yielding the highest mass activity and maximum surface area. However, structural disintegration at low potential raised concerns. Surface-active sites, where P-block metals substitute, were identified due to oxygen's non-stoichiometric behavior. Unfilled d orbitals in P-block metals demonstrated dual characteristics impacting metallicity and non-metallicity. The non-crystalline electrocatalyst exhibited lower Tafel slope, low onset potential, and low BET surface area showcasing stability and activity in OER. The top–down-generated LNF, with dual Ni and Fe sites, outperformed LN or pure FeOOH, despite lower conductivity, establishing a potential nanostructured water oxidation electrocatalyst through a straightforward top–down approach [36].

3.4.2 Nanostructured Perovskites

The extensive bulk of perovskites exhibits a limited catalytic capability owing to their colossal dimensions and minimal surface area. Alongside showcasing a remarkably pure phase and a molecular diversity that is highly active, refining the catalyst at the nanoscale enhances its structural attributes by incorporating additional active sites. A considerable amount of research has been dedicated to perovskite featuring La as an A-site element. By simply modifying the B-site transition earth metal, diverse catalytic traits have been generated. $LaNiO_3$ nanocrystals, as demonstrated by Hardin *et al.* [37], exhibited remarkably elevated activity levels for OER and ORRs. This heightened performance can be attributed to the integration of a hydroxyl group into the lattice structure, exceptional phase purity, and a substantial BET surface area. Comparing the activity of a leading benchmark catalyst, 6-nm IrO_2, revealed that it was 2.5 times less efficient than $nsLaNiO_3$/NC (nitrogen-doped carbon; 89 mA mg^{-1} of oxide). The catalyst's exceptionally active structure emerges from the hydrolysis of La^{3+} and Ni^{2+} nitrates during the reverse-phase growth precipitation. A highly active bi-functional catalyst was created as a result of the novel concept of using calcination to create pure-phase nanocrystals. By using Co as the B-site element, the structural features of perovskite were further improved. Kim *et al.* proposed the hydrothermal technique, which forms an amorphous surface [38], concentrating on creating porous nanostructures and hollow spheres of $LaCoO_3$.

3.5 Perovskite-Based Electrodes for Water Splitting

Creating cost-effective, immensely efficient, and enduring electrocatalysts for both the OER and HER stands as a pivotal challenge in the pursuit of renewable energy through water splitting. Since oxide-based catalysts have distinct electrical, chemical, and physical properties, they have attracted a lot of interest as water splitting electrocatalysts. The greatest emphasis has been focused on perovskites (ABO_3), pyrochlore-type oxides ($A_2B_2O_7$), and Ruddlesden–Popper perovskite (A_2BO_4). In this study, an exploration into potential forthcoming breakthroughs entails a scientometric examination of the progression of various catalysts based on oxides employed in the process of water splitting. Thus far, a total of 29,761 publications (from 1990 to 2020) have been found using a range of linking keywords when searching for these written papers on this topic on the Web of Science (WoS) website. A scientometric analysis was conducted on the resulting papers examining the top citation bursts, topic categories, contributing authors, publications, and contributing nations. According to the data gathered, publications in this field of study have been steadily rising since 2012. It was discovered that when it comes to the production of research findings concerning catalysts based on oxides for the process of water splitting, China and the United States have taken the forefront globally.

3.6 Density-Functional Theory-Based Calculations for Perovskite Electrocatalysts

Numerical simulation has shown to be quite beneficial in explaining experimental findings and providing a concise overview of interatomic interaction by using the microscopic model. Car and Parrinello [38] presented a method for simulating condensed matter dynamics and carried out computations using density theory. This approach is recognized as the most straightforward for explaining the electrical structure of an atom, solid, or molecule based solely on electron density [39].

$$\mu = \frac{h2}{2m}\left(3\pi^2\right)\frac{2}{3}\rho\frac{2}{3}(r) + V(r) \tag{3.12}$$

Here, μ is the chemical potential, m is the electron mass, V(r) is the common potential, and (r) is the inhomogeneous electron density. The identical spinless fermions in the system utilized in the theorem have a ground state energy that changes with particle density. The functional value is minimized in terms of variation in particle density. The one shortcoming of this theory is eliminated by the confined search theorem, which asserts that there will be no degeneracy at ground state. Derivations for spin density functional theory and non-interacting systems are given by the approach. Equation 3.12 asserts that individual moving particles can be thought of as the motion of interacting electrons and is the most significant equation in DFT [40].

3.6.1 Designing of Bi-Functional Perovskite Electrocatalysts by DFT

The rigorous investigation of the electrocatalytic performance of ORR/OER and invention of extremely successful bi-functional electrocatalysts have both benefited greatly from theoretical investigations. When assessing the essential phases of the OER/ORR process, the DFT analysis has grown increasingly important. The rate of the OER heavily depends on the binding-energy of the oxygen intermediate; optimal bonding exhibits greatest activity. DFT calculations performed by Rossmeisl and colleagues [25] distinguished the oxygen intermediates, which are considered a universal descriptor for OER and ORR on surfaces of perovskite oxides. This was undertaken to gain a deeper comprehension of the process. The identification of active perovskite catalysts based on these activity characteristics poses an experimental challenge. The term (GO*–GOH*), representing the redox potential of the transition metal cation before catalyzing OER, is crucial. Analyzing the electronic structure of diverse perovskites through DFT simulations could elucidate the impact of altering electronic structures on surface stability and OER/ORR activity. The addition of oxygen (+) can tune anti-bonding (eg) electrons; however, this might result in a reduction of anti-bonding (eg) electrons. The presence of oxygen vacancies (–) corresponds to an increase in the number of anti-bonding (eg) electrons. Conversely, the emergence of A-site oxygen vacancies leads to a decrease in antibonding (eg) electrons, thereby elevating the oxidation state of B-site cation. Introducing A-site cations causes anti-bonding (eg) electrons to rise subsequently lowering the oxidation state of B-site cation. Unfortunately, errors in B-site placement are easily made. By utilizing the overall and restricted density of states generated from DFT measurements, one can ascertain the valence band maxima (VBM) and minima conduction band (depicted in Figure 3.5) of related compounds with conductive performances. The stability of OER/ORR catalysts can be determined by assessing the oxygen 2p band center in relation to the Fermi level, a value ascertainable through DFT calculations. For certain Co-based perovskites, it is anticipated that OER activity will rise with the oxygen 2p band center bringing the Fermi level into nearer proximity [41]. On the other hand, metal–oxygen covalency serves as a valuable predictor of ORR activity. Interestingly, for Mn-based perovskites, the O 2p band does not significantly indicate ORR activity. DFT calculations also yield the free energy of specific constituent stages through electric potential and thermodynamic energy corrections [42]. DFT's capacity to calculate material properties is primarily responsible for its widespread use. The Kohn–Sham equation resolution is the typical method used to apply this theory. Both in-band computations and the determination of cohesive properties heavily rely on DFT. It also contributed to the creation of the theory of linear response. Rose *et al.* extended the existing framework to encompass electrons within a lattice of stationary point charges during a phase transition. Although the model has advanced to a stage where computations for both ground and excited states are feasible, there exists a disparity between experimental measurements and calculated outcomes, particularly concerning the nickel band structure. Moreover, it failed to ascertain suitable band gaps for semiconducting and nonconducting materials. By employing density functional theory, an examination of the electronic configurations of $LaMO_3$ and Ce-doped LMO was conducted. The results indicate that eg filling of Mn ions in LSM-CZ is 0.93, approaching 1 more closely than in LSM, implying a favorable interface with oxygenated species. Additionally, a revelation was made that the reduced CeO_{2-x} (110) enhances the oxygen absorbability of CeO_2 showcasing a significantly reactive and crucial storage capacity within the realm of electrochemistry. A similar research was conducted by Lee *et al.* [43] utilizing the DFT-GGA + U technique to forecast the catalytic efficacy of $LaBO_3$ in an alkaline environment. Here, B represents elements, such as Mn, Ni, Co, Fe, and Cr, on the (001) surface concerning OERs and ORR. Due to band bending, the binding affinity of the unmodified (001) AO surface toward oxidizing species surpasses that of the unmodified BO_2 surface.

Generated through charge doping on opposing surfaces, the resulting effect was scrutinized. Further exploration involved delving into the HO* adsorption energy of the unadorned AO and BO_2 surfaces revealing a linear correlation with the O 2p band centers of a subsurface layer. The stability assessment of the $LaBO_3$ structure, concerning applied voltage vs. RHE, hinged on the adsorbate coverage on the exposed surfaces of AO and BO_2. This stability was placed in tandem with the experimental characterization process demonstrating comparable outcomes. Figure 3.5 displays the free energies of unadorned surfaces (001), AO, and BO_2 accompanied by surfaces adorned with consistent adsorbate coverage.

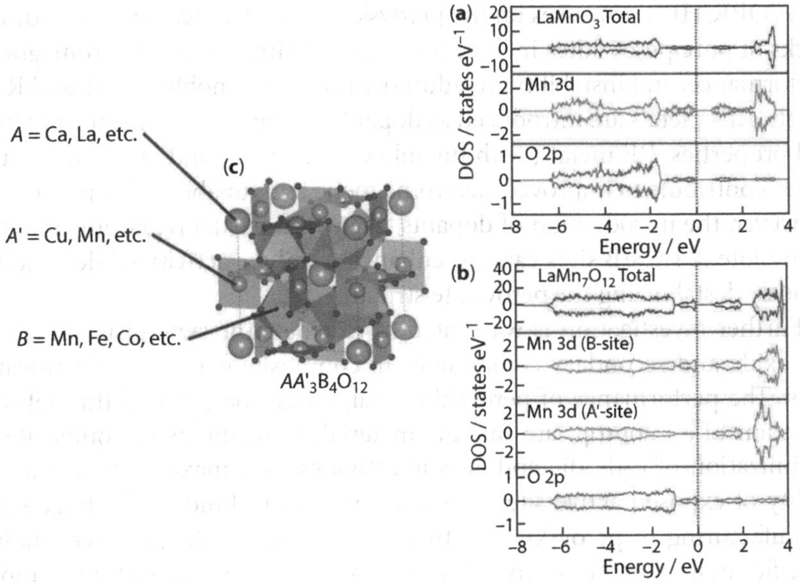

Figure 3.5 Complete and fractional density of states for (a) $LaMnO_3$ and (b) $LaMn_7O_{12}$ ascertained through DFT computations. Moreover, the perovskites C, Quadruple $AA_30B_4O_{12}$, are presented in accordance with the authorization granted in Reference [44], Copyright © 2016 WILEY-VCH Verlag GmbH & Co. KGaA, Weinheim. The principles of density functional theory (DFT) were employed in these calculations. Refer to the colorful figure available at wileyonlinelibrary.com for visual representation.

3.7 Challenges for Water Splitting in Bi-Functional Perovskite Catalysts

While significant progress has been made in the exploration of design principles aimed at enhancing catalytic activity, there has been a notable lack of emphasis on stability, a critical factor for applications involving devices. A considerable number of perovskites undergo amorphization during the OER resulting in the leakage of A-site/B-site cations into the electrolyte. The highly active surface and the bulk exhibit similar amorphization patterns [45, 46] resulting the structural transition from corner sharing octahedra to edge sharing octahedra. It has been observed that no catalyst, including those based on iron and nickel, exhibits long-lasting stability. Despite extensive research on the HER/ORR/OER mechanisms, the intricate process of optimizing perovskite catalysts persists. To gain a better understanding of the surface electrochemistry and electronic structure of perovskite oxides, a well-defined perovskite model is essential facilitating comprehension of the

OER/ORR/HER processes on the perovskite oxide surface. The main drawbacks of pure perovskites include low conductivity, limited electromagnetic performance, and instability. To address these issues, noble metals and Rare Earth (RE) metals are introduced as dopants to enhance chemical and physical properties. RE metals, with tunable valence states and electronic structures, contribute to improved electromagnetic and tunable redox properties. However, the introduction of dopants, even with partial replacement at the perovskite A-site's/B site's cations, enhances catalytic activity while simultaneously destabilizing the perovskite structure.

Further investigation is warranted to elucidate the synergistic effects of the oxide and secondary components in composite perovskite electrocatalysts. The performance of perovskite catalysts can be elevated through the selection of an appropriate support material, adjustments in composition, optimization of cathodic and anodic activities, and maximizing the availability of exposed active sites. The conventional techniques for large-scale manufacturing of perovskite electrocatalysts pose challenges necessitating specific equipment. These investigations are crucial for practical applications aiming to mitigate structural changes in perovskite oxides during catalysis.

3.8 Future Outlook

The widespread application of perovskite catalysts as electrocatalysts faces a significant hurdle in terms of their operational stability. Operational stability poses a notable challenge for perovskite catalysts, as highlighted in current literature. Stability tests primarily employ short-term electrochemical testing revealing swift and pronounced changes in electrode surfaces within the initial hours of operation. To comprehend the performance decline of perovskite electrocatalysts, long-term electrochemical stability studies should be devised aiming to develop a stable and economically feasible catalyst. It is worth noting that the parameters employed for stability tests in research institutions diverge from the industry's requirements. Genuine electrolyzers are capable of operating at various voltages, and investigating catalyst stability under different voltage conditions can offer insights into the prolonged operability of perovskite electrocatalysts. Therefore, a more comprehensive, application-oriented research approach is necessary involving extended testing hours and consideration of industrially relevant operating conditions. This includes exploring a current density range of 0.5 to 2.0 A/cm^2 and elevated temperatures (80°C). The amalgamation of fundamental and practical research contributions is anticipated to expedite the availability of an operational water electrolyzer utilizing perovskite electrocatalysts.

3.9 Conclusion

This study extensively investigates the utilization of perovskite oxides as an electrocatalyst for ORR/HER/OER. The primary focus of this chapter was to identify cost-effective, non-precious metal catalysts with robust catalytic performance and stability. In response to this imperative objective, perovskite oxides have emerged as highly advantageous due to their effective catalytic capabilities. The considerable interest in these oxides stems from their adaptable properties derived from a flexible crystal structure allowing for the substitution of oxygen defects and cations at the A- and B-sites. The results of the investigation highlight the pivotal role of B-site elements in oxygen electrocatalysis. Catalytic activity is notably influenced by substitutions at the A-site and the careful selection of A-site elements. Additionally, the introduction of a modest quantity of noble metal significantly enhances catalytic activity, while the combination of two distinct elements at the B-site produces a synergistic effect. Although considerable research has been dedicated to perovskites as electrocatalysts for water splitting, further exploration is required, particularly concerning the hydrogen evolution reaction. Achieving enhanced activity for the bifunctionality of perovskites proves challenging owing to the distinct properties of HER and OER at the perovskite surface. Nevertheless, it is plausible that diligent research efforts addressing this challenge could effectively realize the intended functions of perovskite-based bifunctional catalysts. A more profound understanding of the OER/ORR/HER mechanisms on the surfaces of perovskites is imperative to advance this field of study. Thorough validation through well-conducted research is necessary for the mechanisms outlined in the literature. Conversely, perovskite catalyst design tactics must be optimized. It may be possible to build active perovskite HER catalysts fast by integrating insights from reaction processes and optimized catalyst design. The reliance on very expensive and cutting-edge characterization techniques hinders further investigation of mechanisms. As a result, the creation of novel materials may result in the formation of new structures and, hence, new functionalities, some of which have been theoretically anticipated using DFT analysis. As a result of the perovskite catalyst's decreased HER activity, there might also be a chance to look into the usage of hybrid catalysts—that is, different catalysts for HER and OER. A more practical solution would be to use perovskite OER with Pt HER catalysts to produce a water electrolyzer.

AI disclosure statement: The author acknowledges the use of AI-based tools for minor language editing or grammatical correction. However, all intellectual contributions, analyses, and conclusions presented in this chapter are entirely the author's own.

References

1. Cheng, F. and Chen, J., Metal–air batteries: from oxygen reduction electrochemistry to cathode catalysts. *Chem. Soc. Rev.*, 41, 6, 2172–2192, 2012.
2. Dusastre, V. (Ed.), *Materials for sustainable energy: a collection of peer-reviewed research and review articles from Nature Publishing Group*, World Scientific, Nature Publishing Group, UK, 2010.
3. Jabeen, N., Hussain, A., Ali, J., Transition Metal Oxides with Broaden Potential Window for High-Performance Supercapacitors, in: *Electroceramics for High Performance Supercapacitors*, pp. 203–231, 2023.
4. Mustafa, Z., Jabeen, N., Hassan, N.U., Jelani, M., Muhammad, A.Q., Sehar, F., Bajaber, M.A., Realization of room temperature exchange bias effect in Co/NiO bilayer via all-solid-state Li-ion redox capacitor. *Ceram. Int.*, *49*, 2, 2115–2120, 2023.
5. Hu, C. and Dai, L., Multifunctional carbon-based metal-free electrocatalysts for simultaneous oxygen reduction, oxygen evolution, and hydrogen evolution. *Adv. Mater.*, 29, 9, 1604942, 2017.
6. Jabeen, N., Hussain, A., Ali, J., Metal Phosphates/Phosphonates for Supercapacitor Applications, in: *Metal Phosphates and Phosphonates: Fundamental to Advanced Emerging Applications*, pp. 245–266, Springer International Publishing, Cham, 2023.
7. Chen, D., Chen, C., Baiyee, Z.M., Shao, Z., Ciucci, F., Nonstoichiometric oxides as low-cost and highly-efficient oxygen reduction/evolution catalysts for low-temperature electrochemical devices. *Chem. Rev.*, 115, 18, 9869–9921, 2015.
8. Hussain, A., Jabeen, N., Hassan, N.U., Hussain, W., Qaiser, M.A., Khan, S., Javed, M.S., High-entropic relaxor ferroelectric perovskites ceramics with A-site modification for energy storage applications. *Ceram. Int.*, *49*, 23, 39419–39427, 2023.
9. Hua, B., Li, M., Sun, Y.-F., Zhang, Y.-Q., Yan, N., Chen, J., Thundat, T., Li, J., Luo, J.-L., A coupling for success: controlled growth of Co/CoOx nanoshoots on perovskite mesoporous nanofibres as high-performance trifunctional electrocatalysts in alkaline condition. *Nano Energy*, 32, 247–254, 2017.
10. Hua, B., Li, M., Luo, J.-L., A facile surface chemistry approach to bifunctional excellence for perovskite electrocatalysis. *Nano Energy*, 49, 117–125, 2018.
11. Sun, Y.-F., Yan-Ling, Y., Chen, J., Li, M., Zhang, Y.-Q., Li, J.-H., Hua, B., Luo, J.-L., Toward a rational photocatalyst design: a new formation strategy of

co-catalyst/semiconductor heterostructures via *in situ* exsolution. *Chem. Commun.*, 54, 12, 1505–1508, 2018.
12. Guo, Y., Yuan, P., Zhang, J., Xia, H., Cheng, F., Zhou, M., Li, J., Qiao, Y., Mu, S., Xu, Q., Co2P–CoN double active centers confined in N-doped carbon nanotube: heterostructural engineering for trifunctional catalysis toward HER, ORR, OER, and Zn–air batteries driven water splitting. *Adv. Funct. Mater.*, 28, 51, 1805641, 2018.
13. Xu, X., Su, C., Shao, Z., Fundamental understanding and application of $Ba_{0.5}Sr_{0.5}Co_{0.8}Fe_{0.2}O_{3-\delta}$ perovskite in energy storage and conversion: past, present, and future. *Energy Fuels*, 35, 17, 13585–13609, 2021.
14. Zhu, Y., Dai, J., Zhou, W., Zhong, Y., Wang, H., Shao, Z., Synergistically enhanced hydrogen evolution electrocatalysis by *in situ* exsolution of metallic nanoparticles on perovskites. *J. Mater. Chem. A*, 6, 28, 13582–13587, 2018.
15. Lasia, A., Hydrogen evolution reaction. *Handb. Fuel Cells*, 2, 416–4405, 2010.
16. Horiuti, J., Matsuda, A., Enyo, M., Kita, H., The mechanism of the hydrogen evolution reaction, in: *Electrochemistry*, pp. 750–779, Pergamon, Proceedings of the First Australian Conference on Held in Sydney, and Hobart, 1965. https://doi.org/10.1016/B978-1-4831-9831-6.50063-1.
17. Bockris, J.O.M., Ammar, I.A., Huq, A.K.M.S., The mechanism of the hydrogen evolution reaction on platinum, silver and tungsten surfaces in acid solutions. *J. Phys. Chem.*, 61, 7, 879–886, 1957.
18. Xu, X., Chen, Y., Zhou, W., Zhu, Z., Su, C., Liu, M., Shao, Z., A perovskite electrocatalyst for efficient hydrogen evolution reaction. *Adv. Mater.no.*, 28, 30, 6442–6448, 2016.
19. Benck, J.D., Hellstern, T.R., Kibsgaard, J., Chakthranont, P., Jaramillo, T.F., Catalyzing the hydrogen evolution reaction (HER) with molybdenum sulfide nanomaterials. *ACS Catal.*, 4, 11, 3957–3971, 2014.
20. Parsons, R., The rate of electrolytic hydrogen evolution and the heat of adsorption of hydrogen. *Trans. Faraday Soc.*, 54, 1053–1063, 1958.
21. Arul, N.S. and Nithya, V.D., *Revolution of Perovskite*, Springer, Singapore, 2020.
22. Trasatti, S., Electrocatalysis by oxides—attempt at a unifying approach. *J. Electroanal. Chem. Interfacial Electrochem.*, 111, 1, 125–131, 1980.
23. Trasatti, S.J.E.A., Electrocatalysis in the anodic evolution of oxygen and chlorine. *Electrochim. Acta*, 29, 11, 1503–1512, 1984.
24. Bockris, J.O'M. and Otagawa, T., The electrocatalysis of oxygen evolution on perovskites. *J. Electrochem. Soc.*, 131, 2, 290, 1984.
25. Man, I.C., Su, H.-Y., Calle-Vallejo, F., Hansen, H.A., Martínez, J., II, Inoglu, N.G., Kitchin, J., Jaramillo, T.F., Nørskov, J.K., Rossmeisl, J., Universality in oxygen evolution electrocatalysis on oxide surfaces. *Chem. Cat. Chem.*, 3, 7, 1159–1165, 2011.
26. Kuznetsov, D.A., Han, B., Yu, Y., Rao, R.R., Hwang, J., Román-Leshkov, Y., Shao-Horn, Y., Tuning redox transitions via inductive effect in metal oxides

and complexes, and implications in oxygen electrocatalysis. *Joule*, 2, 2, 225–244, 2018.
27. Suntivich, J., Gasteiger, H.A., Yabuuchi, N., Nakanishi, H., Goodenough, J.B., Shao-Horn, Y., Design principles for oxygen-reduction activity on perovskite oxide catalysts for fuel cells and metal–air batteries. *Nat. Chem.*, 3, 7, 546–550, 2011.
28. Suntivich, J., May, K.J., Gasteiger, H.A., Goodenough, J.B., Shao-Horn, Y., A perovskite oxide optimized for oxygen evolution catalysis from molecular orbital principles. *Science*, 334, 6061, 1383–1385, 2011.
29. Tulloch, J. and Donne, S.W., Activity of perovskite La1– xSrxMnO3 catalysts towards oxygen reduction in alkaline electrolytes. *J. Power Sources*, 188, 2, 359–366, 2009.
30. Cho, J., Zarnowiecki, D., Bogomolova, S., Wilson, A.L., Villani, A., O'Dea, K., Parletta, N., Consideration of nutritional value and food labels are associated with food intake in adults with depression. *J. Nutr. Intermediary Metab.*, 4, 10–11, 2016.
31. Yuasa, M., Sakai, G., Shimanoe, K., Teraoka, Y., Yamazoe, N., Reverse Micelle-Based Preparation of Carbon-Supported La1– x Sr x Mn1– y Fe y O 3+ δ for Oxygen Reduction Electrode. *J. Electrochem. Soc.*, 151, 10, A1690, 2004.
32. Singh, R.N., Singh, J.P., Singh, A., Electrocatalytic properties of new spinel-type MMoO4 (M= Fe, Co and Ni) electrodes for oxygen evolution in alkaline solutions. *Int. J. Hydrogen Energy*, 33, 16, 4260–4264, 2008.
33. Zhou, W. and Sunarso, J., Enhancing bi-functional electrocatalytic activity of perovskite by temperature shock: A case study of LaNiO3– δ. *J. Phys. Chem. Lett.*, 4, 17, 2982–2988, 2013.
34. Du, J., Zhang, T., Cheng, F., Chu, W., Wu, Z., Chen, J., Nonstoichiometric perovskite CaMnO3– δ for oxygen electrocatalysis with high activity. *Inorg. Chem.*, 53, 17, 9106–9114, 2014.
35. Li, W., Yu, H., Zhang, Z., Hei, W., Liang, K., Yu, H., Electrochemical removal of NOx by La0.8Sr0.2Mn1–xNixO3 electrodes in solid electrolyte cells: role of Ni substitution. *J. Hazard. Mater., 420*, 126640, 2021.
36. Christ, J.M., Ngo, C., Batson, T., Cadigan, C.A., Tong, J., Richards, R.M., O'Hayre, R., Pylypenko, S., Synthesis of high surface area Ca x La (1– x) Al (1– x) Mn x O (3– δ) perovskite oxides for oxygen reduction electrocatalysis in alkaline media. *Catal. Sci. Technol.*, 6, 21, 7744–7751, 2016.
37. Hardin, W.G., Slanac, D.A., Wang, X., Dai, S., Johnston, K.P., Stevenson, K.J., Highly active, nonprecious metal perovskite electrocatalysts for bifunctional metal–air battery electrodes. *J. Phys. Chem. Lett.*, 4, 8, 1254–1259, 2013.
38. Kim, J., Chen, X., Shih, P.-C., Yang, H., Porous perovskite-type lanthanum cobaltite as electrocatalysts toward oxygen evolution reaction. *ACS Sustain. Chem. Eng.*, 5, 11, 10910–10917, 2017.
38. Car, R. and Parrinello, M., Unified approach for molecular dynamics and density-functional theory. *Phys. Rev. Lett.*, 55, 22, 2471, 1985.

39. Callaway, J. and March, N.H., Density functional methods: theory and applications. *Solid State Phys.*, 38, 135–2215, 1984.
40. Finocchi, F., Density Functional Theory for Beginners: Basic Principles and Practical Approaches, in: *Institut des NanoSciences de Paris (INSP) CNRS and University Pierre et Marie Curie*, Paris, 2011.
41. Grimaud, A., May, K.J., Carlton, C.E., Lee, Y.-L., Risch, M., Hong, W.T., Zhou, J., Shao-Horn, Y., Double perovskites as a family of highly active catalysts for oxygen evolution in alkaline solution. *Nat. Commun.*, 4, 1, 2439, 2013.
42. Zhao, B., Zhang, L., Zhen, D., Yoo, S., Ding, Y., Chen, D., Chen, Y., *et al.*, A tailored double perovskite nanofiber catalyst enables ultrafast oxygen evolution. *Nat. Commun.*, 8, 1, 14586, 2017.
43. Lee, Y.-L., Gadre, M.J., Shao-Horn, Y., Morgan, D., Ab initio GGA+ U study of oxygen evolution and oxygen reduction electrocatalysis on the (001) surfaces of lanthanum transition metal perovskites LaBO 3 (B= Cr, Mn, Fe, Co and Ni). *PCCP*, 17, 33, 21643–21663, 2015.
44. Yamada, I., Fujii, H., Takamatsu, A., Ikeno, H., Wada, K., Tsukasaki, H., Kawaguchi, S., Mori, S., Yagi, S., Bifunctional oxygen reaction catalysis of quadruple manganese perovskites. *Adv. Mater.*, 29, 4, 1603004, 2017.
45. Bu, Y., Kim, S., Kwon, O., Zhong, Q., Kim, G., A composite catalyst based on perovskites for overall water splitting in alkaline conditions. *Chem. Electro. Chem.*, 6, 5, 1520–1524, 2019.
46. May, K.J., Carlton, C.E., Stoerzinger, K.A., Risch, M., Suntivich, J., Lee, Y.-L., Grimaud, A., Shao-Horn, Y., Influence of oxygen evolution during water oxidation on the surface of perovskite oxide catalysts. *J. Phys. Chem. Lett.*, 3, 22, 3264–3270, 2012.

4

Design and Engineering of Photoelectrodes

Katarzyna Grochowska, Ameer Nasih, Saiful Islam Khan and Katarzyna Siuzdak*

Centre of Plasma and Laser Engineering, Institute of Fluid-Flow Machinery, Polish Academy of Sciences, Gdańsk, Poland

Abstract
Water is one of the simplest but also one of the most important molecules for all living organisms. This molecule contains only two hydrogen atoms and one oxygen atom that are connected by covalent bonds. On a biological level, water acts as a solvent, carrying nutrients, hormones, antibodies, and oxygen to cells, as well as metabolic waste to the excretory organs. Water is also critical for feeding, breeding, and growing of plants and animals. In the whole ecosystem, water is part of a cycle going through the atmosphere, soil, rivers, lakes, and oceans. At the same time, this molecule is required by most methods of producing hydrogen. It should be noted that hydrogen plays an important role as a versatile energy carrier. This chapter delves into the electrode materials that can be used in solar-driven electrochemical water splitting that can be considered as a ground-breaking field in the green production of hydrogen. Proper selection and engineering of photoelectrodes are critical aspects in designing efficient and stable photoelectrochemical systems.

Keywords: Hydrogen, water splitting, hydrogen evolution reaction, nanomaterials, electrode materials

4.1 Introduction

Nowadays, the world faces escalating climate changes such as rising temperatures, sea level rise, melting ice and glaciers, harsh weather events, changes in precipitation patterns, ocean acidification, shifts in ecosystems,

**Corresponding author*: ksiuzdak@imp.gda.pl

Inamuddin, Tariq Altalhi, Mohammad Luqman and Jorddy Neves Cruz (eds.) Water Splitting: Production of Hydrogen, (75–110) © 2025 Scrivener Publishing LLC

and biodiversity as well as disruption of agriculture. While some natural factors can influence climate, those rapid changes are mainly attributed to human activities, first, through the release of greenhouse gases into the atmosphere. Among the different human activities contributing to climate change, burning of fossil fuels represents one of the most dominant. The combustion of fossil fuels for production of energy is a major source of CO_2, which is the primary greenhouse gas responsible for global warming. Moreover, the human population continues to grow, so the demand for energy also increases. Therefore, the urgency to reduce carbon emissions and transition away from fossil fuels are of key importance, and the shift to renewable energy stands as a critical step in overcoming these issues and in diversification of energy sources.

Renewable energy can be utilized from natural resources, such as sunlight, wind, or water, and provides an ecofriendly alternative to conventional fossil fuels. Solar energy harnesses the power of the Sun through photovoltaic solar cells or solar thermal systems. Wind energy is converted into electrical one by wind turbines. Hydrothermal and geothermal energy can be transformed via hydroelectric power plants and geothermal power plants, respectively. However, one cannot forget also about biomass that can be converted to biofuels and tidal and wave energies, which capture the energy generated by the rise and fall of tides and from ocean waves, respectively. Although hydrogen is not a direct energy source, it can be produced using renewable energy and later used as a clean fuel in fuel cells as only water and heat are byproducts. Therefore, hydrogen can be considered as a secondary energy source or energy carrier.

Nonetheless, the impact of hydrogen production on the environment depends on the fabrication method. One of the most common methods for industrial hydrogen production is steam methane reforming (SMR). In this process, methane reacts with steam at high temperatures to produce hydrogen and CO. Among the advantages of this technique are its high efficiency, versatility, cost effectiveness, and the fact that it is a well-established and mature technology. However, one of the major drawback is the release of carbon dioxide as a byproduct. Other disadvantages are relying on fossil fuels as a feedstock, high energy consumption, catalyst deactivation, and problems with storage and transportation.

Another process of hydrogen production is partial oxidation of hydrocarbons. It differs from SMR by using a partial supply of oxygen instead of steam and involves combustion of hydrocarbons in a limited amount of oxygen or air. As in the case of SMR, this process is also effective in terms of hydrogen production. Moreover, as it is an exothermic process, it can be beneficial with respect to thermal efficiency. Other advantages are high reaction rates, flexibility in feedstock, as a variety of hydrocarbons can be used, and the production of syngas, which is used in the synthesis of many chemicals. Nevertheless, this method also contributes to the generation of carbon dioxide. Also, a mixture of hydrogen and carbon monoxide is produced, which needs further processing depending on the application. Other negative features include high operating temperature and formation of soot and tar, which contributes negatively to the maintenance of equipment.

Hydrogen can be also produced via coal or biomass gasification. When it comes to coal gasification, it relies on abundant, versatile feedstock, and syngas is produced during this process. It can also be integrated into combined heat and power systems that allow the production of not only hydrogen but also electricity. However, it contributes not only to emission of carbon dioxide, but also to sulfur dioxide, nitrogen oxides, and particles polluting the air. It also requires significant amount of water, e.g., for cooling, and generates byproducts such as ash and tars. In contrast to coal gasification, biomass gasification relies on renewable resource. Due to the usage of organic materials, the emission of greenhouse gases is reduced, and it provides a way to utilize organic waste. Nonetheless, this process can be energy intensive, requires sophisticated and expensive solutions, and generates ash. Moreover, the initial costs may be too high to adopt this process on a global scale.

Almost all the abovementioned processes of hydrogen production have one major drawback in common, namely, carbon oxide as a byproduct. The only exception is biomass gasification, though in this case, the byproducts strongly depend on the ongoing biological processes. When it comes to clean and green methods, one can distinguish electrolysis, photoelectrochemical (PEC) water splitting, and thermochemical water splitting.

In the process of electrolysis, water is being split to hydrogen and oxygen by using an electric current. In case of photoelectrochemical water splitting, sunlight is utilized to generate electric current required for water splitting reaction, while thermochemical water splitting requires high temperatures to break water molecules and can rely on concentrated solar power or other heat sources. When electricity and heat are obtained from renewable sources, such as solar energy, not only photoelectrochemical water splitting but also electrolysis and thermochemical water splitting can be considered as environmentally friendly. During this processes, no direct emission of greenhouse gases occurs and produced hydrogen is of high purity (though in the case of the thermochemical method, the separation of hydrogen from other reaction products may be needed if different feedstock other than water is used). Also production systems can be deployed at various scales that leads to decentralized hydrogen generation. Of course, although all those methods are not negatively oriented toward the natural environment, they all face similar challenges. The most crucial ones are related to energy efficiency, initial costs, infrastructure requirements, widespread adoption costs and, in some cases, also stability issues. However, ongoing research and technological advancements aim to address and overcome these challenges and improve the viability of these methods as practical methods for hydrogen production.

Nonetheless, taking into account the current state of technology as well as research and development potential, this chapter focuses on photoelectrochemical water splitting. Following that development of solar technologies is strongly related with development of the new materials and their modifications enabling efficient conversion of light to energy consumed for hydrogen generation, this chapter especially delves into the performance of photoelectrodes.

It should be underlined that semiconductors are preferred in photoelectrochemical systems as their band gap allows for absorption of photons and electron–hole pair generation when the material is illuminated by light. Moreover, due to their intrinsic band structure, they are able to efficiently separate photoexcited charge carriers for driving desired chemical reactions. Additionally, many semiconductors are stable in a wide range of environmental conditions, especially in aqueous electrolytes that are typically used in photoelectrochemical systems.

Historically, one of the first semiconductors that exhibit photoconductivity, understood as a change in resistance when the material is exposed to light, was selenium. In 1873, W. Smith [1] made this discovery, and 10 years later, selenium was demonstrated as the very first photovoltaic material with a power-conversion efficiency below 1% [2]. Nonetheless, the Se band gap is too large, and thus, not much efforts were put to develop such photoelectrodes. Therefore, other semiconductors were explored, including cadmium sulfide and zinc oxide. However, it was titanium dioxide that gained significant attention in photoelectrochemical systems, and due to this, in this chapter, special focus will also be put on this semiconductor.

4.2 Modifications of Semiconductors—A Short Overview

As photoelectrodes play a crucial role in chemical reactions driven by light illumination, the design and engineering of photoelectrodes are also critical. Therefore, to obtain the best photoactive material, one should consider the material selection, surface morphology, interface, device architecture, optimization based on experimental and theoretical data, as well as durability and stability.

When it comes to the selection of material, commonly used materials include silicon, titanium dioxide, zinc oxide, and many others whose performances will be shown further in this chapter. Nonetheless, it should be kept in mind that the band gap of the semiconductor determines the wavelength of absorbed light. As natural solar light is preferable in real applications, the semiconductor should possess the band gap that corresponds to solar spectrum absorption, especially in the visible region. Moreover, the photoactive material should be chemically stable in the operating environment.

Another important matter that should be taken into consideration is the development of the surface area, for example, through nanostructuring, which will influence light absorption as well as charge transport properties. Surface modifications can also affect the ability of the material to trap light and therefore increase light absorption.

The charge transfer efficiency at the semiconductor/electrolyte interface can be also modified by decorating the surface with cocatalysts, e.g., metal or metal oxide nanoparticles. Such photoelectrode engineering may facilitate water oxidation or influence on photocurrent density enhancement, which will be presented later on. Moreover, such functionalization may also lead to an improvement in material stability and, as in the case of nanostructurization, an improvement in light absorption properties. Additionally, it can create active sites for specific catalytic reactions improving overall photoelectrochemical performance.

Charge carrier transfer, charge separation, conductivity, and optical properties of semiconductors can also be changed by the introduction of dopants, i.e., introduction of impurities into the semiconductor lattice. Such a modification changes the material's electronic properties and can change the band gap of the semiconductor. Moreover, decoration and doping can be combined to achieve enhanced electrode performance.

Going further, interfaces between different materials or phases are of key importance in efficiently separating photogenerated charges. Additionally, engineering of the energy band alignment at interfaces influences not only on the charge transfer but also can prevent recombination. To reduce surface recombination and improve charge collection efficiency, one can also consider the application of passivation layers onto the electrode material. The passivation layer can also protect photoelectrodes that are exposed to harsh environments from corrosion over time.

As it comes to whole device architecture, one can use tandem or multilayer composition to enhance overall efficiency and capture light from a broader spectrum. The electrode configuration may also play an important role in charge collection. Depending on the device architecture, photoelectrodes can be connected in series, parallel, or in combination of both. A parallel connection increases the current output, while a series connection increases the voltage output. Therefore, the arrangement of layers as well as interfaces should be carefully designed to minimize losses and maximize overall device efficiency.

In the case when the water splitting process is performed with the assistance of a nanomaterial, it is hard to predict which synthesis or modification method will be the most efficient one. When it comes to the nanoscale properties of a material, it sometimes depends on a very small amount of dopant, formation of some vacancies, or change even in one dimension or shape from a sphere-like particle to a rod. Also changing the concentration of the surface groups can bring totally new properties of the material and, in consequence, affecting its activity under irradiation and its ability to split water. Due to that, we can find in the literature so many approaches that, at first glance, seem to be very similar, but differ in one stage, e.g., calcination in other conditions or the shape of the photocatalyst (sphere particles, nanorods, or nanotubes).

Realizing how many different strategies have been undertaken to reach efficient photocatalytic water splitting, there are multiple challenging issues that should be considered. For example, when the surface area of the photoelectrode increases, also the amount of water molecules that can be adsorbed on this surface increases. Therefore, for more developed surface area, one can expect enhancement of the reaction of light-excited electron–hole pairs on the substrate. However, at the same time, the increase in surface area leads to formation of more defect centers in the photocatalytic material where the already separated charges can accumulate and undergo recombination. Following that, the performance of the photoelectrocatalytic material can be improved by a suitable surface area, configuration of active sites, and crystallinity. To increase the efficiency of hydrogen production, the active sites or the cocatalyst support should be easily accessible for water molecules facilitating its adsorption and then desorption of the reaction products.

In summary, the following features should be met by the materials on which photoelectrochemical process takes place:

a) Absorption of light in the broad range of wavelengths, mostly in the visible range.
b) Long-term stability of the material during its immersion in the electrolyte and under irradiation.
c) Band gap energy higher than 1.7 eV to cope with overpotential loss and the energy required during water splitting processes.
d) The appropriate alignment of the band edge position to catalyze the oxidation and/or reduction of water. Appropriate alignment means that the conduction band should be placed at the more negative energy level than the H_2 generation energy level, while the valence band should be more positive compared to the oxidation potential of water in the semiconductor enabling O_2 release.
e) Efficient separation and then undisruptive transport of charges to make the whole reaction fast.
f) Low charge transfer resistance present at the electrolyte/electrode interface.
g) Material should be easily accessible; synthesis should also be simple providing low environmental impact to make the material attractive for commercial production.

When it comes to the modification of semiconductors that exhibit high stability, the first approaches were based on the noble metals. Such selection is due to their efficient activity toward the hydrogen evolution reaction (HER), as these materials have very low ΔG_{H^*} (see Figure 4.1), being close to zero [3], indicating the most optimized hydrogen adsorption free energy. In Figure 4.1 one can find the hydrogen evolution reaction exchange currents and metal–hydrogen bond strengths of different metals. As can be observed, the Pt-group metals exhibit the most promising catalytic activity toward HER [4].

However, to limit the usage of noble elements, they should act as dopants or just decorative nanoparticles reducing the overall amount of the precious metal. On the other hand, it is not so simple just to reduce the size and amount of the noble metal, but one has to find such interaction between the surface atom and reactive molecule that will make the whole material highly photoactive.

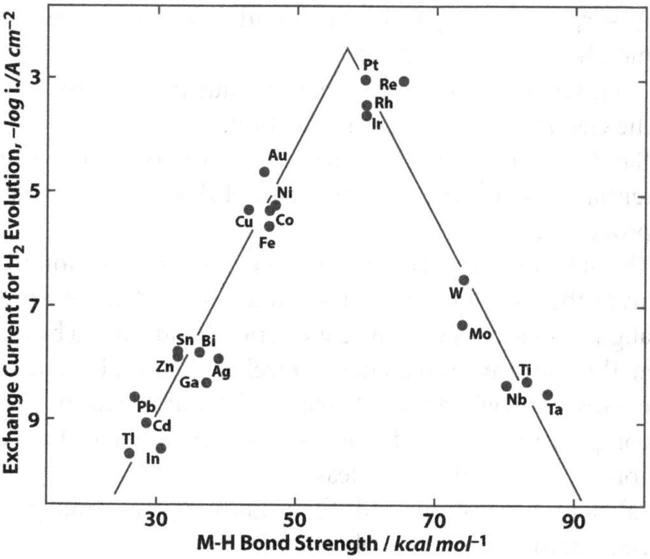

Figure 4.1 Density of exchange current for H_2 evolution for different metals corresponding to adsorption energy of hydrogen (ΔG_H). Reprinted with the permission from ref. [5].

Another approach toward reduction of the costs is based on the formation of bimetallic nanostructures, like PtPd, PtNi, PtCo, and FePt [6]. Since some of the platinum is replaced with other non-metals, these alloys frequently have superior qualities than pure platinum, such as increased catalytic activity, enhanced durability, and reduced cost. They are frequently found in electrochemical devices such as fuel cells.

Another crucial issue is the selection of platform for catalyst loading to make them isolated and keep them stable for a long time. That is why, for example, Pt nanoparticles are frequently deposited onto carbon or other metal-based support [7]. Their selection as the substrate is due to tunable electrical conductivity, abundant defects, and ability to modify chemical nature and structure. The main strategies and the challenges related to them concerning Pt-based electrode materials are shown in Figure 4.2.

As was indicated by Baek *et al.* [4], platinum, together with other noble metals, should be regarded as the most active in hydrogen evolution; however, researchers have designed a wide range of low-cost catalysts [8]. Regarding electrode materials based on molybdenum, one can differentiate three most promising compounds as follows: (a) 1T′-MoS_2, (b) Co-MoS_2, and (c) β-Mo_2C [9]. In an acidic environment, the listed materials show high stability including also a low value of HER reaction

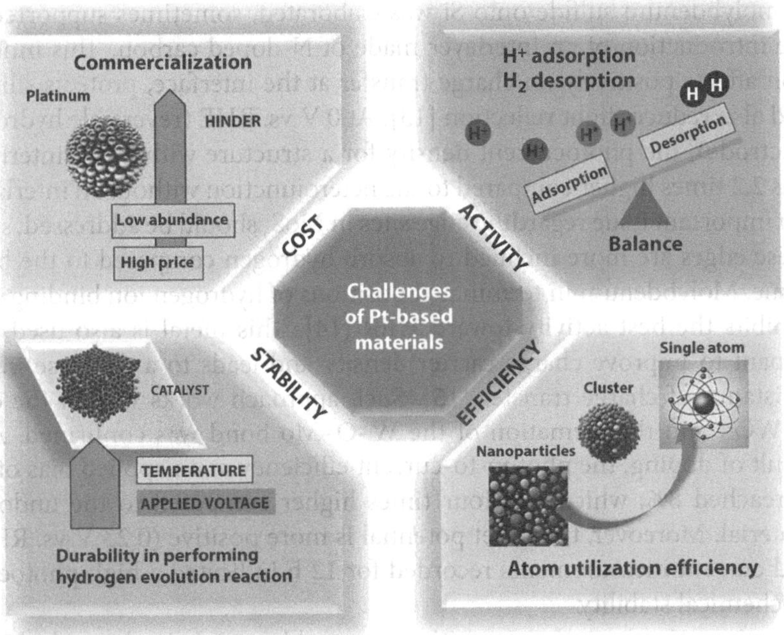

Figure 4.2 Main challenges for Pt-based materials.

and a low Tafel slope coefficient. 1T′-MoS_2 is a transition-metal dichalcogenide characterized by a layered nanostructure where the trigonal phase of MoS_2 is distorted. Molybdenum sulfide, as an n-type semiconductor, can be considered for application in sunlight-supported electrochemical processes due to the narrow band gap (1.9 eV) and direct transitions [10]. The 1T′ phase of MoS_2 exhibits almost optimal hydrogen adsorption energy. Moreover, the layered structure of this compound is supplemented by the semi-metallic behavior. These, together with the activity toward the HER basal plane, favor the HER kinetics [11]. However, for PEC water splitting applications, the recombination of electrons and holes occurs. Therefore, many studies are devoted to combine MoS_2 with various materials such as CdS@$CoMo_2S_4$/MoS_2 and MnS/MoS_2 [10].

In general, there are two known processes enabling the synthesis of 2D molybdenum dichalcogenides and silicon heterojunctions as follows: (a) formation of the material on SiO_2 proceeds by transfer to the desired substrates and (b) direct synthesis of dichalcogenide on the desired substrate. Initially, silicon substrate was used as a light-absorbing material, and MoS_2 was deposited using thermolysis [12]. Then, apart from the challenging chemical vapor deposition process, light-assisted deposition of molybdenum sulfide onto Si was elaborated, sometimes supported by the introduction of an interlayer made of N-doped carbon. This innovation affects positively on charge transfer at the interface, protects silicon, and also reduces light reflection [13]. At 0 V vs. RHE (reversible hydrogen electrode), the photocurrent density for a structure with a CN interlayer was 2.2 times higher compared to the heterojunction without an interlayer. An important issue regarding edge sites in MoS_2 should be addressed, since these edges are more inclined to absorb hydrogen compared to the basal plane. Molybdenum in terminated positions of hydrogen ion binding sites exhibits the best activity toward HER [14]. This metal is also used as a dopant to improve charge carrier density and leads to a decrease in the resistance of charge transfer [15]. Such approach works well in the case of WO_3, and the formation of the W–O–Mo bond was confirmed. As a result of doping, the photon-to-current efficiency at an applied bias of 0.4 V reached 8%, which was four times higher compared to the undoped material. Moreover, the onset potential is more positive (0.23 V vs. RHE), and current measurements recorded for 12 h indicated a high photoelectrochemical stability.

By investigating other transition non-noble metals in the early 2000s, it was found that cobalt oxide can be considered as a promising material for PEC. The usage of cobalt oxide faces some objections since it exhibits indirect transition within its band gap. The presence of such transitions indicates that this material can be rather unsuitable for applications where light capture is crucial. However, there were published results showing that very high solar-to-hydrogen (STH) efficiency can be achieved for CoO [16, 17]. In 2014, Liao *et al.* [18] showed that CoO can even reach 5% solar-to-hydrogen efficiency. It should be noted that CoO is not the only oxide form, and there are also other ones such as Co_3O_4, CoO_2, and Co_2O_3. Among them, Co_3O_4 and CoO are considered the most stable. The stoichiometry of the oxide depends on the annealing temperature and atmosphere, and when the temperature is high enough (900°C), CoO is obtained. Many reports indicated that Co_3O_4 is the most photoactive oxide, and as was shown by Moridon *et al.* [19], the material calcined at 450°C generated a photocurrent of 1.15 mA cm^{-2} at −0.4 vs. Ag/AgCl producing 34.4 μmol L^{-1} of hydrogen. Compared to other fabricated cobalt oxides, Co_3O_4 was the best as a photocathode. Following that, there were proposed different approaches toward the fabrication of oxygen-deficient cobalt oxide, such as hydrogen treatment, solution reduction using different reducing agents ($NaBH_4$ and N_2H_4), thermal annealing in an oxygen-deficient environment, plasma treatment, and heteroatom doping [20]. Regarding hydrogen treatment, annealing under a hydrogen atmosphere was carried out in mild temperature regimes, 150°C–300°C [21]. As a result of this process, amorphous cobalt oxide can remain enriched by hydroxyl groups playing an important role in hydrogen evolution. Such core-shell structure of Co/Co_3O_4 in an alkaline environment achieved 20 mA cm^{-2} at 0.129 mV, and improved performance was justified by its multiphase composition: a metallic core and an amorphous shell that positively affect activity at the electrode/electrolyte interface and bulk conductivity.

Within transition metals, one cannot forget to mention electrodes containing nickel, which is also an Earth-abundant element. Nickel, in the form of phosphides or double hydroxides, is one of the most prominent electrode material for HER. According to theoretical studies, Ni_2P exhibits analogous active sites as hydrogenase enzymes. Such a performance is the result of the synergistic effect occurring due to the presence of donor-acceptor and hydride-acceptor centers on the (001) plane of Ni_2P. The photocathode based on the Ni_2P deposited onto the n^+p-Si-modified thin titanium film reached 12 mA cm^{-2} at 0.41 V. Regarding hydroxides, Ni–Fe-layered double hydroxide shows activity in alkaline solution. It acts as a HER catalyst to facilitate the kinetics of surface reaction and serves as a protective layer to maintain stability, while the presence of the Ti/p-Si junction at the interface supports charge separation. It was shown that the Ni-Fe-layered double hydroxide is also characterized by long-term stability, and the photocurrent density can reach 7 mA cm^{-2} at 0 V [22].

Nickel combined with molybdenum is another alloy that is considered as an active HER electrocatalyst with alkaline stability. After its deposition onto the n^+p-Si, together with titania as an antireflection film, it showed a photocurrent density of 14.3 mA cm^{-2} at 0 V, while the onset potential was only 0.42 V [23]. Such an alloy was also used to form a layer onto the Cu_2O substrate with a TiO_2 protective film. In the case of such configuration, the density of the photocurrent was 6.3 mA cm^{-2} at 0 V in 1 M KOH electrolyte [24]. The level of the photocurrent remained stable for 10 h, with some small decay due to the increased number of the recombination Ti^{3+} centers, then lowering of the activity of the NiMo alloy. Moreover, nickel-based cocatalysts, in general, do not suffer from lack of stability, especially nickel phosphides, which are still considered promising in electrolytes in a wide range of pH.

In the next parts of the chapter, we show the performances of photoelectrodes based on common oxides, non-oxides, and organic compounds. As titanium dioxide is still widely used, we focused more on some strategies to modify this semiconductor to achieve better performance.

4.3 Oxide-Based Semiconductors for Water Splitting

4.3.1 Titania-Based Photoelectrodes

Briefly recalling, by taking advantage of solar energy and considering water as a source of hydrogen, we can split the water molecule via photoelectrochemical water splitting. This process is composed of the following three major steps [25]:

a) Absorption of photons with energy greater than the band gap of the photocatalyst that enables the formation of an electron-hole pair and its separation into single charges.
b) Transport of generated charges to the interface where electrochemical reaction can occur.
c) Charges are taking part in the reactions: oxygen evolution (OER) occurring at the anodic and hydrogen evolution (HER) occurring at the cathodic part of the cell.

In the case when the electron-hole pair is not separated, the recombination processes occur, and the undesirable release of photon or heat energy can happen. Due to that, effective splitting of the e-h pair is also a key factor. If, finally, separated charges are formed, they can take part in the reduction and oxidation process at the electrode surface. The free energy of the formation of H_2 and $1/2O_2$ out of the water molecule is $\Delta G = 237.2$ kJ/mol under standard conditions [26]. Taking into account the Nernst equation, ΔE for this process is 1.23 V vs. NHE (normal hydrogen electrode). Following that, if the standard potential for the reduction of hydrogen ions to hydrogen gas is 0 volts, for the reduction of oxygen gas to water, it equals 1.23 V.

The energy and wavelength are related as follows: Energy (eV) = 1240/λ (nm). Therefore, the semiconductor should receive energy higher than 1.23 eV to make the process possible. To perform water splitting without any additional energy, the bottom of the conduction band (CB) should have energy more negative than the reduction potential of H^+ to H_2. In the case of the valence band (VB), the edge of this band should be more positive than the potential of water molecule oxidation to O_2. Overall, photoactive materials acting as electrodes for sunlight-driven water splitting should satisfy the redox potential gap of 1.23 V as mentioned above. This can be achieved through interaction between light and matter; however, in terms of practical applications, it means that charge carriers must have enough overpotential. As it is known, semiconductor materials used in

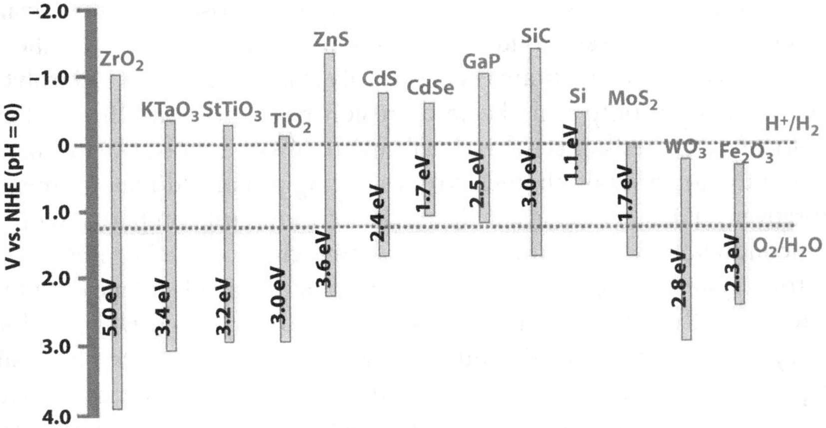

Figure 4.3 The correlation between the redox potentials involved in water splitting and the band structure of certain chosen semiconductor materials, both oxide and non-oxide. Si is given as the reference.

PEC systems can generate a photovoltage of approximately 400–500 mV below their band gap [27]. As a consequence, to efficiently drive the OER and HER across the potential barrier, a semiconductor should exhibit a band gap range of 1.6–2.6 eV [28]. The variety of oxide semiconducting materials used in PEC systems are given in Figure 4.3.

The requirements given above were considered by Fujishima and Honda [29], two Japanese scientists who utilized solar energy and were the first who achieved light-assisted water splitting using titanium dioxide as a photocatalyst. In 1967, Fujishima was working on his PhD under the supervision of professor Kenichi Honda. At that time, he put titania under intense ultraviolet light and observed the generation of gas bubbles, and this phenomenon is now well recognized as the Honda–Fujishima effect. After this pioneering work, titanium dioxide gained a lot of attention, and the scientific community undertook many efforts to boost the performance of this photocatalyst, to make the process more and more efficient under natural solar radiation. When Japanese scientists performed their experiment, they used strong UV light from an artificial source. However, as it is known, the spectrum of solar radiation contains only 5% of UV light, so performing water splitting supported by a natural light source is not so effective. It should be evoked here that in this first experiment, the selection of UV light was justified by the absorption spectrum of titanium dioxide that fits the UV range. This limitation is the consequence of a wide energy band gap being equal to 3.2 eV.

Regardless of its wide energy band gap, titania possesses many advantages that make this semiconductor commonly used in photoelectrochemical systems. First of all, titania is chemically stable in aqueous electrolytes and exhibits not only remarkable corrosion resistance but also excellent photocorrosion resistance. These features are of key importance for maintaining the integrity of photoelectrodes during prolonged measurements. Moreover, TiO_2 is widely available at relatively low cost compared to other semiconductors, which influence the commercial potential of PEC water splitting systems based on titania. Another important advantage of titania materials is their biocompatibility, which can be exploited when biological systems are considered. Additionally, titania electrodes can be easily prepared using a variety of methods, from sol–gel methods, to electrodeposition and anodization. This particular feature allows for fabrication of titania electrodes with different sets of properties such as morphology, structure, and surface properties. Moreover, these properties can be further tuned through, e.g., doping, surface modification, and nanostructuring. Overall, this may lead to improving efficiency of photoelectrochemical water splitting through enhancing of light absorption in the visible range, charge carrier mobility, and catalytic activity of titania electrodes.

The most suitable geometry of titania dedicated especially for photoelectrochemical water splitting depends on different factors, namely, surface area, charge transfer efficiency, and stability. Although many nanostructures, such as nanowires or nanoparticles (see Figure 4.4), can be

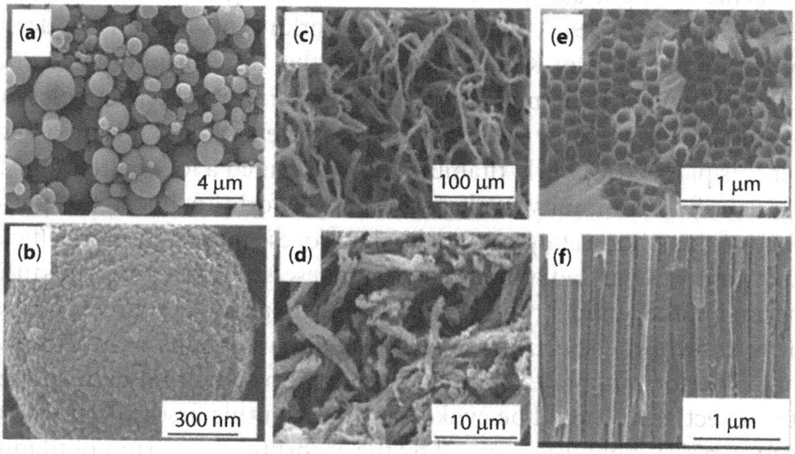

Figure 4.4 SEM images of different TiO_2 nanostructures that can be used in photoelectrochemical water splitting. Reprinted with permission from ref. [30] under the terms of the Creative Commons Attribution License (CC BY).

optimized, in general, nanotubes are believed to exhibit favorable properties due to their high surface area and efficient charge transport pathways.

One cannot also forget about the fact that electrodes for solar-driven water splitting should exhibit enhanced optical absorption. Figure 4.5a provides the mechanism of the light-trapping process in spherical nanoparticle-based electrodes. It was demonstrated that nano-embossed hollow spherical nanoparticles possess higher light-harvesting capabilities in comparison to the small nanobead-based materials [30]. Various high-aspect nanostructures, e.g., nanowires, nanotubes, and nanofibers, have been proposed to improve the light-trapping phenomenon. Nanotubular TiO_2 electrodes showed increased light harvesting efficiency up to 20% in comparison to the nanoparticulated layer. Nevertheless, most publications lack discussion and quantitative analysis of how the morphology of such structures influence their light-trapping properties. However, Leem *et al.* [31] performed three-dimensional finite-difference time domain (FDTD) simulations and compared light propagation in TiO_2 nanorods and nanotubes. The authors proved that the simulated optical output of TiO_2 nanotubes is enhanced compared to nanorods (see Figure 4.5b) and depends on the nanotube geometry, namely, the wall thickness in this case.

When discussing the photoelectrode design for electrochemical water splitting, one should consider the electron diffusion in the material. First of all, the electron transport in nanotubes follows one-dimensional pathway, while in the case of the nanoparticles, the electrons have to cope with the charge transfer resistance occurring at the particle/particle interface. In consequence, transport is less disruptive for nanotubular array. It should also be highlighted here that traps play a crucial role in limiting transport

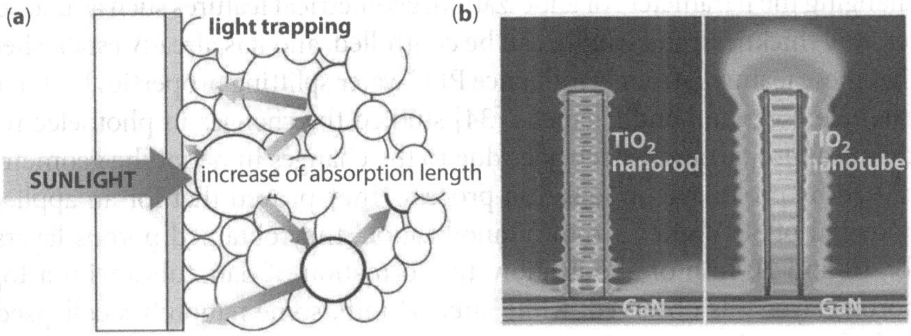

Figure 4.5 (a) Schematic of light trapping phenomenon in nanoparticulated film. (b) FDTD simulations to investigate optical properties of titania nanorods and nanotubes. (a) Reprinted with permission from ref. [30] under the terms of the Creative Commons Attribution License (CC BY). (b) Reprinted with permission from ref. [31].

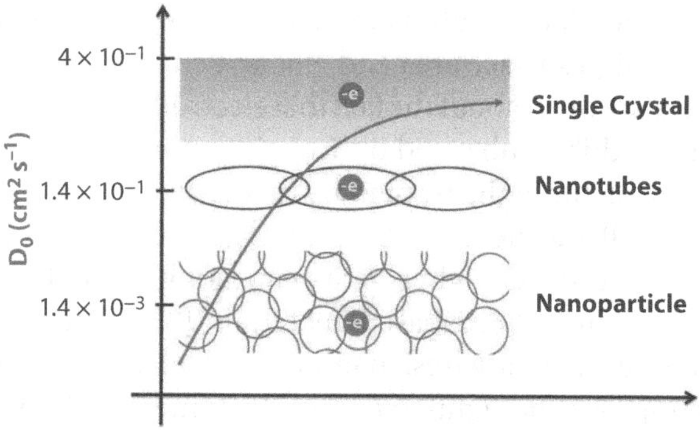

Figure 4.6 Trap-free electron diffusion (D_0) for TiO_2 nanoparticles, nanotubes, and single crystal. Reprinted with permission from ref. [32]. Copyright 2014 American Chemical Society.

in the electrode, and the material architecture, with, e.g., different impurities and local vacancies, influences this transport [32]. It was proven that the TiO_2 nanotubes exhibit two orders of magnitude larger free electron diffusion coefficient compared to nanoparticle films as electron scattering at the grain boundaries of particles (see Figure 4.5) inhibits trap-free diffusion. Additionally, the value of trap-free electron diffusion coefficient for tubular layer approaches the one for the single crystal (Figure 4.6), which indicates that structural disorder in the nanotubes is lower than that in the nanoparticles. This, in turn, affects the overall conversion efficiency.

It was already explained why titania nanotubes have attracted so much interest in photoelectrochemical application. However, going further, titania nanotubes are stable, cost effective, and can be easily fabricated, e.g., via the anodization process both on metallic and semi-transparent substrate. By changing the parameters of anodization, geometrical features such as diameter, wall thickness, and length, can be controlled, and it is already established that these features strongly influence PEC water splitting properties [33]. For example, Regonini and Clemens [34] studied the changes in photoelectrochemical properties of electrodes due to the changes in nanotube geometry caused by prolonged anodization process. They proved that for an applied potential of 30 V after 5 min of anodization, they obtained porous layers; for 10 and 20 min of anodization, the formation of nanotubes with a top porous layer was observed; while after 60 min, some nanotubes collapsed; and after another hour, the prepared layers were composed mainly of clusters of nanotubes. The photoelectrochemical analysis of the obtained electrodes suggested that the optimized length of the nanotubes is 1.1 μm. On the other hand, Paulose *et al.* [35] showed that 45-μm-long TiO_2 nanotubes with 120-nm pore diameter and 15-nm wall thickness exhibited photoconversion efficiency of 16.25%. Although for decades, there has been an extensive research on TiO_2 nanotubes, it has been suggested that still further investigation is required to quantitatively explore the impact of tube geometry on their photoelectrochemical properties as during anodization, all the features, namely, wall thickness, tube diameter, and length, are changed with processing parameters [33]. The area of titania synthesis procedure is still explored since the final morphology depends also on the substrate quality and the arrangement of the anodization setup, namely, the distance between electrodes. Nonetheless, in 2023, Inoue *et al.* [36] also reported on the impact of titania nanotube length on the PEC water splitting efficiency. This time, the pore size and wall thickness were fixed allowing for precise comparison between the prepared materials of lengths from 0.47 to 21 μm. To make the diameter and wall thickness constant, the authors applied a procedure, which they called "post anodization treatment," which relied on keeping the as-anodized samples in the electrolyte at the same temperature after anodization for some time, depending on the initial geometry, without stirring. From electrochemical impedance spectroscopy (EIS) measurements, nanotubes with a length of 9.6 μm were expected to be the most promising in water splitting. Interestingly, the tubes of 1.5-μm length exhibited the best photocurrent response under simulated solar irradiation, while 4.9-μm-long tubes showed the best photoresponse under UV irradiation. The possible explanation given by researchers considered the electrolyte convection inside and outside nanotubes resulting from the generation of bubbles on the electrode surface. It should be noted here that no bubbles were generated during EIS measurements, a small amount of bubbles was formed during measurements under solar illumination, and a large amount of bubbles was produced during measurements under UV irradiation. Nonetheless, it can be easily seen that discussion on the effect of particular nanotube dimensions on PEC performance is not trivial.

To enhance PEC performance of TiO_2 photoelectrodes, different approaches were undertaken as follows: a) morphology tuning, b) doping, and c) interface engineering. The variety of proposed approaches is shown in Figure 4.7.

For example, in 2023. Lin *et al.* [37] reported on H_2 production for pure titania nanotubes and titania nanotubes modified with CeO_2 nanoparticles. For modified titania, a PEC evolution rate of 17.86 μmol h^{-1} was obtained at 1.55 V vs. RHE, while for pristine material, this rate was 4.5 times smaller. A modified electrode showed also a higher photocurrent density (2.11 mA cm^{-2}) than that of pure titania (0.49 mA cm^{-2}) at 1.23 V vs. RHE. Different rare earth oxides, like Y_2O_3, were also used to decorate titania nanotubes

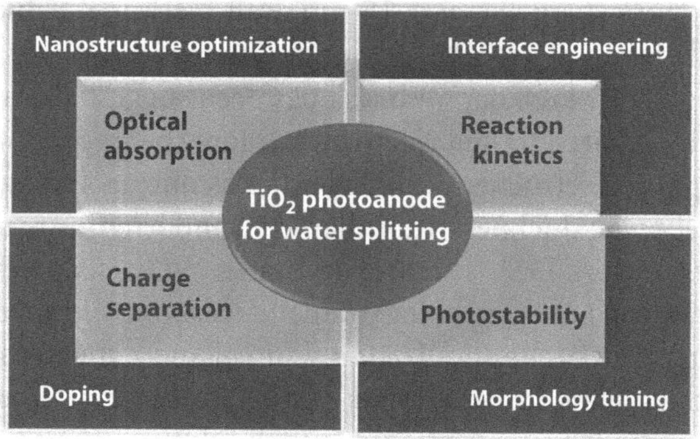

Figure 4.7 Schematic representation of the proposed modification approaches to change TiO_2 properties.

[38]. The modified electrode exhibited 2.75 higher photoconversion than the unmodified one, though in this case, the best photoresponse was of 609 μA cm^{-2} at 1.23 V vs. RHE. Another approach to improve the properties of titania nanotubes was presented by Meng *et al.* [39]. The authors fabricated 3D nanotube arrays with an anatase–rutile phase junction and oxygen vacancies, and such structure achieved 1.5 mA cm^{-2} at 0.22 V vs. Ag/AgCl, being almost two times higher than that of pristine TiO_2 nanotubes. Tong *et al.* [40] reported on increased visible light photoelectrochemical water splitting using ultra-thin carbon doped TiO_2 nanotubes. Such material exhibited 41.3 μmol cm^{-2} h^{-1} H_2 evolution rate under simulated solar irradiation and 1.3 μmol cm^{-2} h^{-1} under visible light, which is 2 and 10 times higher, respectively, compared to that of pure titania. Other examples of PEC performance based on ordered TiO_2 nanotubes are given in Table 4.1.

Nonetheless, in some cases, to achieve significant enhancement in photoconversion efficiency, one has to scarify the stability of the electrode material, and fabrication of a stable material with a solar-to-hydrogen efficiency above 10% remains challenging [41]. Zhang *et al.* [41] proposed possible future routes ensuring both high stability and photoconversion efficiency as follows: (i) interface engineering to develop effective and efficient light absorption, passivation, and catalytic layers and (ii) tuning the structure and crystal morphology of vertically oriented highly ordered titania arrays as this may result in increased electron lifetime.

Table 4.1 Performance of TiO_2 nanotube-based electrodes toward photoelectrochemical water splitting.

Electrode material	Photocurrent density	Conditions	References
Si–TiO_2NT[(a)]	0.83 mA cm^{-2} at 0 V vs. Ag/AgCl	1.0 M KOH	[42]
Black Si–TiO_2NT	1.78 mA cm^{-2} at 0 V vs. Ag/AgCl	1.0 M KOH	[43]
g-C_3N_4/TiO_2NT	0.86 mA cm^{-2} at 1.23 V vs. RHE	0.1 M Na_2SO_4	[44]
Au@A-B-TiO_2NT[(b)]	2.82 mA cm^{-2} at 1.23 V vs. RHE	1 M KOH	[45]
Mixed-phase TiO_2 NT	0.11 mA cm^{-2} at 0.5 V vs. Ag/AgCl	1.0 M NaOH	[46]
AN–TiO_2 NT[(c)]	1 mA cm^{-2} at 1.18 V vs. RHE	1.0 M KOH	[47]
CdS/TiO_2NT	4.56 mA cm^{-2} at 1.23 V vs. RHE	0.2 M Na_2SO_3 (pH = 10.2)	[48]
$Co(OH)_2$/NT	0.6 mA cm^{-2} at 1.23 V vs. RHE	0.5 M Na_2SO_4	[49]
Cu-TiO_2NT	4.43 mA cm^{-2} at 2.0 V vs. Ag/AgCl	0.5 KOH under visible light	[50]
$SrTiO_3$/TiO_2NT	4.05 mA cm^{-2} at 0.6 V vs. SCE[(d)]	1.0 M NaOH under visible light	[51]

Measurements were conducted under AM 1.5G, 100 mW cm^{-2} unless stated otherwise. Adapted from ref. [41] under the terms of the Creative Commons Attribution License (CC BY). (a) NT, nanotube arrays; (b) A, amorphous, B, black; (c) AN, argon–ammonia; (d) SCE, standard calomel electrode.

4.3.2 Iron Oxide-Based Photoelectrodes

Iron oxide, as a photoelectrode material, was introduced shortly after titanium dioxide as it was in 1976 [52]. To its advantages belong good chemical stability, narrow band gap, and Earth abundance, while low absorption coefficient, short carrier lifetime, high recombination rate, and low conductivity are considered as major drawbacks. It should be underlined that hematite is one of the most studied photoelectrodes based on iron for water oxidation as it has the ability to reach the maximum STH efficiency of 15% [53], which is much higher than that of 10% required for real applications. However, the abovementioned limitations inhibit the PEC performance of hematite, and some strategies to overcome this issue have been applied. For example, hematite in the form of nanofibers with Sn-doping and FeOOH as cocatalyst achieved a photocurrent density of 1.86 mA cm^{-2} [54]. 1D hematite nanobundle arrays with gradient doping by P and Co–Pi as cocatalyst reached 2 mA cm^{-2} photoresponse [55]. Much higher photocurrents of 4.68 mA cm^{-2} and 6 mA cm^{-2} were registered for hematite nanosheets with plasmonic Ag nanoparticles and Co-Pi cocatalysts [56] and hematite nanorods doped by hydrogen and with titanium dioxide overlayer [57], respectively. All the mentioned above photocurrent densities are reported for 1.23 V vs. RHE and for simulated solar light (AM 1.5 G, intensity of 100 mW cm^{-2}) and alkaline 1 M electrolyte.

In contrast to hematite electrodes, the one based on non-hematite iron oxide structures are not so widely explored. The probable reason for this is related to the lack of these polymorphs in nature and the necessity to be synthetically fabricated. The β-Fe_2O_3 thin films reached a photocurrent density of 1.03 mA cm^{-2}, while after modification with Sb_2Se_3, it reached 3.07 mA cm^{-2} at 1.23 V vs. RHE [58]. $ZnFe_2O_4$ nanorods achieved a photoresponse of 1.7 mA cm^{-2} at 1.6 V vs. RHE [59], whereas β-Fe_2O_3/$ZnFe_2O_4$ 3D nanospikes exhibited 2.19 mA cm^{-2} photocurrent density at 1.23 V vs. RHE [60]. One of the highest photoresponses, namely, 3.23 and 4.1 mA cm^{-2} were reported for nonporous $BiVO_4$/Fe_2TiO_5 [61] and Co_{Ox}/Fe_2TiO_5/TiO_2 nanotubes [62], respectively, at 1.23 V vs. RHE. As in the case of presented data for hematite photoelectrodes, the measurements were also conducted in alkaline 1 M electrolyte under simulated solar irradiation.

Although many different materials are considered for the cathodic and anodic sides of PEC systems, Kment *et al.* [63] anticipated that the iron oxide-based electrode material will be used in practical devices. However, one cannot deny that much better materials in terms of performance are studied, and thus, the obvious limitations of iron oxides need to be further addressed.

4.3.3 Tungsten Oxide-Based Photoelectrodes

In 1976, tungsten oxides were first introduced as photoelectrode materials [64]. In the case of tungsten oxides, good stability in acid environments, tunable composition, and good electron transport are the main advantages. The numerous limitations of such oxide semiconductors cover a large band gap, corrosion due to peroxospecies, and low stability at pH above 4. Moreover, they still suffer from PEC water splitting performance. Therefore, solar-to-chemical energy conversion needs to be improved to prevent recombination, ensure sufficient carrier lifetime, and establish proper valence band edge [65]. Moreover, it should be mentioned that tungsten oxides are mainly used for oxygen evolution reaction, and only some publications mention eminent photocatalytic activity for $CdS–WO_3$ and WO_{3-x}/CdS for efficient hydrogen generation [66]. Therefore, although tungsten-based photocatalytic systems can be used in effective water splitting applications, and there are some developments in the field of photoelectrochemical hydrogen evolution, we will not focus further on this oxide semiconductor in this chapter.

4.3.4 Zinc Oxide-Based Photoelectrodes

Another semiconductor that was verified as a photoelectrode was zinc oxide, and first attempts were initiated in 1977 [67]. This material is characterized by a large band gap and thus low visible-light absorption, high recombination rate, susceptibility to photocorrosion, high carrier mobility, and low costs of fabrication. It seems like the mentioned limitations may be the cause of hindering the use of ZnO in PEC research. Nonetheless, like many other materials, zinc oxide-based photoelectrodes can show some improvement when appropriate modifications, already stated in this chapter, are applied.

So far, N-doped ZnO nanorods have been shown to exhibit 0.160 mA cm^{-2} photoresponse at 1.1 V vs. SCE and 17.5% incident photon to electron efficiency (IPCE; gives insight into the contribution of electrodes in the conversion of incident photons into charge carriers) [68], while N doped ZnO nanowires show a photoresponse of 0.4 mA cm^{-2} at 1.0 V vs. Ag/AgCl [69]. Better results were obtained for ZnO–Cu photoelectrodes for which 0.35 mA cm^{-2} photocurrent density at 1.0 V vs. SCE and 3% IPCE were registered [70]. Even higher photocurrent densities of 0.7 mA cm^{-2} at 0.4 V vs. Ag/AgCl and 4.6 mA cm^{-2} at 1.23 V vs. RHE were measured for Ag/Co-doped ZnO nanowire/flake-like composite [71] and Ni-doped ZnO nanostructures [72], respectively. However, the most pronounced photoresponse of 8 mA cm^{-2} at 1.0 V vs. Ag/AgCl was reported for Pb-doped ZnO, and for this electrode, IPCE was approximately 30% at 390 nm [73]. Moreover, the applied bias photon-to-current-efficiency, which represents the development of the electrode performance as a function of the applied potential, was 0.636 at 0.5 V. The authors stated that reported efficiencies were higher than for those of other reported zinc oxide-based photoactive materials. Nonetheless, despite promising results, still, zinc oxide is believed to be challenging in terms of using it for large-scale applications as light absorption in visible range, recombination rate, and charge carrier transfer are not as efficient as the PEC systems demand them to be [74].

4.3.5 $BiVO_4$-Based Photoelectrodes

The last oxide semiconductor that we would like to mention was introduced as a photoelectrode relatively late as it was in 1998 [75]. This material possesses good chemical stability, suitable band gap, and low onset potential. However, among the limitations, one can find poor photocurrent stability, low carrier mobility, high recombination rate, and sluggish kinetics of oxygen evolution. Nonetheless, as in the case of tungsten oxide, $BiVO_4$ can be used as a photoanode, but to achieve high solar-to-hydrogen efficiency, integration with an appropriate photocathode is required. So far, reported STH efficiencies for such systems vary between 1.5% and 3%, which is considered low [76]. Moreover, it should be remembered that theoretical photocurrent density of $BiVO_4$ is 7.5 mA cm^{-2}, while solar-to-hydrogen efficiency is 9.2% [77]. Therefore, we will not focus more on this semiconductor.

4.4 Non-Oxide Semiconductors for Photoelectrochemical Water Splitting

Another class of semiconductors that can be used for photoelectrochemical water splitting is non-oxide semiconductors—one that includes silicon, groups II–V and II–VI materials, though it should be already mentioned that such electrodes suffer from PEC/chemical corrosion and passivation under operation conditions. The problem of non-oxide semiconductors with self-corrosion reaction is caused by the fact that this reaction is more thermodynamically favorable and, therefore, less energetic than hydrogen and oxygen evolution reactions. This leads to disintegrating and dissolving of electrodes in the electrolyte over time and thus to gradual decrease in photocurrent. To overcome this problem, protection layers or a cocatalyst should be introduced to inhibit self-corrosion to some extent. Nevertheless, a small band gap that allows for harvesting of light in the wide range of solar spectrum [78] is considered as a huge advantage. The variety of different non-oxide semiconductors, with their relative positions of the band structures both toward HER and OER activity, are given in Figure 4.3.

Going to exemplary PEC performances of non-oxide semiconductor photoelectrodes, CdSe material verified in sulfide/polysulfide electrolyte and doped with Sb showed a photocurrent density of 0.2 mA cm^{-2} vs. SCE, and the reported system efficiency was 0.2 % [79]. CdSe with an NiO protection layer exhibited a photoresponse of 2 mA cm^{-2} vs. RHE in 0.1. M KCl with an incident photon-to-electron efficiency of 7.8% at 450 nm [80]. CdS electrodes tested in 0.1 M Na_2S and 0.1 M Na_2SO_3 with TiO_2 as cocatalyst showed 1.2 mA cm^{-2} vs. SCE and an IPCE of 21% at 475 nm as well as stability for 9 h [81]. GaAs and Si electrodes, both with a Ni cocatalyst and TiO_2 protection layers, exhibited 14.6 and 34 mA cm^{-2} vs. SCE photocurrent density in 1 M KOH, respectively [82]. However, while for the modified GaAs electrode no stability was reported, the modified Si electrode was stable for 100 h. CdTe electrodes measured in 0.1 M KH_2PO_4 showed a photoresponse of 1.2 mA cm^{-2} vs. RHE with a solar energy conversion of 1.5% and a stability of 1,500 s [83], whereas CdTe with Pt cocatalyst and a CdS protection layer exhibited a photocurrent density of 6.0 mA cm^{-2} vs. RHE, 12-h stability, an IPCE of 30%, solar energy conversion of 0.2%, and 100% Faradaic efficiency [84]. One of the highest photocurrent density of 24 mA cm^{-2} vs. Ag/AgCl was achieved for CIGS with Pt cocatalyst and ZnS as protection coverage in 0.5 M H_2SO_4. This electrode was also stable for 10 h and showed 40% of IPCE at 800 nm [85]. Another high photoresponse was registered for $CuGaSe_2$ with a CdS protection layer in 0.1 M Na_2SO_4. The electrode reached 15 mA cm^{-2} vs. RHE, 42% IPCE at 520 nm, and 48-h stability [86]. Another remarkable photocurrent density of 30 mA cm^{-2} vs. RHE in 0.5 H_2SO_4 and 60% of IPCE at 400–600 nm were reported for $SbSe_3$ photoelectrode with Pt cocatalyst and TiO_2 protection film [87].

Nonetheless, elemental Si still reaches 39 mA cm^{-2} when used as a photoanode and 38.5 mA cm^{-2} when used as a photocathode. In here, it should be remembered that theoretical maximum photocurrents are limited due to the band gap value. For example, a material with a band gap of 2.5 eV should achieve 10 mA cm^{-2} photocurrent density under solar light illumination. However, high recombination rate, incomplete light absorption, and high overpotential losses also influence the maximum photocurrents. Therefore, photoactive materials are not able to reach their theoretical values of photoresponse [88].

As can be seen in the given examples, even different strategies to enhance the photoresponse and extend the stability of non-oxide electrodes still do not guarantee satisfactory outcomes. When taking into account even complex fabrication methods, it is not surprising that oxide semiconductors dominate in the research works. Looking forward, we should focus on improving the stability of non-oxide semiconductors, and approaches such as appropriate electrolyte selection, applying protective coatings, loading with cocatalysts, and controlling the hole oxidative capacity should be considered. Another possible route involves exploring a new class of non-oxides including metal-free ceramics, polymers, and ferroelectrics. This may open a new direction in this field and opportunity for commercialization.

4.5 Organic Semiconductors for Photoelectrochemical Water Splitting

Although the research on photoelectrodes for water splitting is mainly focused on the inorganic materials, organic photoelectrodes exhibit also some promising properties that can be utilized in this field, namely, broad absorption spectrum, tunability properties through different approaches of molecular design, efficient processing, and high charge carrier mobility. Photoelectrodes that rely on organic semiconductors have been neglected as it was challenging to achieve both effective light harvesting and drive redox half reaction. Unfortunately, the reported photocurrent values did not reach more than 10 $\mu A\ cm^{-2}$ [89]. One of the first used organic semiconductors, polyacetylene film, achieved approximately 5 $\mu A\ cm^{-2}$ at -1.0 V vs. RHE with formation of hydrogen bubbles at the cathode [90]. Till now, a variety of different organic semiconductors have been proposed for use in photoelectrochemical water splitting (see Figure 4.8).

Linear polymers are large molecules composed of repeating units, which are linked together forming a chain-like structure. These materials can be engineered in such a way to absorb a broad range of light wavelengths as well as be functionalized with catalytic groups or nanoparticles that play a crucial role in facile solar-driven water splitting reaction. For nitrogen-containing materials, the STH efficiency of 1.16% has been reported as the best one (for boron-doped nitrogen-deficient carbon nitride) [91]. Metal–organic (MOFs) and covalent organic frameworks (COFs) are porous materials whose structures can be tailored for specific applications. MOFs are composed of metal ions or clusters coordinated

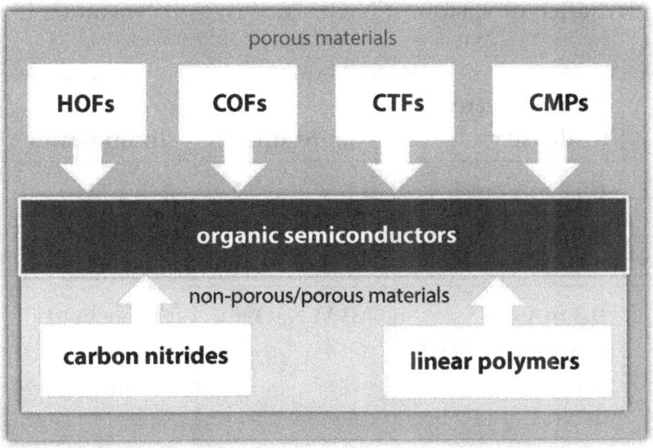

Figure 4.8 Schematic representation of different classes of organic semiconductors. HOFs, hydrogen-bonded organic; COFs, covalent organic frameworks; CTFs, covalent triazine frameworks; CMPs, conjugated microporous polymers.

with organic ligands, while COFs of light elements, i.e., carbon, hydrogen, nitrogen, and oxygen, with covalent bonds forming the framework structures. When it comes to solar-to-hydrogen efficiency for COFs, 0.40% value was reported, whereas for MOFs, no data were published [91]. Covalent triazine frameworks (CTFs) are a class of organic materials forming porous nanostructures that are considered a part of the COFs group and distinguished by triazine-based building blocks consisting of nitrogen-containing heterocycles. Other classes of porous materials are represented by conjugated microporous polymers (CMPs) composed of organic polymer with extended π-conjugated structures and hydrogen-bonded organic frameworks (HOFs) in which compounds are assembled through hydrogen bonds. What is unique, in comparison with MOFs and COFs, is the fact that HOFs are soluble in organic solvents and therefore can be regenerated in the solution [92]. Finally, carbon nitrides that are a group of materials composed of carbon and nitrogen atoms arranged in two- or three-dimensional networks have gained significant attention in the field of photoelectrochemical hydrogen production due to good thermal stability, appropriate band gap of 2.7 eV, non-toxicity, and photocorrosion resistance [93], though to boost their PEC performance, one needs to construct composite electrodes. A summary of CN materials and the obtained results are shown in Table 4.2.

Nonetheless, organic semiconductors still represent a small group of materials that are used as photoelectrodes in solar-driven water splitting.

Table 4.2 Performance of various CN electrodes; g-C_3N_4 stands for graphitic carbon nitride.

Electrode material	Photocurrent density	Conditions	Illumination	References
MoS_2/S-doped g-C_3N_4	0.12 mA cm^{-2} at 0.5 V vs. Ag/AgCl	0.1 M Na_2SO_4	AM 1.5G	[94]
TiO_2/g-C_3N_4/FeP	0.3 mA cm^{-2} at 1.23 V vs. RHE 1.1 mA cm^{-2} at 2.0 V vs. RHE	1.0 M NaOH	Visible light (above 420 nm)	[95]
Y:ZnO@ polymeric carbon nitride	0.4 mA cm^{-2} at 1.23 V vs. RHE	0.5 M Na_2SO_4	AM 1.5G	[96]
ZnO/g-C_xN_y	0.25 mA cm^{-2} at 1.23 V vs. RHE	0.1 M KOH	AM 1.5G	[97]
g-C_3N_4/$BiVO_4$	0.44 mA cm^{-2} at 0.56 V vs. SCE	0.5 M KH_2PO_4 containing 1 M Na_2SO_3	Visible light (above 420 nm)	[98]
g-C_3N_4/SnO_2	1.8 mA cm^{-2} at 0.2 V vs. Ag/AgCl	0.1 M NaOH	AM 1.5G	[99]
Fe_2O_3/carbon nitride/Co-Pi	0.7 mA cm^{-2} at 1.23 V vs. RHE	1 M NaOH	AM 1.5G	[100]
SnO_{2-x}/g-C_3N_4	0.05 mA cm^{-2} at 1.23 V vs. RHE	0.2 M Na_2SO_4	Visible light	[101]
$NiTiO_3$/g-C_3N_4	0.3 mA cm^{-2} at 0.6 V vs. Ag/AgCl 0.75 mA cm^{-2} at 1.0 V vs. Ag/AgCl	0.1 M Na_2SO_4	Visible light	[102]
g-C_3N_4/CdS	5.4 mA cm^{-2} at 0 V vs. Ag/AgCl	0.5 M Na_2S + 0.5 M Na_2SO_3	Visible light (above 420 nm)	[103]

Additionally, their optical and electronic properties have to be optimized to improve their STH efficiency and to make them viable in large-scale hydrogen production. It can be deliberately said that organic materials, although they are promising candidates due to feasible and facile structural modification possibilities and Earth abundance, should be more intensively studied.

4.6 Conclusions

Water splitting for hydrogen production represents an important turning point in the pursuit of sustainable and clean energy solutions. In light of the pressing issues raised by climate change and the widespread endeavor to reduce dependency on fossil fuels, the possibility of utilizing water splitting to extract hydrogen presents a glimmer of hope for a more sustainable future.

While the perspectives on water splitting for hydrogen production are undoubtedly optimistic, challenges remain. Overcoming issues related to cost, efficiency, and infrastructure development will require collaborative efforts from the scientific community, industry stakeholders, and policymakers. Nevertheless, the potential benefits—from a cleaner environment to energy security—underscore the significance of continuing and accelerating research in the field.

The design and engineering of photoelectrodes represent a critical area of research regarding solar-driven water splitting. Methodic exploring of materials, nanostructures, and device architectures affects the advancements in the development of efficient and cost-effective photoelectrochemical systems. Nonetheless, it is safe to say that we are on the right track as materials with excellent STH efficiencies have already been fabricated (see Figure 4.9).

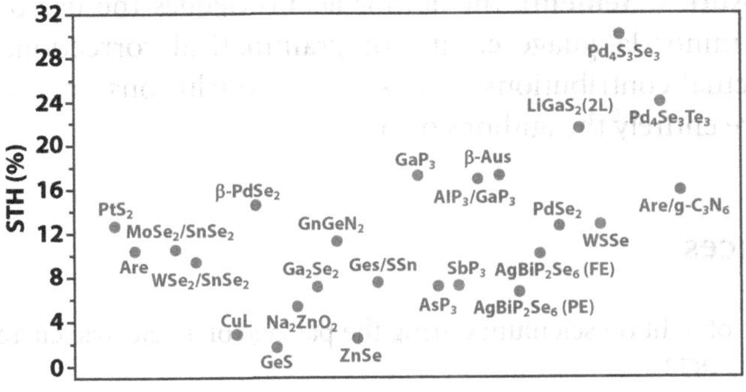

Figure 4.9 STH efficiency for selected materials. Reprinted with permission from ref. [104] under the terms of the Creative Commons Attribution License (CC BY).

In terms of efficiency of photoelectrodes, a rational design of such materials involves careful consideration of several key factors, such as light absorption, charge separation and transport, surface reactions, and stability. By tailoring the abovementioned properties of photoactive materials, one can optimize the performance of photoelectrodes for enhanced solar energy conversion efficiency.

Specifically, the incorporation of new materials—e.g., metal oxides and carbon-based nanomaterials—has demonstrated significant potential in surmounting current obstacles and expanding the frontiers of photoelectrochemical water splitting. These compounds are ideal for use in photoelectrodes because of their special qualities, which include tunable band gaps, high charge carrier mobility, and superior chemical stability. Additionally, the precise control over the morphology and nanostructure of photoelectrodes has been made possible by advancements in manipulation technologies at the nanoscale. This allows for increasing surface catalytic activity and light-harvesting capacities. Changing the material morphology and structure may also lead to fabrication of more stable photoelectrodes with enhanced performance.

Ultimately, the successful realization of efficient and scalable photoelectrochemical systems holds great promise for addressing global energy challenges. Although there is no known existing material that satisfies all the requirements for photoelectrodes, ongoing research efforts are bringing us closer to achieve this goal by modifications of already existing materials or exploration of new ones. With continued research realized between materials scientists, chemists, physicists and engineers, and development efforts, photoelectrodes can play a pivotal role in unlocking the full potential of solar energy conversion and facilitate the transition to a clean and renewable hydrogen economy.

AI disclosure statement: The author acknowledges the use of AI-based tools for minor language editing or grammatical correction. However, all intellectual contributions, analyses, and conclusions presented in this chapter are entirely the author's own.

References

1. Effect of light on selenium during the passage of an electric current. *Nature*, 7, 303, 1873.
2. Fritts, C.E., On a new form of selenium cell, and some electrical discoveries made by its use. *Am. J. Sci.*, 465, s3–26, 1883.

3. Niu, H., Wang, Q., Huang, C., Zhang, M., Yan, Y., Liu, T., Zhou, W., Noble metal-based heterogeneous catalysts for electrochemical hydrogen evolution reaction. *Appl. Sci.*, 13, 2177, 2023.
4. Li, C. and Baek, J.-B., Recent advances in noble metal (Pt, Ru, and Ir)-based electrocatalysts for efficient hydrogen evolution reaction. *ACS Omega*, 5, 31, 2020.
5. Trasatti, S., Work function, electronegativity, and electrochemical behaviour of metals. *J. Electroanal. Chem. Interf. Electrochem.*, 39, 163, 1972.
6. Hang, N.T.N., Yang, Y., Nam, N.Q.T., Nogami, M., Phuc, L.H., Long, N.V., Pt-based multimetal electrocatalysts and potential applications: recent advancements in the synthesis of nanoparticles by modified polyol methods. *Crystals*, 12, 375, 2022.
7. Guo, F., Macdonald, T.J., Sobrido, A.J., Liu, L., Feng, J., He, G., Recent advances in ultralow-Pt-loading electrocatalysts for the efficient hydrogen evolution. *Adv. Sci.*, 10, 2301098, 2023.
8. Liu, M., Wang, J.-A., Klysubun, W., Wang, G.-G., Sattayaporn, S., Li, F., Cai, Y.-W., Zhang, F., Yu, J., Yang, Y., Interfacial electronic structure engineering on molybdenum sulfide for robust dual-pH hydrogen evolution. *Nat. Commun.*, 12, 5260, 2021.
9. Piñeiro García, A., Perivoliotis, D., Wu, X., Gracia-Espino, E., Benchmarking molybdenum-based materials as cathode electrocatalysts for proton exchange membrane water electrolysis: can these compete with Pt? *ACS Sustain. Chem. Eng.*, 11, 7641–7654, 2023.
10. Sitara, E., Nasir, H., Mumtaz, A., Ehsan, M.F., Sohail, M., Iram, S., Bukhari, S.A.B., Efficient photoelectrochemical water splitting by tailoring MoS_2/CoTe heterojunction in a photoelectrochemical cell. *Nanomaterials*, 10, 2341, 2020.
11. Ekspong, J., Sandström, R., Rajukumar, L.P., Terrones, M., Wågberg, T., Gracia-Espino, E., Stable sulfur-intercalated 1T′ MoS_2 on graphitic nanoribbons as hydrogen evolution electrocatalyst. *Adv. Funct. Mater.*, 28, 1802744, 2018.
12. Kwon, K.C., Choi, S., Hong, K., Moon, C.W., Shim, Y.-S., Kim, D.H., Kim, T., Sohn, W., Jeon, J.-M., Lee, C.-H., Nam, K.T., Han, S., Kim, S.Y., Jang, H.W., Wafer-scale transferable molybdenum disulfide thin-film catalysts for photoelectrochemical hydrogen production. *Energy Environ. Sci.*, 9, 2240, 2016.
13. Zhang, S., Zhao, H., Li, X., Li, Y., Jin, Y., Liu, X., Shi, G., Wong, P.K., A hierarchical Si_{PN}/CN/MoS_x photocathode with low internal resistance and strong light-absorption for solar hydrogen production. *Appl. Catal., B*, 300, 120758, 2022.
14. Yu, Y., Nam, G.-H., He, Q., Wu, X.-J., Zhang, K., Yang, Z., Chen, J., Ma, Q., Zhao, M., Liu, Z., Ran, F.-R., Wang, X., Li, H., Huang, X., Li, B., Xiong, Q., Zhang, Q., Liu, Z., Gu, L., Du, Y., Huang, W., Zhang, H., High phase-purity 1T′-MoS_2- and 1T′-$MoSe_2$-layered crystals. *Nat. Chem.*, 10, 638, 2018.

15. Ghadge, S.D., Datta, M.K., Velikokhatnyi, O.I., Kumta, P.N., Molybdenum doped bilayer photoanode nanotubes for enhanced photoelectrochemical water splitting. *Int. J. Hydrogen Energy*, 47, 993, 2022.
16. Barakat, N.A.M., Khil, M.S., Sheikh, F.A., Kim, H.Y., Synthesis and optical properties of two cobalt oxides (CoO and Co_3O_4) nanofibers produced by electrospinning process. *J. Phys. Chem. C*, 112, 12225, 2008.
17. Ebadi, M., Mat-Teridi, M.A., Sulaiman, M.Y., Basirun, W.J., Asim, N., Ludin, N.A., Ibrahim, M.A., Sopian, K., Electrodeposited p-type Co_3O_4 with high photoelectrochemical performance in aqueous medium. *RSC Adv.*, 5, 36820, 2015.
18. Liao, L., Zhang, Q., Su, Z., Zhao, Z., Wang, Y., Li, Y., Lu, X., Wei, D., Feng, G., Yu, Q., Cai, X., Zhao, J., Ren, Z., Fang, H., Robles-Hernandez, F., Baldelli, S., Bao, J., Efficient solar water-splitting using a nanocrystalline CoO photocatalyst. *Nat. Nanotech.*, 9, 69, 2014.
19. Moridon, S.N.F., Salehmin, M.I., Mohamed, M.A., Arifin, K., Minggu, L.J., Kassim, M.B., Cobalt oxide as photocatalyst for water splitting: Temperature-dependent phase structures. *Int. J. Hydrogen Energy*, 44, 25495, 2019.
20. Badreldin, A., Abusrafa, A.E., Abdel-Wahab, A., Oxygen-deficient cobalt-based oxides for electrocatalytic water splitting. *ChemSusChem*, 14, 10, 2021.
21. Yan, X., Tian, L., Atkins, S., Liu, Y., Murowchick, J., Chen, X., Converting $CoMoO_4$ into CoO/MoO_x for overall water splitting by hydrogenation. *ACS Sustain. Chem. Eng.*, 4, 3743, 2016.
22. Zhao, J., Cai, L., Li, H., Shi, X., Zheng, X., Stabilizing silicon photocathodes by solution-deposited Ni–Fe layered double hydroxide for efficient hydrogen evolution in alkaline media. *ACS Energy Lett.*, 2, 1939, 2017.
23. Shaner, M.R., McKone, J.R., Gray, H.B., Lewis, N.S., Functional integration of Ni–Mo electrocatalysts with Si microwire array photocathodes to simultaneously achieve high fill factors and light-limited photocurrent densities for solar-driven hydrogen evolution. *Energy Environ. Sci.*, 8, 2977, 2015.
24. Morales-Guio, C.G., Liardet, L., Mayer, M.T., Tilley, S.D., Grätzel, M., Hu, X., Photoelectrochemical hydrogen production in alkaline solutions using Cu_2O coated with earth-abundant hydrogen evolution catalysts. *Angew. Chem. Int. Ed.*, 54, 664, 2015.
25. Chang, J.-H., Kumar, M., Nayak, A.K. (Eds.), *Nanostructured Materials for Photoelectrochemical Water Splitting*, IOP Publishing, Bristol, UK, 2021.
26. Li, X., Zhao, L., Yu, J., Liu, X., Zhang, X., Liu, H., Zhou, W., Water splitting: from electrode to green energy system. *Nano Micro Lett.*, 12, 131, 2020.
27. Polman, A. and Atwater, H.A., Photonic design principles for ultrahigh-efficiency photovoltaics. *Nat. Mater.*, 11, 174, 2012.
28. Kumar, M., Meena, B., Subramanyam, P., Suryakala, D., Subrahmanyam, C., Recent trends in photoelectrochemical water splitting: the role of cocatalysts. *NPG Asia Mater.*, 14, 88, 2022.
29. Marschall, R., 50 years of materials research for photocatalytic water splitting. *Eur. J. Inorg. Chem.*, 2021, 2435, 2021.

30. Gagliardi, S., Rondino, F., Paoletti, C., Falconieri, M., On the morphology of nanostructured TiO_2 for energy applications: the shape of the ubiquitous nanomaterial. *Nanomaterials*, 12, 2608, 2022.
31. Leem, Y.-C., Seo, O., Jo, Y.-R., Kim, J.H., Chun, J., Kim, B.-J., Noh, D.Y., Lim, W., Kim, Y.-I., Park, S.-J., Titanium oxide nanotube arrays for high light extraction efficiency of GaN-based vertical light-emitting diodes. *Nanoscale*, 8, 10138, 2016.
32. Villanueva-Cab, J., Jang, S.-R., Halverson, A.F., Zhu, K., Frank, A.J., Trap-free transport in ordered and disordered TiO_2 nanostructures. *Nano Lett.*, 14, 2305, 2014.
33. Liang, S., He, J., Sun, Z., Liu, Q., Jiang, Y., Cheng, H., He, B., Xie, Z., Wei, S., Improving photoelectrochemical water splitting activity of TiO_2 nanotube arrays by tuning geometrical parameters. *J. Phys. Chem. C*, 116, 9049, 2012.
34. Regonini, D. and Clemens, F.J., Anodized TiO_2 nanotubes: effect of anodizing time on film length, morphology and photoelectrochemical properties. *Mater. Lett.*, 142, 97, 2015.
35. Paulose, M., Shankar, K., Yoriya, S., Prakasam, H.E., Varghese, O.K., Mor, G.K., Latempa, T.A., Fitzgerald, A., Grimes, C.A., Anodic growth of highly ordered TiO_2 nanotube arrays to 134 μm in length. *J. Phys. Chem. B*, 110, 16179, 2006.
36. Inoue, K., Matsuda, A., Kawamura, G., Tube length optimization of titania nanotube array for efficient photoelectrochemical water splitting. *Sci. Rep.*, 13, 103, 2023.
37. Lin, S.-W., Tong, M.-H., Chen, Y.-X., Chen, R., Zhao, H.-P., Jiang, X., Yang, K., Lu, C.-Z., CeO_2/TiO_2 heterojunction nanotube arrays for highly efficient visible-light photoelectrochemical water splitting. *ACS Appl. Energy Mater.*, 6, 1093, 2023.
38. Sharma, S., Shanmugam, R., Harikrishna, R.B., Prasad, U., Ranga Rao, G., Kannan, A.M., Y_2O_3 electrodeposited TiO_2 nanotube arrays as photoanode for enhanced photoelectrochemical water splitting. *Int. J. Hydrogen Energy*, 52, 1415, 2024.
39. Meng, M., Li, C., Li, J., Wu, J., Feng, Y., Sun, L., Yuan, H., Liu, K., 3D TiO_2 nanotube arrays with anatase-rutile phase junction and oxygen vacancies for promoted photoelectrochemical water splitting. *J. Phys. D: Appl. Phys.*, 56, 055502, 2023.
40. Tong, M.-H., Wang, T.-M., Lin, S.-W., Chen, R., Jiang, X., Chen, Y.-X., Lu, C.-Z., Ultra-thin carbon doped TiO_2 nanotube arrays for enhanced visible-light photoelectrochemical water splitting. *Appl. Surf. Sci.*, 623, 156980, 2023.
41. Zhang, X., Zhang, S., Cui, X., Zhou, W., Cao, W., Cheng, D., Sun, Y., Recent advances in TiO_2-based photoanodes for photoelectrochemical water splitting. *Chem. Asian J.*, 17, e202200668, 2022.
42. Dong, Z., Ding, D., Li, T., Ning, C., Facile fabrication of Si-doped TiO_2 nanotubes photoanode for enhanced photoelectrochemical hydrogen generation. *Appl. Surf. Sci.*, 436, 125, 2018.

43. Dong, Z., Ding, D., Li, T., Ning, C., Black Si-doped TiO_2 nanotube photoanode for high-efficiency photoelectrochemical water splitting. *RSC Adv.*, 8, 5652, 2018.
44. Liu, C., Wang, F., Zhang, J., Wang, K., Qiu, Y., Liang, Q., Chen, Z., Efficient photoelectrochemical water splitting by g-C_3N_4/TiO_2 nanotube array heterostructures. *Nano Micro Lett.*, 10, 37, 2018.
45. Shi, L., Li, Z., Dao, T.D., Nagao, T., Yang, Y., A synergistic interaction between isolated Au nanoparticles and oxygen vacancies in an amorphous black TiO_2 nanoporous film: toward enhanced photoelectrochemical water splitting. *J. Mater. Chem. A*, 6, 12978, 2018.
46. Yalavarthi, R., Naldoni, A., Kment, S., Mascaretti, L., Kmentová, H., Tomanec, O., Schmuki, P., Zbořil, R., Radiative and non-radiative recombination pathways in mixed-phase TiO_2 nanotubes for PEC water-splitting. *Catalysts*, 9, 204, 2019.
47. Xu, Y., Ahmed, R., Klein, D., Cap, S., Freedy, K., McDonnell, S., Zangari, G., Improving photo-oxidation activity of water by introducing Ti^{3+} in self-ordered TiO_2 nanotube arrays treated with Ar/NH_3. *J. Power Sources*, 414, 242, 2019.
48. Wang, L., Han, J., Feng, J., Wang, X., Su, D., Hou, X., Hou, J., Liang, J., Dou, S.X., Simultaneously efficient light absorption and charge transport of CdS/TiO_2 nanotube array toward improved photoelectrochemical performance. *Int. J. Hydrogen Energy*, 44, 30899, 2019.
49. Liu, C., Wang, F., Qiu, Y., Liang, Q., Mitsuzak, N., Chen, Z., Facile electrodeposition of cobalt hydroxide on anodic TiO_2 nanotubes arrays for enhanced photoelectrochemical application. *JPPA*, 353, 200, 2018.
50. Grochowska, K., Molenda, Z., Karczewski, J., Bachmann, J., Darowicki, K., Ryl, J., Siuzdak, K., Laser induced formation of copper species over TiO_2 nanotubes towards enhanced water splitting performance. *Int. J. Hydrogen Energy*, 45, 19192, 2020.
51. Cheng, X., Zhang, Y., Hu, H., Shang, M., Bi, Y., High-efficiency $SrTiO_3$/TiO_2 hetero-photoanode for visible-light water splitting by charge transport design and optical absorption management. *Nanoscale*, 10, 3644, 2018.
52. Hardee, K.L. and Bard, A.J., Semiconductor electrodes: V . The application of chemically vapor deposited iron oxide films to photosensitized electrolysis. *J. Electrochem. Soc.*, 123, 1024, 1976.
53. Farhoosh, S., Eftekharinia, B., Tayebi, M., Lee, B.-K., Naseri, N., Newly designed ternary hematite-based heterojunction for PEC water splitting. *Appl. Surf. Sci.*, 550, 149374, 2021.
54. Ling, Y., Wang, G., Wheeler, D.A., Zhang, J.Z., Li, Y., Sn-doped hematite nanostructures for photoelectrochemical water splitting. *Nano Lett.*, 11, 2119, 2011.
55. Luo, Z., Li, C., Liu, S., Wang, T., Gong, J., Gradient doping of phosphorus in Fe_2O_3 nanoarray photoanodes for enhanced charge separation. *Chem. Sci.*, 8, 91, 2017.

56. Peerakiatkhajohn, P., Yun, J., Chen, H., Lyu, M., Butburee, T., Wang, L., Stable hematite nanosheet photoanodes for enhanced photoelectrochemical water splitting. *Adv. Mater.*, 28, 6405, 2016.
57. Jeon, T.H., Moon, G., Park, H., Choi, W., Ultra-efficient and durable photoelectrochemical water oxidation using elaborately designed hematite nanorod arrays. *Nano Energy*, 39, 211, 2017.
58. Liao, A., Zhou, Y., Xiao, L., Zhang, C., Wu, C., Asiri, A.M., Xiao, M., Zou, Z., Direct Z scheme-fashioned photoanode systems consisting of Fe_2O_3 nanorod arrays and underlying thin Sb_2Se_3 layers toward enhanced photoelectrochemical water splitting performance. *Nanoscale*, 11, 109, 2019.
59. Zhu, X., Guijarro, N., Liu, Y., Schouwink, P., Wells, R.A., Le Formal, F., Sun, S., Gao, C., Sivula, K., Spinel structural disorder influences solar-water-splitting performance of $ZnFe_2O_4$ nanorod photoanodes. *Adv. Mater.*, 30, 1801612, 2018.
60. Hussain, S., Hussain, S., Waleed, A., Tavakoli, M.M., Yang, S., Rauf, M.K., Fan, Z., Nadeem, M.A., Spray pyrolysis deposition of $ZnFe_2O_4/Fe_2O_3$ composite thin films on hierarchical 3-D nanospikes for efficient photoelectrochemical oxidation of water. *J. Phys. Chem. C*, 121, 18360, 2017.
61. Gao, Y., Li, Y., Yang, G., Li, S., Xiao, N., Xu, B., Liu, S., Qiu, P., Hao, S., Ge, L., Fe_2TiO_5 as an efficient co-catalyst to improve the photoelectrochemical water splitting performance of $BiVO_4$. *ACS Appl. Mater. Interfaces*, 10, 39713, 2018.
62. Liu, Q., He, J., Yao, T., Sun, Z., Cheng, W., He, S., Xie, Y., Peng, Y., Cheng, H., Sun, Y., Jiang, Y., Hu, F., Xie, Z., Yan, W., Pan, Z., Wu, Z., Wei, S., Aligned Fe_2TiO_5-containing nanotube arrays with low onset potential for visible-light water oxidation. *Nat. Commun.*, 5, 5122, 2014.
63. Kment, S., Sivula, K., Naldoni, A., Sarmah, S.P., Kmentová, H., Kulkarni, M., Rambabu, Y., Schmuki, P., Zbořil, R., FeO-based nanostructures and nanohybrids for photoelectrochemical water splitting. *Prog. Mater Sci.*, 110, 100632, 2020.
64. Hodes, G., Cahen, D., Manassen, J., Tungsten trioxide as a photoanode for a photoelectrochemical cell (PEC). *Nature*, 260, 312, 1976.
65. Zhang, M., Yang, C., Zhang, Z., Tian, W., Hui, B., Zhang, J., Zhang, K., Tungsten oxide polymorphs and their multifunctional applications. *Adv. Colloid Interface Sci.*, 300, 102596, 2022.
66. Shabdan, Y., Markhabayeva, A., Bakranov, N., Nuraje, N., Photoactive tungsten-oxide nanomaterials for water-splitting. *Nanomaterials*, 10, 1871, 2020.
67. Gerischer, H., On the stability of semiconductor electrodes against photodecomposition. *J. Electroanal. Chem. Interf. Electrochem.*, 82, 133, 1977.
68. Wang, M., Ren, F., Zhou, J., Cai, G., Cai, L., Hu, Y., Wang, D., Liu, Y., Guo, L., Shen, S., N doping to ZnO nanorods for photoelectrochemical water splitting under visible light: Engineered impurity distribution and terraced band structure. *Sci. Rep.*, 5, 12925, 2015.

69. Yang, X., Wolcott, A., Wang, G., Sobo, A., Fitzmorris, R.C., Qian, F., Zhang, J.Z., Li, Y., Nitrogen-doped ZnO nanowire arrays for photoelectrochemical water splitting. *Nano Lett.*, 9, 2331, 2009.
70. Dom, R., Baby, L.R., Kim, H.G., Borse, P.H., Enhanced solar photoelectrochemical conversion efficiency of ZnO:Cu electrodes for water-splitting application. *Int. J. Photoenergy*, 2013, 1, 2013.
71. Sreedhar, A., Reddy, I.N., Hoai Ta, Q.T., Namgung, G., Cho, E., Noh, J.-S., Facile growth of novel morphology correlated Ag/Co-doped ZnO nanowire/flake-like composites for superior photoelectrochemical water splitting activity. *Ceram. Int.*, 45, 6985, 2019.
72. Neelakanta Reddy, I., Venkata Reddy, C., Sreedhar, A., Shim, J., Cho, M., Yoo, K., Kim, D., Structural, optical, and bifunctional applications: Supercapacitor and photoelectrochemical water splitting of Ni-doped ZnO nanostructures. *J. Electroanal. Chem.*, 828, 124, 2018.
73. Aboud, A.A., Shaban, M., Revaprasadu, N., Effect of Cu, Ni and Pb doping on the photo-electrochemical activity of ZnO thin films. *RSC Adv.*, 9, 7729, 2019.
74. Kegel, J., Povey, I.M., Pemble, M.E., Zinc oxide for solar water splitting: A brief review of the material's challenges and associated opportunities. *Nano Energy*, 54, 409, 2018.
75. Kudo, A., Ueda, K., Kato, H., Mikami, I., Photocatalytic O_2 evolution under visible light irradiation on $BiVO_4$ in aqueous $AgNO_3$ solution. *Catal. Lett.*, 53, 229, 1998.
76. Gaikwad, M.A., Suryawanshi, U.P., Ghorpade, U.V., Jang, J.S., Suryawanshi, M.P., Kim, J.H., Emerging surface, bulk, and interface engineering strategies on $BiVO_4$ for photoelectrochemical water splitting. *Small*, 18, 2105084, 2022.
77. Chen, S., Huang, D., Xu, P., Xue, W., Lei, L., Cheng, M., Wang, R., Liu, X., Deng, R., Semiconductor-based photocatalysts for photocatalytic and photoelectrochemical water splitting: will we stop with photocorrosion? *J. Mater. Chem. A*, 8, 2286, 2020.
78. Feng, J., Huang, H., Yan, S., Luo, W., Yu, T., Li, Z., Zou, Z., Non-oxide semiconductors for artificial photosynthesis: Progress on photoelectrochemical water splitting and carbon dioxide reduction. *Nano Today*, 30, 100830, 2020.
79. Masumdar, E.U., Deshmukh, L.P., Mane, S.H., Karande, V.S., Pujari, V.B., Bhosale, P.N., CdSe : Sb electrode for photoelectrochemical applications. *J. Mater. Sci.: Mater. Electron.*, 14, 43, 2003.
80. Ruberu, T.P.A., Dong, Y., Das, A., Eisenberg, R., Photoelectrochemical generation of hydrogen from water using a CdSe quantum dot-sensitized photocathode. *ACS Catal.*, 5, 2255, 2015.
81. Pareek, A., Gopalakrishnan, A., Borse, P.H., Efficiency and stability aspects of CdS photoanode for solar hydrogen generation technology. *J. Phys. Conf. Ser.*, 755, 012006, 2016.

82. Hu, S., Shaner, M.R., Beardslee, J.A., Lichterman, M., Brunschwig, B.S., Lewis, N.S., Amorphous TiO_2 coatings stabilize Si, GaAs, and GaP photoanodes for efficient water oxidation. *Science*, 344, 1005, 2014.
83. Waldiya, M., Bhagat, D., Mukhopadhyay, I., Photoelectrochemical study of electrochemically synthesized CdTe thin films from acetate-anion based ionic liquid bath. *Electrochim. Acta*, 331, 135437, 2020.
84. Su, J., Minegishi, T., Katayama, M., Domen, K., Photoelectrochemical hydrogen evolution from water on a surface modified CdTe thin film electrode under simulated sunlight. *J. Mater. Chem. A*, 5, 4486, 2017.
85. Chae, S.Y., Park, S.J., Han, S.G., Jung, H., Kim, C.-W., Jeong, C., Joo, O.-S., Min, B.K., Hwang, Y.J., Enhanced photocurrents with ZnS passivated $Cu(In,Ga)(Se,S)_2$ photocathodes synthesized using a nonvacuum process for solar water splitting. *J. Am. Chem. Soc.*, 138, 15673, 2016.
86. Moriya, M., Minegishi, T., Kumagai, H., Katayama, M., Kubota, J., Domen, K., Stable hydrogen evolution from CdS-modified $CuGaSe_2$ photoelectrode under visible-light irradiation. *J. Am. Chem. Soc.*, 135, 3733, 2013.
87. Yang, W., Kim, J.H., Hutter, O.S., Phillips, L.J., Tan, J., Park, J., Lee, H., Major, J.D., Lee, J.S., Moon, J., Benchmark performance of low-cost Sb_2Se_3 photocathodes for unassisted solar overall water splitting. *Nat. Commun.*, 11, 861, 2020.
88. Kurnia, F., Scott, J.A., Valanoor, N., Hart, J.N., A review of non-oxide semiconductors for photoelectrochemical water splitting. *J. Mater. Chem. C*, 11, 802, 2023.
89. Otep, S., Michinobu, T., Zhang, Q., Pure organic semiconductor-based photoelectrodes for water splitting. *Sol. RRL*, 4, 1900395, 2020.
90. Shirakawa, H., Ikeda, S., Aizawa, M., Yoshitake, J., Suzuki, S., Polyacetylene film: A new electrode material for photoenergy conversion. *Synth. Met.*, 4, 43, 1981.
91. Guo, Y., Zhou, Q., Zhu, B., Tang, C.Y., Zhu, Y., Advances in organic semiconductors for photocatalytic hydrogen evolution reaction. *EES Catal.*, 1, 333, 2023.
92. Li, P., Ryder, M.R., Stoddart, J.F., Hydrogen-bonded organic frameworks: A rising class of porous molecular materials. *Acc. Mater. Res.*, 1, 77, 2020.
93. Xiong, W., Huang, F., Zhang, R.-Q., Recent developments in carbon nitride based films for photoelectrochemical water splitting. *Sustain. Energy Fuels*, 4, 485, 2020.
94. Ye, L., Wang, D., Chen, S., Fabrication and enhanced photoelectrochemical performance of MoS_2/S-doped g-C_3N_4 heterojunction film. *ACS Appl. Mater. Interfaces*, 8, 5280, 2016.
95. Qin, D., Quan, J., Duan, S., San Martin, J., Lin, Y., Zhu, X., Yao, X., Su, J., Rodríguez-Gutiérrez, I., Tao, C., Yan, Y., High-performance photoelectrochemical water oxidation with phosphorus-doped and metal phosphide cocatalyst-modified g-C_3N_4 formation through gas treatment. *ChemSusChem*, 12, 898, 2019.

96. Fang, Y., Xu, Y., Li, X., Ma, Y., Wang, X., Coating polymeric carbon nitride photoanodes on conductive Y:ZnO nanorod arrays for overall water splitting. *Angew. Chem. Int. Ed.*, 57, 9749, 2018.
97. Hajduk, S., Berglund, S.P., Podlogar, M., Dražić, G., Abdi, F.F., Orel, Z.C., Shalom, M., Conformal carbon nitride coating as an efficient hole extraction layer for ZnO nanowires-based photoelectrochemical cells. *Adv. Mater. Inter.*, 4, 1700924, 2017.
98. Wang, Y., Sun, J., Li, J., Zhao, X., Electrospinning preparation of nanostructured g-C_3N_4/$BiVO_4$ composite films with an enhanced photoelectrochemical performance. *Langmuir*, 33, 4694, 2017.
99. Bian, H., Wang, A., Li, Z., Li, Z., Diao, Y., Lu, J., Li, Y.Y., *g*-C_3N_4-modified water-crystallized mesoporous SnO_2 for enhanced photoelectrochemical properties. *Part Part Syst. Charact.*, 35, 1800155, 2018.
100. An, X., Hu, C., Lan, H., Liu, H., Qu, J., Strongly coupled metal oxide/reassembled carbon nitride/Co–Pi heterostructures for efficient photoelectrochemical water splitting. *ACS Appl. Mater. Interfaces*, 10, 6424, 2018.
101. Li, K., Zeng, X., Gao, S., Ma, L., Wang, Q., Xu, H., Wang, Z., Huang, B., Dai, Y., Lu, J., Ultrasonic-assisted pyrolyzation fabrication of reduced SnO_{2-x}/g-C_3N_4 heterojunctions: Enhance photoelectrochemical and photocatalytic activity under visible LED light irradiation. *Nano Res.*, 9, 1969, 2016.
102. Huang, Z., Zeng, X., Li, K., Gao, S., Wang, Q., Lu, J., Z-scheme $NiTiO_3$/g-C_3N_4 heterojunctions with enhanced photoelectrochemical and photocatalytic performances under visible LED light irradiation. *ACS Appl. Mater. Interfaces*, 9, 41120, 2017.
103. Wang, R., Yan, J., Zu, M., Yang, S., Cai, X., Gao, Q., Fang, Y., Zhang, S., Zhang, S., Facile synthesis of interlocking g-C_3N_4/CdS photoanode for stable photoelectrochemical hydrogen production. *Electrochim. Acta*, 279, 74, 2018.
104. Jakhar, M., Kumar, A., Ahluwalia, P.K., Tankeshwar, K., Pandey, R., Engineering 2D materials for photocatalytic water-splitting from a theoretical perspective. *Materials*, 15, 2221, 2022.

5

MXene Electrocatalysts for Water Splitting

Dujearic-Stephane Kouao and Katarzyna Siuzdak*

Centre of Plasma and Laser Engineering, Institute of Fluid-Flow Machinery, Polish Academy of Sciences, Gdańsk, Poland

Abstract

Electrochemical water splitting can be treated as hope to exploit renewable energy sources and production of alternative fuels, such as hydrogen and oxygen, used then in fuel cells. Recently, special attention has been paid to the development of efficient electrocatalysts capable of driving the water splitting process with low energy consumption. Related to this, two-dimensional transition metal carbides and nitrides, also known as MXenes, are becoming extensively investigated as electrocatalyst materials due to their large specific surface area, outstanding metallic conductivity, high hydrophilicity, and tunable surface functionalities. In this chapter, the recent advances regarding materials engineering of MXene-based electrocatalysts are discussed. First, the synthesis methods targeted to the fabrication of MXenes from their MAX precursor are briefly summarized. Then, strategies based on (i) the regulation of the intrinsic properties of MXenes, namely, the control of surface-terminating groups and structural defects and (ii) the doping of MXenes with heteroatoms to raise the electrocatalytic performance of MXenes for hydrogen evolution reaction and oxygen evolution reaction processes are discussed.

***Keywords*:** MAX phases, 2D materials, MXenes, surface-terminating groups, electrocatalysts, overpotential, tafel slope, water splitting

5.1 Introduction

Electrocatalytic oxygen and hydrogen evolution reactions (i.e., OER and HER) are considered green and sustainable alternatives to traditional fossil fuels. Although electrocatalysts, such as Pt/C (for HER) and RuO_2 (for OER),

**Corresponding author*: ksiuzdak@imp.gda.pl

Inamuddin, Tariq Altalhi, Mohammad Luqman and Jorddy Neves Cruz (eds.) Water Splitting: Production of Hydrogen, (111–138) © 2025 Scrivener Publishing LLC

are already commercially available, and their high cost limits large-scale application [1]. Thus, the fabrication of highly active and cost-effective alternative catalysts to noble metal catalysts is of crucial interest [2]. However, until now, the design and fabrication of such electrocatalysts still remain a major challenge.

Transition metal carbides/carbonitrides/nitrides (MXenes) are a new family of two-dimensional materials, with a general formula of $M_{n+1}X_nT_x$, where M, X, and T_x elements represent early transition, carbon or nitrogen, and surface terminations (i.e., –O, –OH, or –F), respectively. [3] MXenes are synthesized by selective etching of their MAX phase precursor. The MAX phases with the general formula $M_{n+1}AX_n$ (where A is an A group element in the Periodic Table) belong to the family of ternary carbide or nitride ceramics with hexagonal crystal symmetry and exhibit a stacked layered structure [4]. The purity level of the final product as well as their synthesis cost and environmental impact are of key importance in introducing the product into the global market. Related to this, various chemical and physical routes have been rigorously investigated for the production of MXenes from the MAX precursor [5]. Generally, M–X bonds in MAX are covalent bonds, while M–A bonds are metallic bonds with low bond strength. This allows selective removal of "A" layers in MAX phases by appropriate chemical etchants [6]. Numerous strategies targeted to the production of MXenes from their MAX precursors are summarized in this chapter.

MXenes have gained wide application as catalysts in HER and OER process due to their outstanding electrical conductivity and functionalization versatility. The electrocatalytic performances of several types of MXenes, such as $Ti_3C_2T_x$ [7], Ti_2CT_x [8], V_2CT_x [9], Ta_2CS_2 [10], Mo_2CT_x [11], $V_4C_3T_x$ [12], Ti_3CN [13], and Cr_2CT_x [14], have been reported. Although these works have revealed the promising features of MXenes as electrocatalysts, their sluggish kinetics during the water splitting process requires further activation [1]. Indeed, MXenes are highly oxidizable in the aqueous solution leading to the formation of a metal oxide layer on its surface, which affects the electrical conductivity of the MXene-based catalyst to drop [1]. Concurrently, strategies based on (i) control of surface terminating groups, (ii) introduction of structural defects, and (iii) doping with heteroatoms have been widely applied to manipulate their electrochemical activity toward efficient water splitting [15]. This chapter presents recent advances regarding the abovementioned methodologies to regulate the electrocatalytic activity of MXene-based electrocatalysts for HER and OER processes.

5.2 Overview of the Synthesis of MXenes

Since their discovery, the mainly used method for the synthesis of MXenes consists of wet-chemical etching of the aluminum layers in the MAX phase precursors by acid solutions containing fluoride ions (e.g., HF or a mixture HF with HCl). However, this synthesis method leads to considerable safety and environmental concerns resulting from both the high toxicity and corrosive activity of fluorine-containing reagents associated with the exothermic reaction occurring during the Al etching process [16–18]. In recent years, diverse routes have been proposed for the production of MXenes taking into account the trend of green chemistry and clean production and cost-efficient synthesis. The idea is to explore different etchants that support selective removal of aluminum layers present in the MAX phase. In this section, different strategies utilized for the removal of the Al layers during the preparation of MXenes are discussed.

5.2.1 Selective Etching Strategy

Gogotsi and Barsoum *et al.* [19] explored the high reactivity between Al and fluorine ions, which is described by Equations 5.1 to 5.5, and show how the process occurs in the presence of hydrofluoric acid (HF):

$$Ti_3AlC_2 + 3HF \rightarrow AlF_3 + \frac{3}{2}H_2 + Ti_3C_2 \tag{5.1}$$

$$Ti_3C_2 + 2HF \rightarrow Ti_3C_2(F)_2 + H_2 \tag{5.2}$$

$$Ti_3C_2 + 2H_2O \rightarrow Ti_3C_2(OH)_2 + H_2 \tag{5.3}$$

$$Ti_3C_2 + 2H_2O \rightarrow Ti_3C_2(O)_2 + 2H_2 \tag{5.4}$$

Structural characterization of the samples before and after HF treatment was carried out by X-ray diffraction (Figure 5.1a). The result revealed that the intense diffraction peak in the XRD pattern of the Ti_3AlC_2 phase located at 39° completely vanished after the HF etching. In addition, after HF treatment, the signal attributed to the (002) plane shifted toward the lower angle, which indicates a widening of the interlayer spacing. A new 2D transition metal carbide material has just been identified. Figure 5.1b

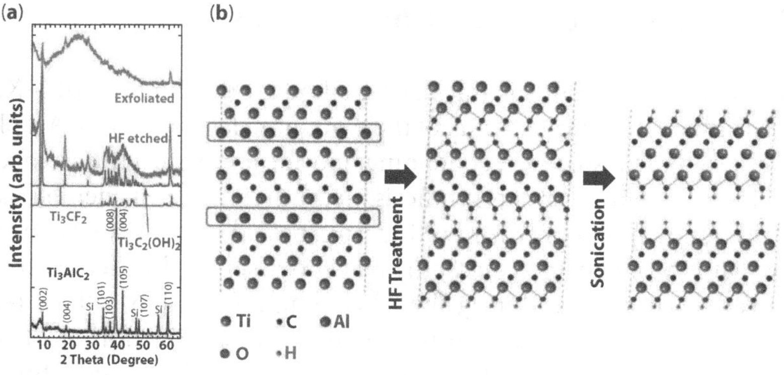

Figure 5.1 (a) XRD pattern of Ti_3AlC_2 before and after HF treatment. (b) Schematic of the exfoliation process for Ti_3AlC_2 (reproduced with the permission from ref. [3]).

shows changes in the composition of the Ti_3AlC_2 MAX phase after the HF treatment and the sonication of the dispersed solution.

Later on, the HF etching method was successfully applied to other MAX phases, including V_2AlC [20], Nb_2AlC [21], Nb_4AlC_3 [22], Ta_4AlC_3 [23], Mo_2TiAlC_2 [24], $Hf_3(AlSi)_4C_6$ [25], $Zr_3Al_3C_5$ [26], and Mo_4VAlC_4 [27], to produce stacked layers with obvious open gaps, like morphology, confirming the effectiveness of HF in the removal of the Al layers in the MAX phases. To date, the mixture of hydrochloric acid (HCl) with fluorine salts, such as LiF [28, 29], NaF [28–30], KF [28, 29], and NH_4F [28, 29] have been also investigated as fluoride etchants. However, the use of HF are of high concern regarding environmental and safety issues [16, 17]. The works of Xie *et al.* [31] and Li *et al.* [32] have revealed that inexpensive alkali salts, such as NaOH and KOH, can be used to substitute for the fluoride compounds conventionally used for MAX phase etching owing to their good reactivity with Al [31–33]. It should be also noted that the use of alkali as etchants leads to the formation of aluminum hydroxides over the MXene surface. To overcome this problem, Xie *et al.* [31] proposed a two-step etching process consisting, first, of soaking the Ti_3AlC_2 MAX powder in 1 M NaOH solution for 100 h and subsequently in 1 M H_2SO_4 solution for 2 h at 80°C. In this method, NaOH is used to remove aluminum layers out of Ti_3AlC_2, whereas acid treatment eliminates Al (oxide) hydroxide contaminants. Both the temperature and concentration of the base solution have been also optimized by Li *et al.* [32] to produce $Ti_3C_2T_x$ MXenes with 92 wt% purity only in a one-step process. Their results have shown that high-purity $Ti_3C_2T_x$ MXenes can be synthesized in an alkaline

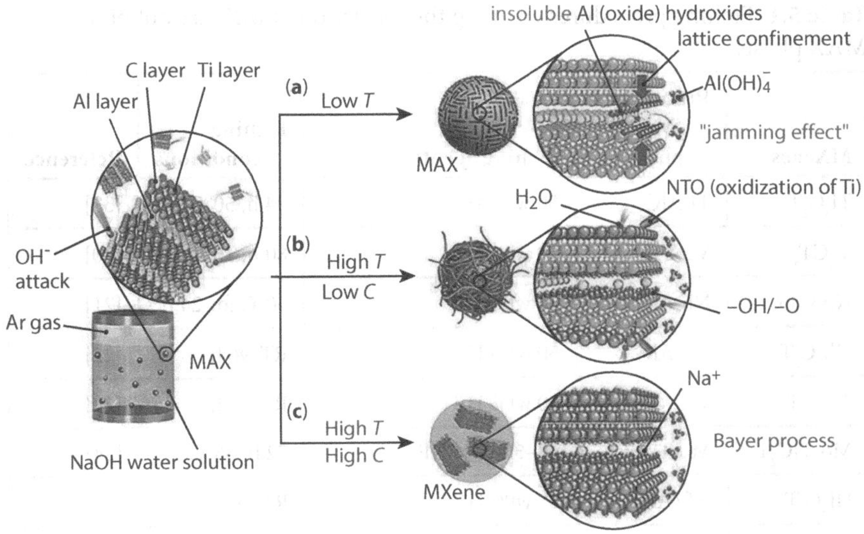

Figure 5.2 Influence of etching conditions, i.e., the temperature and concentration of NaOH on the purity of the MXenes (reproduced with the permission from ref. [32]).

medium by increasing both temperatures and concentration of NaOH as show in Figure 5.2.

Electrochemical etching method has been explored as well. The electrochemical method involves the etching of Al in the MAX phase under an applied voltage. This method can be carried out in both: two- or three-electrode arrangement, where the bulk Ti_3AlC_2 [34] or MAX phase is coated onto a conducting substrate [35] acting as the working electrode. The successful synthesis of MXenes from different MAX phases, like Ti_3AlC_2 [34], Ti_2AlC [36], V_2AlC [37], and Cr_2AlC [37], immersed in 1 M HCl using the electrochemical etching method has already been reported. NaCl or HCl solutions are typically used as electrolytes for this process, which does not necessarily involve the presence of fluorine ions for etching. Electrochemical etching is regarded as a green and safe synthesis method since it can be carried out in fluorine-free electrolytes. However, the formation of carbide-derived carbon from the MAX phase on the surface of produced MXene surface hinders the subsequent etching process resulting in a low MXene yield [35]. Various etching agents and synthesis conditions are set in Table 5.1.

Table 5.1 Etching procedures enabling the formation of MXenes out of the MAX phase.

MXenes	Precursor (MAX phase)	Etching agents	Etching conditions	Reference
$Ti_3C_2T_x$	Ti_3AlC_2	50 wt% HF	24 h, 50°C	[38]
V_2CT_x	V2AlC	50 wt% HF	80 h, 35°C	[20]
Nb_2CTx	Nb_2AlC	50 wt% HF	30°C for 24 h	[21]
$Nb_4C_3T_x$	Nb_4AlC_3	50 wt% HF	RT, 96 h	[22]
$Ta_4C_3T_x$	Ta_4AlC_3	50 wt% HF	RT, 72 h	[23]
$Mo_2TiC_2T_x$	Mo_2TiAlC_2	48–50 wt% HF	72 h, 55°C	[24]
$Hf_3C_2T_x$	$Hf_3(AlSi)_4C_6$	35 wt% HF	RT, 60 h	[25]
$Zr_3C_2T_x$	$Zr_3Al_3C_5$	50 wt% HF	RT, 72 h	[26]
$Mo_4VC_4T_x$	Mo_4VAlC_4	48–51 wt% HF	8 days, 55°C	[27]
Mo_2CT_x	Mo_2Ga_2C	2.00 g LiF + 6 M HCl/2.00 g NaF + 6 M HCl/2.00 g KF + 6 M HCl/2.00 g NH_4F + 6 M HCl	180°C, 24 h	[12]
$Ti_3C_2T_x$/ Ti_2CT_x	Ti_3AlC_2/ Ti_2AlC	2.07 g LiF + 36–38 wt.% HCl/3.35 g NaF + 36–38 wt.% HCl/7.52 g KF·$2H_2O$ + 36–38 wt.% HCl/2.96 g NH_4F + 36–38 wt.% HCl	55°C, 48 h	[29]
$Ta_4C_3T_x$	Ta_4AlC_3	4 M NaF + 12 M HCl	60°C for 48 h	[30]
V_2CT_x	V_2AlC	1.5 g NaF + 6 M HCl	90°C, 5 days	[39]
$Ti_3C_2T_x$	Ti_3AlC_2	1 M NaOH + 1 M H_2SO_4	(80°C, 100 h) + (80°C, 2 h)	[31]
$Ti_3C_2T_x$	Ti_3AlC_2	27.5 M NaOH / 125 M KOH	270°C, 12 h/180°C, 24 h	[32]

(*Continued*)

Table 5.1 Etching procedures enabling the formation of MXenes out of the MAX phase. (*Continued*)

MXenes	Precursor (MAX phase)	Etching agents	Etching conditions	Reference
$Ti_3C_2T_x$	Ti_3AlC_2	1 M NH_4Cl + 0.2 M TMAOH	5 V, 5 h	[34]
Ti_2CT_x	Ti_2AlC	2 M HCl/1 M HCl	0.6 V, 120 h/0.3 V, 9 h, 50°C	[36]
V_2CT_x	V_2AlC	1 M HCl	0.5 V, 9 h, 50°C	[37]
Cr_2CTx	Cr_2AlC	1 M HCl	1 V, 9 h, 50°C	[37]
$Ti_3C_2T_x$	Ti_3AlC_2	$ZnCl_2$	550°C, 5 h, Ar	[40]
V_2CT_x	V_2AlC	$ZnCl_2$	550°C, 5 h, Ar	[40]
Nb_2CT_x	Nb_2AlC	AgCl	700°C, 24 h, Ar	[41]

5.2.2 Delamination of Multilayer MXenes

In general, MXenes obtained after Al etching in the MAX precursor exhibit stacked accordion-like multilayers morphology. However, similar to other two-dimensional (2D) materials, both the single-layer MXenes exhibit high specific surface area, good hydrophilicity, high electrical conductivity, and excellent mechanical properties compared to their multilayered counterparts [42]. Therefore, tremendous research efforts have been dedicated to elaborate a simple method for the efficient delamination of the obtained multilayer MXenes after the etching process into single-layer or few-layer MXenes. In the multilayer MXenes, the individual layers stick together by van der Waals forces [43]. The main purpose of separation present between stacked layers is the intercalation of organic molecules or inorganic ions into this space to weaken the interaction between particular layers and widen the interlayer spacing. Both dimethyl sulfoxide (DMSO) and N,N-dimethylformamide (DMF) were the first compounds used for the delamination of the multilayer $Ti_3C_2T_x$ into single-layer sheets (Figure 5.3) [44]. However, the use of DMSO for the delamination of other types of

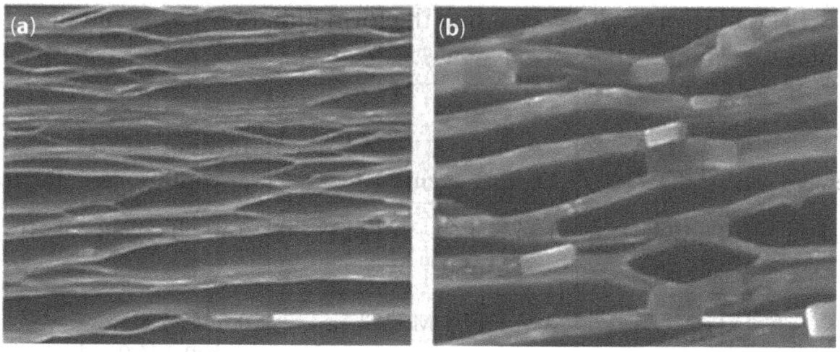

Figure 5.3 Morphology of $Ti_3C_2T_x$ (a) before (b) after the intercalation in DMF (reproduced with permission from ref. [42]).

MXenes, such as V_2CT_x and Mo_2CT_x, was not successful. Different organic alkali, such as tetrabutylammonium hydroxide (TBAOH), hydroxyl choline, and *n*-butylamine, have been investigated as well using a two-step delamination procedure. The first step involves the dispersion of the multilayer MXenes in organic alkali solution. This allows the alkali cations to be intercalated between the MXenes sheets, thus swelling the multilayer structure. In the next step, the delamination of the multilayer MXenes is further achieved by either handshaking or ultrasonic treatment.

5.3 Kinetic and Reaction Mechanism of Electrocatalytic Water Splitting

Hydrogen and oxygen evolutions (HER and OER) through water splitting in the presence of semiconductor catalysts have long been investigated as a potential strategy for greener energy production. The change in the free energy (ΔG) required to split one H_2O molecule to H_2 and $\frac{1}{2}O_2$ under standard conditions (at 25°C and 1 atm) is $\Delta G = 237.2$ kJ mol^{-1}, which corresponds to a potential barrier of 1.23 V to drive the splitting reaction [45]. In general, the water splitting systems suffers from sluggish kinetics of the electrochemical reactions at the surface of the electrodes associated with the unbalanced production of H_2 and O_2 [46]. Therefore, the use of electrocatalysts is essential to speed up the sluggish HER and OER kinetics. The performance evaluation of the electrocatalyst is based on the analysis of the pivotal electrocatalytic kinetic parameters such as overpotential and Tafel slope. In this section, these key parameters are provided and discussed.

5.3.1 Key Kinetic Parameters Characterizing Electrocatalytic Activity

5.3.1.1 Overpotential

The applied potential to the electrode is defined according to Nernst equation as follows:

$$E = E^0 + \frac{RT}{nF} ln \frac{C_{Ox}}{C_{Red}} \quad (5.5)$$

where C_{Ox} and C_{Red} are the concentrations of the oxidized and reduced species, respectively, E^0 and E represents the formal potential of the overall reaction and the potential of the electrode. R stands for the gas constant, T for the absolute temperature, and F for Faraday's constant. In the ideal situation, the potential hold at the working electrode to drive the electrochemical reaction should be equal to the equilibrium reaction potential. However, in the most of the cases, a much higher potential than that at equilibrium should be applied to the electrode to drive the process. The overpotential (η) is defined as the potential difference between the applied potential E (i.e., when a given electric current flows through the system, $I \neq 0$) and the potential E_{eq} under equilibrium condition ($I = 0$) [47].

$$\eta = E - E_{eq} \quad (5.6)$$

Notably, this potential is commonly referred to in the literature as the value that has to be applied to the working electrode to achieve a current density of 10 mA cm^{-2}. The electrocatalytic performance of an electrode in an electrochemical system is defined by the value of the overpotential. A lower value of the overpotential is indicative of a superior electrocatalytic ability.

5.3.1.2 Tafel Slope

The current at the anode/electrolyte interface and at the cathode/electrolyte interface are related to the overpotential (η) according to the Butler–Volmer equation [48]:

$$i_A = i_o \left(e^{\frac{\alpha_A nF}{RT} \times \eta} \right) \quad (5.7)$$

$$i_C = i_o \left(e^{-\frac{\alpha_C nF}{RT} \times \eta} \right) \tag{5.8}$$

where α_A and α_C are the charge transfer coefficients for the electrochemical reactions occurring at the anode and cathode, respectively, R stands for the gas constant, T for the absolute temperature, F for the Faraday constant, and i_o for the exchange current under equilibrium condition ($\eta = 0$, $E = E_{eq}$).

In the logarithmic coordinates Equations 5.7 and 5.8 can be rewritten in the following form

$$log(i) = \log(i_0) + \left(\frac{1}{b}\right)\eta \tag{5.9}$$

The plot of $\eta\, vs.\, log(i)$ is known as the Tafel plot, and the Tafel slope is defined as follows:

$$b = \frac{\partial\eta}{\partial \log(i)} = \frac{2.303RT}{\alpha F} \tag{5.10}$$

According to Equation 5.10, the Tafel slope (b) is inversely proportional to the charge transfer coefficients (α). This implies that the smaller values of the Tafel slope indicates fast electrochemical kinetics. Since the most important parameters for the electrocatalyst performance analysis have been evoked, the next section is dedicated to the description of the HER and OER mechanisms.

5.3.2 Mechanism of Water Splitting Process

Regarding water splitting process, at the beginning, the reactants are adsorbed on the surface of the electrocatalyst, and it facilitates charge transfer at the electrode/electrolyte interface [49]. The different reactions with intermediates arising during HER and OER strongly depend on the nature of the electrolyte (i.e.. acidic or alkaline media). In general, HER can proceed through two different mechanisms, namely, the Volmer–Heyrovsky and Volmer–Tafel mechanisms [50]. In both cases (i.e., Volmer–Heyrovsky and Volmer–Tafel mechanisms), the first step consists of the adsorption of the proton (H^+) on the surface of the electrocatalyst to form the adsorbed intermediate H^+ site (H*). Then, depending on the amount of adsorbed intermediate H^+ site present on the surface of the catalyst, the next steps will go through the Volmer–Heyrovsky mechanism (if the amount of H* is low) or by the Volmer–Tafel mechanism (if the amount of H* is high) [51]. In the Volmer–Tafel mechanism, owing to the large amount of H*, H_2 is produced by a combination of two H*. In the case of the Volmer–Heyrovsky mechanism, the situation is completely different. In an acidic electrolyte, H* can combine with both a proton and an electron (e^-) to produce an H_2, whereas in a basic medium, H_2 is produced by combining H* with both H_2O and an e^-. Figure 5.4 depicts hydrogen evolution according to both Volmer–Heyrovsky and Volmer–Tafel mechanisms.

Compared to HERs, OER is more complex and involve three intermediate reactions. In an acidic electrolyte, the electrocatalyst first adsorbs the H_2O molecule. Then, water oxidation takes place to form the adsorbed intermediate HO* sites. In the next step, the absorbed HO* couples with the proton (H^+) and electron (e^-) to form the intermediate site O*, which in turn combines with another H_2O to produce the intermediate HOO*. Finally, HOO* is oxidized by electron transfer to form O_2 gas. However, in an alkaline electrolyte, the catalyst adsorbs the hydroxyl (OH^-), and the OH^- oxidizes to form the intermediate HO* sites. Then, HO* reacts with another hydroxyl group to produce O* and H_2O. Next, an intermediate HOO* is formed by the coupling of the adsorbed O* with OH^-. Finally, HOO* reacts with an OH^- to generate O_2 and H_2O. In the OER process, three different intermediaries, i.e., HO*, O*, and HOO* are involved as described in Figure 5.5 [51].

Electrocatalytic water splitting offers the best alternatives for the production of clean, sustainable, and renewable energy. Compared to other

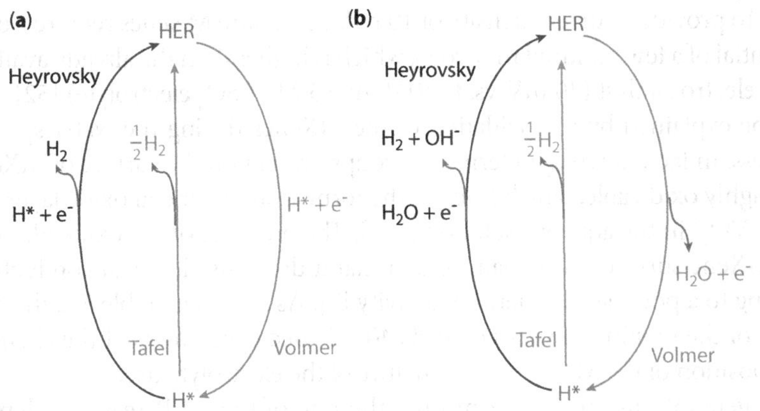

Figure 5.4 Schematic illustration of HER according to Volmer–Heyrovsky and Volmer–Tafel mechanisms (a) in acidic and (b) alkaline media.

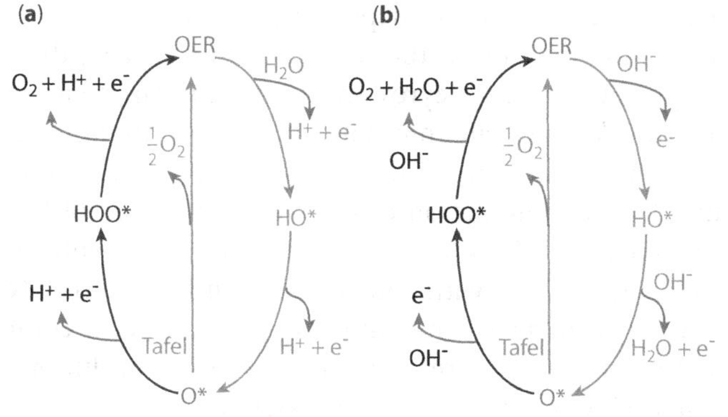

Figure 5.5 Schematic illustration of OER process (a) in acid and (b) alkaline media.

non-commercial electrocatalysts, MXenes are considered as the highly favored catalyst for the water splitting process. This results from their outstanding electrical conductivity associated with their unique structural and electrochemical properties. The role of MXenes as a potential material in promoting electrocatalytic HER and OER process is discussed in the next section.

5.4 MXene-Based Electrocatalyst for Hydrogen Evolution Reaction

Typically, pristine MXenes with different chemical compositions, including $Ti_3C_2T_x$, Ti_2CT_x, V_2CT_x,, Ta_2CS_2, $Nb_4C_3T_x$, $V_4C_3T_x$, Ti_3CN, Ta_4C_3Tx, and Cr_2CT_x, have been studied as electrocatalysts for water splitting (show in Table 5.2). To provide a current density of 10 mA cm^{-2}, pure MXenes require a over-potential of a few hundred millivolts, which is higher than the already available Pt/C electrocatalyst (40 mV vs. to RHE in 0.5 M H_2SO_4 electrolyte) [52]. This can be explained by the oxidation of the MXenes during the water splitting process. In fact, the oxygen terminal groups present on the surface of MXenes are highly oxidizable, which leads to the formation of a metal oxide layer (i.e., TiO_x, VO_x) in the aqueous solution [7, 9]. The presence of the oxide film over the MXene surface is the main reason that it drops its electrical conductivity leading to a poor electrochemical activity [7]. As shown in Table 5.2, the Tafel slope of the pristine MXenes exceeds 40 mV dec^{-1} regardless of the chemical composition of the MXenes or the nature of the electrolyte used.

In general, the steps determining the rate of the HER reaction depend on the value of the Tafel slopes. In this regard, the Tafel slopes of 120, 40, and 30 mV dec^{-1} imply that the steps determining the rate of the HER

Table 5.2 Overpotential and Tafel slope of pristine MXenes or MXenes modified with heteroatoms for HER.

Type of MXenes	Overpotential/ mV vs. RHE	Tafel slope/ mV dec^{-1}	Electrolyte	Reference
$Ti_3C_2T_x$	718	1,063	0.5 M H_2SO_4	[7]
$Ti_3C_2T_x$	843	340	0.5 M H_2SO_4	[53]
$Ti_3C_2T_x$	491	454	0.5 M H_2SO_4	[54]
$Ti_3C_2T_x$	449	350	0.5 M H_2SO_4	[55]
$Ti_3C_2T_x$ (layered/ nanosheets)	265/75	138/100	0.5 M H_2SO_4	[2]
$Ti_3C_2T_x$	Not measured	162	0.5 M H_2SO_4	[56]
N–$Ti_3C_2T_x$	162	69	0.5 M H_2SO_4	[57]
E–$Ti_3C_2O_x$	190	60.7	0.5 M H_2SO_4	[58]
Co–Ti_3C_2	300	147	1 M NaOH	[59]
Pt NCs–$Ti_3C_2T_x$	40	50.8	0.5 M H_2SO_4	[60]
Pt–$Ti_3C_2T_x$	34	29.7	0.5 M H_2SO_4	[61]
Ru–N–$Ti_3C_2T_x$	23	42	0.5 M H_2SO_4	[62]
$Ti_3C_2T_x$	259	268	1 M KOH	[63]
$Ti_3C_2T_x$	443	311	1 M KOH	[64]
Ti_2CT_x	609	124	0.5 M H_2SO_4	[8]
V_2CT_x	618	193	0.5 M H_2SO_4	[9]
Defective V_2CT_x (O-defects)/defective V_2CT_x (O-defects) + stain activation	300/134	137/48.4	0.5 M H_2SO_4	[1]
Ti_3CN	600	284	0.5 M H_2SO_4	[13]
$Nb_4C_3T_x$ (etching time = 180 h/220 h)	263/335	122/140	1 M KOH	[65]
$V_4C_3T_x$	Not measured	236	0.5 M H_2SO_4	[12]

(Continued)

Table 5.2 Overpotential and Tafel slope of pristine MXenes or MXenes modified with heteroatoms for HER. (*Continued*)

Type of MXenes	Overpotential/ mV vs. RHE	Tafel slope/ mV dec^{-1}	Electrolyte	Reference
Mo_2CT_x	283	82	0.5 M H_2SO_4	[8]
Mo_2CT_x	300	Not measured	0.5 M H_2SO_4	[11]
Ta_2CS_2/Ta_4C_3Tx	105/139	73.7/122.9	1 M KOH	[10]
Cr_2CT_x (etching time = 0.5 h/8 h)	212/225	109/110	1 M KOH	[14]

reaction are the Volmer, Heyrovsky, and Tafel steps, respectively (shown in Figure 5.4), described in the following equations [66]:

$$H_2O + e^- \rightarrow H^* + OH^-, \text{Volmer step} \tag{5.11}$$

$$H^* + H_2O + e^- \rightarrow H_2 + OH^-, \text{Heyrovsky step} \tag{5.12}$$

$$H^* + H^* \rightarrow H_2, \text{Tafel step} \tag{5.13}$$

The large Tafel slopes for pristine MXenes (>40 mV dec^{-1}) indicate sluggish kinetics of the electrochemical reactions corresponding to a mixture of Heyrovsky and Volmer steps [9, 66] according to Equations 5.11 and 5.12 compared to the commercial Pt/C electrocatalyst (18.4 mV dec^{-1}) [9], which undergoes the Tafel pathway. Therefore, it is necessary to accelerate the electrochemical reaction rate occurring at MXene-based materials to achieve better electrocatalytic performance attractive for commercialization. Following that, several methods were explored to improve the electrochemical activity of metal carbides, namely, the (i) control of surface termination groups, (ii) introduction of structural defects, and (iii) doping with heteroatoms, which are the most reported in the literature.

5.4.1 Effects of Surface-Terminating Groups on the Catalytic Performance of MXenes in the HER Process

As has been already discussed, the MXenes can be easily oxidized leading to the formation of the metal oxide layer on the MXene surface. This significantly dropped the electrical conductivity of the material. Hence, various chemical or thermal treatments (under reactive gas) of the as-synthesized MXene-based materials have been undertaken to improve its conductivity before its further utilization as electrocatalyst. These treatments are not only used to make defects in MXenes (i.e., generation of O, T or C-vacancy) or introduction of heteroatoms into the MXene matrix but also allow the modification of the terminated groups developed on the surface. Those terminated groups, such as –O, –OH, –F, –Cl, and –S, participate in the charge transfer process occurring at the MXene/electrolyte interface, thus contributing to the improvement of the HER activity. For example, the introduction of ordered sulfur-terminated group in tantalum carbide MXenes (Ta_2CS_2) by Wu *et al.* [10] has increased the electrical conductivity by approximately 22.5 times with respect to the pristine $Ta_4C_3T_x$. It results from the high charge carrier concentration found for Ta_2CS_2, which is almost 80 times higher than that of $Ta_4C_3T_x$. Cl-terminated MXenes were prepared by thermal treatment of Ti_3AlC_2 MAX phase loaded with copper chloride salt at a high temperature of 550°C for 5 h under argon gas flows [64]. For HER, the obtained Cl-terminated titanium carbides ($Ti_3C_2Cl_2$) as electrocatalysts has demonstrated significant reduction in both overpotential (259 mV) at 10 mA cm^{-2} and 259 mV, and Tafel slope (92 mV dec^{-1}) compared to the prepared MXenes (444 mV, 311 mV dec^{-1}) obtained simply by HF etching, as shown in Figure 5.6 [64].

The modification of the surface termination can also be realized through alkalization process of the already synthesized MXenes or by employing appropriate thermal treatment. For example, fluorine-terminated MXenes were prepared by annealing of the Ti_2CT_x MXenes at 200°C for 2 h in vacuum conditions [2]. This process allows the removal of –OH and –O surface termination, while the fluorine-terminated group content remains

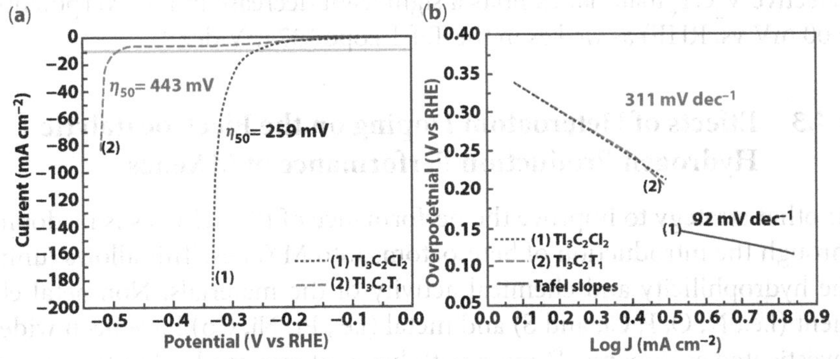

Figure 5.6 (a) Linear sweep voltammetry curves and (b) Tafel slope of pristine $Ti_3C_2T_x$ and $Ti_3C_2Cl_2$ (reproduced with permission from ref. [62]).

nearly unchanged. In this way, the F-rich surface of Ti_2CT_x was fabricated by Li *et al.* [2] and further used as an electrocatalyst for the water splitting process. The as-prepared F-terminated Ti_2CT_x was then tested as a catalyst [2] and exhibits a small overpotential of 75 mV toward HER.

5.4.2 Effects of the Structural Defects on the Electrocatalytic Activity

Generally, in the field of materials science, a defect is defined as any localized disruption of a perfect crystalline order [67]. This includes point defects (e.g., vacancy, heteroatom substitution, and presence of interstitial atoms), linear defects (e.g., edge and screw dislocations), two-dimensional defects (e.g. crystal boundaries), and bulk defects [68]. It is well known that the structural defects strongly influence the electrical and mechanical properties of the materials [69]. Therefore, the precise control of the defects are explored to tune the electrical conductivity of MXenes, thus improving the electrochemical activity of the material. Related to this, the choice of the point defect that has to be generated in the material has paramount importance. For example, Gan *et al.* [69] pointed out that the presence of carbon vacancy in titania carbide MXenes strongly enhances the electrical conductivity of the materials compared to titanium vacancy. Jin *et al.* [1] showed, through density functional theory calculation, that optimizing the gradient amount of O defects in the surface of V_2CT_x MXenes can modulate the outermost orbitals of oxygen atoms to be at half occupied. O defects are generated by heat treatment in a mixture of Ar/H_2 atmosphere, and the defect amount is controlled by adjusting the heating time [1]. Compared to previously reported overpotential and Tafel slope recorded in acid electrolyte V_2CT_x by Wang *et al.* [9] prepared without heat treatment (i.e., 618 mV vs. RHE, 193 mV dec^{-1}), the defective V_2CT_x material exhibits a significant decrease in the overpotential (300 mV vs. RHE) as well as in the Tafel slope 137 mV dec^{-1}.

5.4.3 Effects of Heteroatom Doping on the Electrocatalytic Hydrogen Production Performance of MXenes

Another strategy to improve the performance of the MXenes is its doping through the introduction of heteroatoms into MXenes. This allows tuning the hydrophilicity and chemical activity of the materials. Nonmetal element (i.e., N, O, P, Cl, and S) and metal (i.e., Pt, Ni, Co) have been widely investigated as dopants. Doping with heteroatoms can lead to the modification of electrical and structural properties of the materials compared to their undoped counterparts. In general, MXenes exhibit a strong hydrogen adsorption feature. As a consequence, poor electrocatalytic activity results from the strong binding between the adsorbed H and -O terminates in the MXene surface. It has been demonstrated that the presence of heteroatoms in the structure can effectively weaken the H–O interaction, thereby enhancing efficient hydrogen evolution [70]. The work of Gan *et al.* [70] revealed that after doping of Mo_2CO_2 with Fe, Co, and Ni, the material exhibits better activity than pristine MXenes. Phosphorization of Mo_2CT_x MXenes was performed by Qu *et al.* [11] to achieve P-doped Mo_2CT_x. A remarkable decrease in the overpotential for the doped material (100 mV) with respect to the bare Mo_2CT_x (300 mV) was observed. Han *et al.* [57] reported a drop in the overpotential approximately 3.5 times with respect to that of pristine MXenes after N doping. An electrocatalyst with outstanding activity comparable to that of the commercial Pt/C electrocatalyst has been synthesized through ruthenium and nitrogen co-doping into $Ti_3C_2T_x$ [62]. The fabricated material exhibits an overpotential of 23, 27, and 81 mV in 0.5 M H_2SO_4, 1 M KOH, and 1 M PBS electrolytes, respectively. This work have demonstrated that the synergistic effect between metal and nonmetal elements can contribute in promoting the HER catalytic activity.

Table 5.3 Overpotential and Tafel slope of pristine MXenes or MXenes modified with heteroatoms for OER.

MXenes	Overpotential/ mV vs. RHE	Tafel slope/ mV dec^{-1}	Electrolyte	Reference
$Ti_3C_2T_x$	570	420	0.5 M H_2SO_4.	[71]
Ag–$Ti_3C_2T_x$	250	180	0.5 M H_2SO_4	
Pt–V_S–Ti_2CSBr	790	Not measured	0.5 M H_2SO_4	[72]
Pt–V_O–Ti_2CO_2	310	Not measured	0.5 M H_2SO_4	[73]
Ta_2CS_2	287	108.2	1 M KOH	[10]
Co@V_2CT_x	242	90.4	1 M KOH	[74]
Co@Nb_2CT_x	310	100.2	1 M KOH	
Co@$Ti_3C_2T_x$	388	109.2	1 M KOH	
Cl–$Ti_3C_2T_x$	Not measured	48	1 M KOH	[64]
$Ti_3C_2T_x$	Not measured	136	1 M KOH	

5.5 MXene-Based Electrocatalyst for Oxygen Evolution Reaction

MXene materials are also widely studied for the oxygen evolution reaction occurring in the anodic regime of the potential range. Similar to the HER process, pristine MXenes exhibit slow kinetics for the OER process, as shown in Table 5.3. Recent advances in optimizing the OER process via defect generation, heteroatom doping, and modification of surface terminations of electrocatalyst based on MXenes are discussed in this section.

5.5.1 Effects of Surface-Terminating Groups on the Catalytic Performance of MXenes in the OER Process

The –O, –F, and –OH termination groups present on the MXene surface also can greatly affect the OER performance of the family of MXene materials. In fact, the high electrical conducting sulfur-terminated tantalum carbide MXenes (Ta_2CS_2) prepared by Wu *et al.* [10] demonstrated an overpotential of 287 mV vs. RHE at 10 mA cm^{-2} and faster OER kinetics (a Tafel slope of 108.2 mV dec^{-1}) compared to the commercial RuO_2 (327 mV vs. RHE, 115.7 mV dec^{-1}). The introduction of Cl terminated on the surface of titanium carbide MXenes, as prepared by Sarfraz *et al.* [63], improved the OER performance (overpotential of 364 mV and a Tafel slope of 50 mV dec^{-1}) of the material compared to the pristine MXenes used as an electrocatalyst (overpotential of 418 mV and a Tafel slope of 176 mV dec^{-1}) as shown in Figure 5.7a, b.

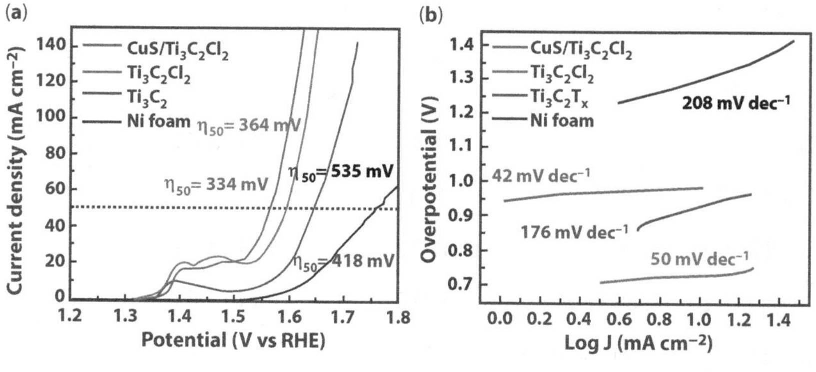

Figure 5.7 (a) Linear sweep voltammetry curves and (b) Tafel slope of electrocatalyst based on MXenes (reproduced from ref. [61] under CC BY-NC Licence).

5.5.2 Effects of Heteroatom Doping on the Electrocatalytic Oxygen Production Performance of MXenes

The doping of MXenes with transition metals not only weakens the binding between hydrogen and the surface termination group in the MXene surface but also affects the interaction strength of the intermediates appearing during the oxygen evolution reaction as indicated in Figure 5.5. In this way, it promotes kinetics of the OER reaction [75]. Investigation on the effect of Pt-doped Ti_2CNL (N = O, S and Se, L = F, Cl, Br, and I) on the OER performance by Ma *et al.* [72] revealed that especially the Pt–V_S–Ti_2CSBr electrocatalyst (V_S for sulfur vacancy) exhibits very low overpotential of 340 mV vs. RHE. They have also pointed out of that the density of states (DOS) of the electrocatalysts is one of the key parameters that affect the OER performance of the electrodes. In fact, the visible small peaks on the

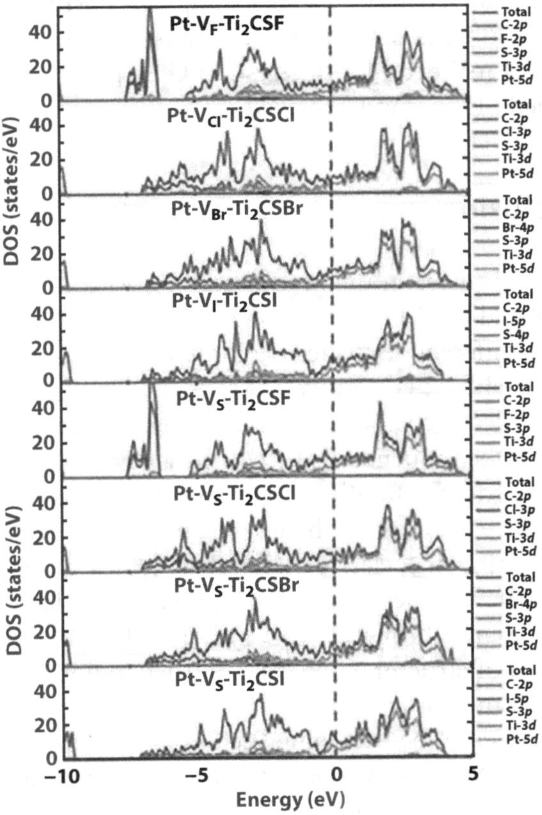

Figure 5.8 The density of states (DOS) of the Pt-doped Ti_2CNL (N = O, S and Se, L = F, Cl, Br, and I) (reproduced with permission from ref. [70]).

Pt-doped Ti_2CNL catalyst, shown in Figure 5.8 at the Fermi level, indicate the presence of some charge carriers that contribute to the increase in the conductivity of the catalyst. In general, the DOS is sensitive to the electronegativity and atomic size of the heteroatom—dopant or heteroatom—defect introduced in the MXene matrix [72]. Indeed, the electronegativity of the heteroatoms will lead to the redistribution charge carriers in the modified materials, whereas the size of the heteroatom will lead to the change in the lattice constants [72]. The increase in the lattice constants will obviously decrease the electrical conductivity of the materials and, as a consequence, will drop its electrocatalytic performance. Electrocatalysts based on titanium carbide MXenes containing different nitrogen concentrations were investigated for OER by Tang *et al.* [76]. The introduction of the non-metal dopant has led to a remarkable decrease in the water contact angle from 38.5° for pristine $Ti_3C_2T_x$ to 9° for the N-doped material. This improvement in the hydrophilicity of MXenes after nitrogen doping indicates an increase in active sites participating in the REL reaction, thereby raising the OER kinetics of the electrocatalysts. In fact, a drop in the Tafel slope from 243.4 mV dec^{-1} for the fresh Ti_3C_2 MXenes to 216.4 mV dec^{-1} for the N-doped MXenes was achieved.

5.6 Conclusion

Herein, recent advances in the modification of MXene-based electrocatalysts for HER and OER processes are presented. Particular attention was devoted to the different strategies explored for the etching of the MAX precursor to produce high-purity MXenes. MXenes are widely studied as catalysts for electrochemical water splitting. In general, those studies mainly focus on the influence of the chemical composition of MXenes or their functionalization with other chemical compounds on the electrocatalytic performance toward both hydrogen and oxygen evolutions. Herein, the factors affecting the electrochemical activities of pristine MXenes are discussed. Then, achievements in improving the electrocatalytic performance of MXenes through (i) control of surface-terminating groups, (ii) introduction of structural defects, and (iii) doping with heteroatoms are presented. Although encouraging results have been reported, several challenges still need to be overcome such as long-term stability in water-based electrolytes. Suitable synthesis methods enabling effective control of the structural defect and surface termination groups are still extensively explored. Preparing MXene materials that are highly crystalline and free from metal oxide contamination while the whole process is environmentally friendly is also challenging. From an electrocatalytic perspective, more computational studies using DFT calculations could be beneficial for new variations of MXenes and for better understanding of the HER and OER processes with their assistance. Overall, MXene-based materials show significant potential in the field of electrocatalysis and could be candidates to compete with platinum-based materials. Therefore, it will be worth establishing the conformational relationship between MXene electrocatalysts and their electrochemical performances to benefit water splitting applications.

Acknowledgments

The work was financially supported by the National Science Centre (Poland) with project no. 2020/02/Y/ST8/00030 and Czech Science Foundation with grant no. 21-05030K.

AI disclosure statement: The author acknowledges the use of AI-based tools for minor language editing or grammatical correction. However, all intellectual contributions, analyses, and conclusions presented in this chapter are entirely the author's own.

References

1. Wan, J., Wang, Y., Zhang, H., Wang, Y., Local electrons perturbation of V_2CT_x via defect and strain for efficient hydrogen production. *Chem. Eng. J.*, 470, 144151, 2023.
2. Li, S., Tuo, P., Xie, J., Zhang, X., Xu, J., Bao, J., Pan, B., Xie, Y., Ultrathin MXene nanosheets with rich fluorine termination groups realizing efficient electrocatalytic hydrogen evolution. *Nano Energy*, 47, 512–518, 2018.
3. Huang, J., Feng, M., Peng, Y., Huang, C., Yue, X., Huang, S., Encapsulating Ni Nanoparticles into Interlayers of Nitrogen-Doped Nb_2CT_x MXene to Boost Hydrogen Evolution Reaction in Acid. *Small*, 19, 2206098, 2023.
4. Sokol, M., Natu, V., Kota, S., Barsoum, M.W., On the Chemical Diversity of the MAX Phases. *Trends Chem.*, 1, 210–223, 2019.
5. Zhan, X., Si, C., Zhou, J., Sun, Z., MXene and MXene-based composites: Synthesis, properties and environment-related applications. *Nanoscale Horiz.*, 5, 235–258, 2020.
6. Lim, K.R.G., Shekhirev, M., Wyatt, B.C., Anasori, B., Gogotsi, Y., Seh, Z.W., Fundamentals of MXene synthesis. *Nat. Synth.*, 1, 601–614, 2022.

7. Wang, Z., Yu, K., Feng, Y., Qi, R., Ren, J., Zhu, Z., Stabilizing Ti3C2Tx-MXenes with TiOF2 nanospheres intercalation to improve hydrogen evolution reaction and humidity-sensing performance. *Appl. Surf. Sci.*, 496, 143729, 2019.
8. Seh, Z.W., Fredrickson, K.D., Anasori, B., Kibsgaard, J., Strickler, A.L., Lukatskaya, M.R., Gogotsi, Y., Jaramillo, T.F., Vojvodic, A., Two-Dimensional Molybdenum Carbide (MXene) as an Efficient Electrocatalyst for Hydrogen Evolution. *ACS Energy Lett.*, 1, 589–594, 2016.
9. Wang, J., Guan, Y., Zhang, Q., Zhu, H., Li, X., Li, Y., Dong, Z., Yuan, G., Cong, Y., Well-dispersed ultrafine Pt nanoparticles anchored on oxygen-rich surface of V_2CT_x (MXene) for boosting hydrogen evolution reaction. *Appl. Surf. Sci.*, 582, 152481, 2022.
10. Wu, T., Pang, X., Zhao, S., Xu, S., Liu, Z., Li, Y., Huang, F., One-Step Construction of Ordered Sulfur-Terminated Tantalum Carbide MXene for Efficient Overall Water Splitting. *Small Struct.*, 3, 2100206, 2022.
11. Qu, G., Zhou, Y., Wu, T., Zhao, G., Li, F., Kang, Y., Xu, C., Phosphorized MXene-Phase Molybdenum Carbide as an Earth-Abundant Hydrogen Evolution Electrocatalyst. *ACS Appl. Energy Mater.*, 1, 7206–7212, 2018.
12. Tran, M.H., Schäfer, T., Shahraei, A., Dürrschnabel, M., Molina-Luna, L., Kramm, U.I., Birkel, C.S., Adding a New Member to the MXene Family: Synthesis, Structure, and Electrocatalytic Activity for the Hydrogen Evolution Reaction of $V_4C_3T_x$. *ACS Appl. Energy Mater.*, 1, 3908–3914, 2018.
13. Uwadiunor, E., Johnson, D., Hansen, K., Djire, A., Controlling the Surface Reactivity of Hybrid Ti_3CN MXene via *In-situ* Electrocatalysis. *ChemCatChem*, 14, e202200702, 2022.
14. Zou, X., Liu, H., Xu, H., Wu, X., Han, X., Kang, J., Reddy, K.M., A simple approach to synthesis Cr_2CT_x MXene for efficient hydrogen evolution reaction. *Mater. Today Energy*, 20, 100668, 2021.
15. Wei, Y., Soomro, R.A., Xie, X., Xu, B., Design of efficient electrocatalysts for hydrogen evolution reaction based on 2D MXenes. *J. Energy Chem.*, 55, 244–255, 2021.
16. Xue, N., Li, X., Zhang, M., Han, L., Liu, Y., Tao, X., Chemical-combined ball-milling synthesis of fluorine-free porous MXene for high-performance lithium ion batteries. *ACS Appl. Energy Mater.*, 3, 10234–10241, 2020.
17. Gentile, A., Marchionna, S., Balordi, M., Pagot, G., Ferrara, C., Di Noto, V., Ruffo, R., Critical Analysis of MXene Production with *In-Situ* HF Forming Agents for Sustainable Manufacturing. *ChemElectroChem*, 9, e202200891, 2022.
18. Nagtode, V.S., Cardoza, C., Yasin, H.K.A., Mali, S.N., Tambe, S.M., Roy, P., Singh, K., Goel, A., Amin, P.D., Thorat, B.R., Cruz, J.N., Pratap, A.P., Green Surfactants (Biosurfactants): A Petroleum-Free Substitute for Sustainability—Comparison, Applications, Market, and Future Prospects. *ACS Omega*, 8, 11674–11699, 2023.

19. Naguib, M., Kurtoglu, M., Presser, V., Lu, J., Niu, J., Heon, M., Hultman, L., Gogotsi, Y., Barsoum, M.W., Two-dimensional nanocrystals produced by exfoliation of Ti_3AlC_2. *Adv. Mater.*, 23, 4248–4253, 2011.
20. Ye, M., Shen, J., Zhu, X., Cao, Z., Wang, W., Li, H., Dong, J., Gao, S., Xu, D., Li, L., Superior-performance aqueous zinc-ion batteries based on the *in situ* growth of mno2 nanosheets on V_2CT_X MXene. *ACS Nano*, 15, 2971–2983, 2021.
21. Li, Z., Dall'Agnese, Y., Guo, J., Huang, H., Liang, X., Xu, S., Flexible free-standing all-MXene hybrid films with enhanced capacitive performance for powering a flex sensor. *J. Mater. Chem. A*, 8, 16649–16660, 2020.
22. Ghidiu, M., Naguib, M., Shi, C., Mashtalir, O., Pan, L.M., Zhang, B., Yang, J., Gogotsi, Y., Billinge, S.J.L., Barsoum, M.W., Synthesis and characterization of two-dimensional Nb_4C_3 (MXene). *Chem. Commun.*, 50, 9517–9520, 2014.
23. Liu, M.C., Zhang, Y.S., Zhang, B.M., Zhang, D.T., Tian, C.Y., Kong, L.B., Hu, Y.X., Large interlayer spacing 2D $Ta4C_3$ matrix supported 2D MoS_2 nanosheets: A 3D heterostructure composite towards high-performance sodium ions storage. *Renew. Energy*, 169, 573–581, 2021.
24. Hakim, M.W., Fatima, S., Tahir, R., Iqbal, M.Z., Li, H., Rizwan, S., Ni-intercalated Mo_2TiC_2Tx free-standing MXene for excellent gravimetric capacitance prepared via electrostatic self-assembly. *J. Energy Storage*, 61, 106662, 2023.
25. Zhou, J., Zha, X., Zhou, X., Chen, F., Gao, G., Wang, S., Shen, C., Chen, T., Zhi, C., Eklund, P., Du, S., Xue, J., Shi, W., Chai, Z., Huang, Q., Synthesis and Electrochemical Properties of Two-Dimensional Hafnium Carbide. *ACS Nano*, 11, 3841–3850, 2017.
26. Zhou, J., Zha, X., Chen, F.Y., Ye, Q., Eklund, P., Du, S., Huang, Q., A Two-Dimensional Zirconium Carbide by Selective Etching of Al_3C_3 from Nanolaminated $Zr_3Al_3C_5$. *Angew. Chem. - Int. Ed.*, 55, 5008–5013, 2016.
27. Deysher, G., Shuck, C.E., Hantanasirisakul, K., Frey, N.C., Foucher, A.C., Maleski, K., Sarycheva, A., Shenoy, V.B., Stach, E.A., Anasori, B., Gogotsi, Y., Synthesis of Mo_4VAlC_4 MAX Phase and Two-Dimensional Mo_4VC_4 MXene with Five Atomic Layers of Transition Metals. *ACS Nano*, 14, 204–217, 2020.
28. Guo, Y., Jin, S., Wang, L., He, P., Hu, Q., Fan, L.Z., Zhou, A., Synthesis of two-dimensional carbide Mo_2CT_x MXene by hydrothermal etching with fluorides and its thermal stability. *Ceram. Int.*, 46, 19550–19556, 2020.
29. Liu, F., Zhou, A., Chen, J., Jia, J., Zhou, W., Wang, L., Hu, Q., Preparation of Ti_3C_2 and Ti_2C MXenes by fluoride salts etching and methane adsorptive properties. *Appl. Surf. Sci.*, 416, 781–789, 2017.
30. Rafieerad, A., Yan, W., Alagarsamy, K.N., Srivastava, A., Sareen, N., Arora, R.C., Dhingra, S., Fabrication of Smart Tantalum Carbide MXene Quantum Dots with Intrinsic Immunomodulatory Properties for Treatment of Allograft Vasculopathy. *Adv. Funct. Mater.*, 31, 2106786, 2021.

31. Xie, X., Xue, Y., Li, L., Chen, S., Nie, Y., Ding, W., Wei, Z., Surface Al leached Ti_3AlC_2 as a substitute for carbon for use as a catalyst support in a harsh corrosive electrochemical system. *Nanoscale*, 6, 11035–11040, 2014.
32. Li, T., Yao, L., Liu, Q., Gu, J., Luo, R., Li, J., Yan, X., Wang, W., Liu, P., Chen, B., Zhang, W., Abbas, W., Naz, R., Zhang, D., Fluorine-Free Synthesis of High-Purity $Ti_3C_2T_x$ (T=OH, O) via Alkali Treatment. *Angew. Chem. - Int. Ed.*, 57, 6115–6119, 2018.
33. Sarfraz, M.H., Zubair, M., Aslam, B., Ashraf, A., Siddique, M.H., Hayat, S., Cruz, J.N., Muzammil, S., Khurshid, M., Sarfraz, M.F., Hashem, A., Dawoud, T.M., Avila-Quezada, G.D., Abd_Allah, E.F., Comparative analysis of phyto-fabricated chitosan, copper oxide, and chitosan-based CuO nanoparticles: antibacterial potential against Acinetobacter baumannii isolates and anti-cancer activity against $HepG_2$ cell lines. *Front. Microbiol.*, 14, 1188743, 2023.
34. Yang, S., Zhang, P., Wang, F., Ricciardulli, A.G., Lohe, M.R., Blom, P.W.M., Feng, X., Fluoride-Free Synthesis of Two-Dimensional Titanium Carbide (MXene) Using A Binary Aqueous System. *Angew. Chem.*, 130, 15717–15721, 2018.
35. Lukatskaya, M.R., Halim, J., Dyatkin, B., Naguib, M., Buranova, Y.S., Barsoum, M.W., Gogotsi, Y., Room-Temperature Carbide-Derived Carbon Synthesis by Electrochemical Etching of MAX Phases. *Angew. Chem.*, 126, 4977–4980, 2014.
36. Sun, W., Shah, S.A., Chen, Y., Tan, Z., Gao, H., Habib, T., Radovic, M., Green, M.J., Electrochemical etching of Ti_2AlC to Ti_2CTX (MXene) in low-concentration hydrochloric acid solution. *J. Mater. Chem. A*, 5, 21663–21668, 2017.
37. Pang, S.Y., Wong, Y.T., Yuan, S., Liu, Y., Tsang, M.K., Yang, Z., Huang, H., Wong, W.T., Hao, J., Universal Strategy for HF-Free Facile and Rapid Synthesis of Two-dimensional MXenes as Multifunctional Energy Materials. *J. Am. Chem. Soc.*, 141, 9610–9616, 2019.
38. Bae, Y.H., Park, S., Noh, J.S., Control of electrical properties of $Ti_3C_2T_x$ mediated by facile alkalization. *Surf. Interfaces*, 41, 103258, 2023.
39. Wu, M., He, Y., Wang, L., Xia, Q., Zhou, A., Synthesis and electrochemical properties of V_2C MXene by etching in opened/closed environments. *J. Adv. Ceram.*, 9, 749–758, 2020.
40. Li, M., Lu, J., Luo, K., Li, Y., Chang, K., Chen, K., Zhou, J., Rosen, J., Hultman, L., Eklund, P., Persson, POA., Du, S., Chai, Z., Huang, Z., Huang, Q., Element Replacement Approach by Reaction with Lewis Acidic Molten Salts to Synthesize Nanolaminated MAX Phases and MXenes. *J. Am. Chem. Soc.*, 141, 4730–4737, 2019.
41. Li, Y., Shao, H., Lin, Z., Lu, J., Liu, L., Duployer, B., Persson, POA., Eklund, P., Hultman, L., Li, M., Chen, K., Zha, X.H., Du, S., Rozier, P., Chai, Z., Raymundo-Piñero, E., Taberna, P.L., Simon, P., Huang, Q., A general Lewis acidic etching route for preparing MXenes with enhanced electrochemical performance in non-aqueous electrolyte. *Nat. Mater.*, 19, 894–899, 2020.

42. VahidMohammadi, A., Kayali, E., Orangi, J., Beidaghi, M., Techniques for MXene delamination into single-layer flakes, in: *2D Metal Carbides and Nitrides (MXenes): Structure, Properties and Applications*, pp. 177–195, Springer International Publishing, Cham, 2019.
43. Inman, A., Šedajová, V., Matthews, K., Gravlin, J., Busa, J., Shuck, C.E., VahidMohammadi, A., Bakandritsos, A., Shekhirev, M., Otyepka, M., Gogotsi, Y., Shear delamination of multilayer MXenes. *J. Mater. Res.*, 37, 4006–4016, 2022.
44. Mashtalir, O., Naguib, M., Mochalin, V.N., Dall'Agnese, Y., Heon, M., Barsoum, M.W., Gogotsi, Y., Intercalation and delamination of layered carbides and carbonitrides. *Nat. Commun.*, 4, 1716, 2013.
45. Marquez, F., Masa, A., Cotto, M., Garcia, A., Duconge, J., Campo, T., Elizalde, E., Morant, C., Photocatalytic hydrogen production by water splitting using novel catalysts under UV-vis light irradiation. *Adv. Energy Res.*, 2, 33–45, 2014.
46. Wang, Z., Li, C., Domen, K., Recent developments in heterogeneous photocatalysts for solar-driven overall water splitting. *Chem. Soc. Rev.*, 48, 2109–2125, 2019.
47. Vidal-Iglesias, F.J., Solla-Gullón, J., Rodes, A., Herrero, E., Aldaz, A., Understanding the Nernst equation and other electrochemical concepts: An easy experimental approach for students. *J. Chem. Educ.*, 89, 936–939, 2012.
48. Anantharaj, S., Ede, S.R., Karthick, K., Sam Sankar, S., Sangeetha, K., Karthik, P.E., Kundu, S., Precision and correctness in the evaluation of electrocatalytic water splitting: Revisiting activity parameters with a critical assessment. *Energy Environ. Sci.*, 11, 744–771, 2018.
49. Li, Z., Gao, R., Feng, M., Deng, Y.P., Xiao, D., Zheng, Y., Zhao, Z., Luo, D., Liu, Y., Zhang, Z., Wang, D., Li, Q., Li, H., Wang, X., Chen, Z., Modulating Metal–Organic Frameworks as Advanced Oxygen Electrocatalysts. *Adv. Energy Mater.*, 11, 2003291, 2021.
50. Lasia, A., Mechanism and kinetics of the hydrogen evolution reaction. *Int. J. Hydrogen Energy*, 44, 19484–19518, 2019.
51. Tang, Y., Yang, C., Xu, X., Kang, Y., Henzie, J., Que, W., Yamauchi, Y., MXene Nanoarchitectonics: Defect-Engineered 2D MXenes towards Enhanced Electrochemical Water Splitting. *Adv. Energy Mater.*, 12, 2103867, 2022.
52. Kang, Z., Cai, J., Ye, D., Zhao, H., Luo, J., Zhang, J., Three-dimensional nitrogen-doped MXene as support to form high-performance platinum catalysts for water-electrolysis to produce hydrogen. *Chem. Eng. J.*, 446, 137443, 2022.
53. Bat-Erdene, M., Batmunkh, M., Sainbileg, B., Hayashi, M., Bati, A.S.R., Qin, J., Zhao, H., Zhong, Y.L., Shapter, J.G., Highly Dispersed Ru Nanoparticles on Boron-Doped $Ti_3C_2T_x$ (MXene) Nanosheets for Synergistic Enhancement of Electrocatalytic Hydrogen Evolution. *Small*, 17, 2102218, 2021.

54. Yang, Z., Chen, Y., Chen, G., Wang, J., Li, H., Ultrathin Ti_3C2T_x MXene sheets with high electrochemically active area anchored Pt boosting hydrogen evolution. *Heliyon*, 9, e19197, 2023.
55. Sun, W., Wang, Y., Xiang, K., Bai, S., Wang, H., Zou, J., Arramel, Jiang, J., CoP Decorated on $Ti_3C_2T_x$ MXene Nanocomposites as Robust Electrocatalyst for Hydrogen Evolution Reaction. *Wuli Huaxue Xuebao/Acta Phys. - Chim. Sin.*, 40, 202308015, 2024.
56. Liu, J., Liu, Y., Xu, D., Zhu, Y., Peng, W., Li, Y., Zhang, F., Fan, X., Hierarchical "nanoroll" like MoS_2/Ti_3C_2Tx hybrid with high electrocatalytic hydrogen evolution activity. *Appl. Catal. B Environ.*, 241, 89–94, 2019.
57. Han, M., Yang, J., Jiang, J., Jing, R., Ren, S., Yan, C., Efficient tuning the electronic structure of N-doped Ti-based MXene to enhance hydrogen evolution reaction. *J. Colloid Interface Sci.*, 582, 1099–1106, 2021.
58. Jiang, Y., Sun, T., Xie, X., Jiang, W., Li, J., Tian, B., Su, C., Oxygen-Functionalized Ultrathin $Ti_3C_2T_x$ MXene for Enhanced Electrocatalytic Hydrogen Evolution. *Chem. Sus. Chem.*, 12, 1368–1373, 2019.
59. Wang, J., Liu, Y., Yang, G., Cobalt decorated ultra-thin Ti_3C_2 MXene electrocatalyst for high-efficiency hydrogen evolution reaction. *Mater. Res. Express*, 6, 025056, 2019.
60. Jian, X., Wei, T.-T., Zhang, M.-M., Li, R., Liu, J.-X., Liang, Z.-H., Construction of $Ti_3C_2T_x$ MXene Supported Low-Platinum Electrocatalyst for Hydrogen Evolution Reaction by Direct Electrochemical Strategy. *J. Electrochem. Soc.*, 168, 096504, 2021.
61. Wu, Y., Wei, W., Yu, R., Xia, L., Hong, X., Zhu, J., Li, J., Lv, L., Chen, W., Zhao, Y., Zhou, L., Mai, L., Anchoring Sub-Nanometer Pt Clusters on Crumpled Paper-Like MXene Enables High Hydrogen Evolution Mass Activity. *Adv. Funct. Mater.*, 32, 2110910, 2022.
62. Liu, H., Hu, Z., Liu, Q., Sun, P., Wang, Y., Chou, S., Hu, Z., Zhang, Z., Single-atom Ru anchored in nitrogen-doped MXene ($Ti_3C_2T_X$) as an efficient catalyst for the hydrogen evolution reaction at all pH values. *J. Mater. Chem. A*, 8, 24710–24717, 2020.
63. Sarfraz, B., Mehran, M.T., Shahzad, F., Hussain, S., Naqvi, S.R., Khan, H.A., Mahmood, K., Bifunctional CuS/Cl-terminated greener MXene electrocatalyst for efficient hydrogen production by water splitting. *RSC Adv.*, 13, 22017–22028, 2023.
64. Sarfraz, B., Mehran, M.T., Baig, M.M., Naqvi, S.R., Khoja, A.H., Shahzad, F., HF free greener Cl-terminated MXene as novel electrocatalyst for overall water splitting in alkaline media. *Int. J. Energy Res.*, 46, 10942–10954, 2022.
65. Tan, Y., Zhu, Z., Zhang, X., Zhang, J., Zhou, Y., Li, H., Qin, H., Bo, Y., Pan, Z., $Nb_4C_3T_x$ (MXene) as a new stable catalyst for the hydrogen evolution reaction. *Int. J. Hydrogen Energy*, 46, 1955–1966, 2021.
66. Long, A., Li, W., Zhou, M., Gao, W., Liu, B., Wei, J., Zhang, X., Liu, H., Liu, Y., Zeng, X., MoS_2 nanosheets grown on nickel chalcogenides: Controllable

synthesis and electrocatalytic origins for the hydrogen evolution reaction in alkaline solution. *J. Mater. Chem. A*, 7, 21514–21522, 2019.

67. Spitaler, J. and Estreicher, S.K., Perspectives on the theory of defects. *Front. Mater.*, 5, 2110910, 2018.
68. Wu, H., Huang, Q., Shi, Y., Chang, J., Lu, S., Electrocatalytic water splitting: Mechanism and electrocatalyst design. *Nano Res.*, 16, 9142–9157, 2023.
69. Gan, J., Li, F., Tang, Q., Vacancies-Engineered M_2CO_2 MXene as an Efficient Hydrogen Evolution Reaction Electrocatalyst. *J. Phys. Chem. Lett.*, 12, 4805–4813, 2021.
70. Gan, J., Li, F., Tang, Y., Tang, Q., Theoretical Study of Transition-Metal-Modified Mo_2CO_2 MXene as a Catalyst for the Hydrogen Evolution Reaction. *ChemSusChem*, 13, 6005–6015, 2020.
71. Sharma, V., Dhiman, R., Mahajan, A., Ti^{2+} and Ti^{4+} species enriched MXene electrocatalyst for highly efficient hydrogen evolution and oxygen evolution reaction kinetics. *Appl. Surf. Sci.*, 612, 155883, 2023.
72. Ma, N., Wang, Y., Zhang, Y., Liang, B., Zhao, J., Fan, J., First-principles screening of Pt doped Ti_2CNL (N = O, S and Se, L = F, Cl, Br and I) as high-performance catalysts for ORR/OER. *Appl. Surf. Sci.*, 596, 153574, 2022.
73. Ma, N., Li, N., Zhang, Y., Wang, T., Zhao, J., Fan, J., Strain adjustment Pt-doped Ti_2CO_2 as an efficient bifunctional catalyst for oxygen reduction reactions and oxygen evolution reactions by first-principles calculations. *Appl. Surf. Sci.*, 590, 153149, 2022.
74. Zhao, X., Zheng, X., Lu, Q., Li, Y., Xiao, F., Tang, B., Wang, S., Yu, D.Y.W., Rogach, A.L., Electrocatalytic enhancement mechanism of cobalt single atoms anchored on different MXene substrates in oxygen and hydrogen evolution reactions. *EcoMat*, 5, 2110910, 2023.
75. Zhao, R., Yang, C., Zhang, Q., Xiang, H., Li, Y., Li, X., Transition Metal-Promoted VC(001) for Overall Water Splitting and Oxygen Reduction. *J. Phys. Chem. C*, 125, 14607–14615, 2021.
76. Tang, Y., Yang, C., Tian, Y., Luo, Y., Yin, X., Que, W., The effect of: *In situ* nitrogen doping on the oxygen evolution reaction of MXenes. *Nanoscale Adv.*, 2, 1187–1194, 2020.

6

Inorganic Photocatalysts for Water Splitting

Nogueira, A. E.[1*], Ribeiro, L. S.[2], Torres, J. A.[2,3], Sala, R. L.[3], Pinto F. M.[4], La Porta F. A.[5] and Santos, F. L.[6]

[1]*Department of Chemistry, Division of Fundamental Sciences (IEF), Aeronautics Institute of Technology—ITA, São José dos Campos—SP, Brazil*
[2]*Embrapa Instrumentation, São Carlos—SP, Brazil*
[3]*Department of Chemistry, University of São Carlos—UFSCar, São Carlos—SP, Brazil*
[4]*Engineering Department, Federal University of Lavras, Lavras-MG, Brazil*
[5]*Laboratory of Nanotechnology and Computational Chemistry, Federal Technological University of Paraná, Londrina-PR, Brazil*
[6]*Department of Chemistry, Institute of Exact and Biological Sciences (ICEB), Federal University of Ouro Preto-UFOP, Ouro Preto—MG, Brazil*

Abstract
The advancement of sustainable technologies has been driven by the use of inorganic photocatalysts, with emphasis on their application in the H_2 generation process through water splitting. The crystalline structure of these photocatalysts, characterized by an ordered atomic arrangement, is essential for their performance. Generally composed of a combination of metallic and non-metallic elements, semiconductor materials, such as titanium dioxide (TiO_2), play an indispensable role in catalyzing photochemical reactions through light absorption. Furthermore, they offer remarkable durability and stability and can handle adverse conditions. By exploring inorganic photocatalysts, this chapter focuses on fundamental aspects, from understanding photocatalysis to relevant physical and chemical properties. Some well-known inorganic photocatalysts, like TiO_2, ZnO, $BiVO_4$, perovskites, and others, are shown, along with ways to make them better, like doping and nanostructure optimization. Present obstacles and forthcoming prospects in the exploration and utilization of these photocatalysts are deliberated emphasizing the ongoing necessity for technological innovations to enhance water

**Corresponding author*: andre.nogueira@ita.br

Inamuddin, Tariq Altalhi, Mohammad Luqman and Jorddy Neves Cruz (eds.) Water Splitting: Production of Hydrogen, (139–170) © 2025 Scrivener Publishing LLC

splitting technology and its sustainable applications. Optimizing the efficiency and stability of photocatalysts is essential to boosting sustainable hydrogen production and contributing to the transition to renewable energy sources.

Keywords: Semiconductors, artificial photosynthesis, energy, green hydrogen, photocatalysis

6.1 Introduction

The advancement of sustainable technologies has been driven by the use of inorganic photocatalysts, with an emphasis on their application in the H_2 generation process through water splitting. The crystalline structure of these photocatalysts, characterized by an ordered atomic arrangement, is essential for their performance (Figure 6.1). Generally consisting of a combination of metallic and non-metallic elements, such as titanium dioxide (TiO_2), these semiconductor materials serve a pivotal role in catalyzing photochemical reactions through light absorption. This is due to the metal's ability to provide free electrons and the non-metal's ability to provide electron holes.

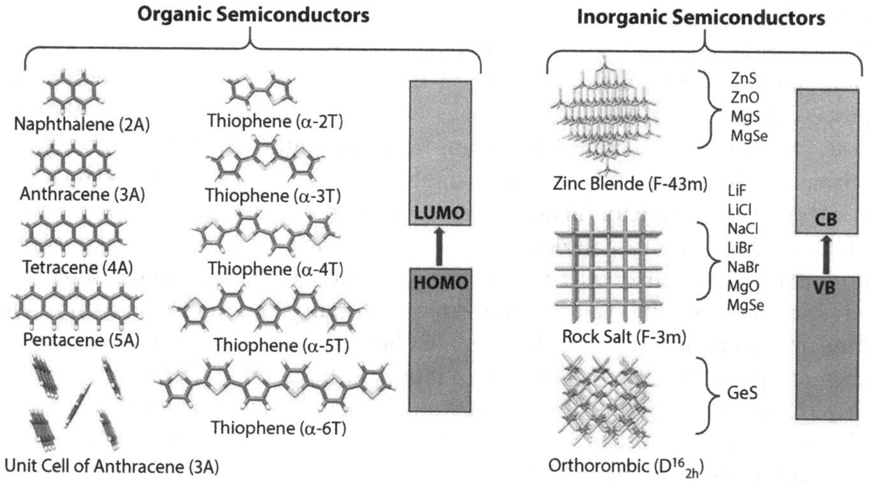

Figure 6.1 Representation of molecular structures encompassing some organic and inorganic semiconductors. HOMO refers to the highest occupied molecular orbital, and LUMO represents the lowest unoccupied molecular orbital, whereas VB stands for valence band, and CB stands for conduction band. Figure adapted with permission from ref. [1]. Copyright© 2019 American Chemical Society. Distributed under a Creative Commons Attribution License 4.0 (CC-BY).

Furthermore, inorganic photocatalysts offer a series of benefits compared to their organic counterparts. The durability and stability of these inorganic materials are remarkable enabling longer lifespans and sustained efficiency in catalytic processes. It is important to note that inorganic photocatalysts are strong and can handle bad environmental conditions like changes in temperature and humidity. This makes them better for use in a wide range of settings. These extra attributes further underscore the significance of inorganic photocatalysts as essential components in the advancement of sustainable and environmentally friendly technologies.

By delving into the exploration of inorganic photocatalysts, this chapter focuses on fundamental aspects. At first, it focuses on understanding how photocatalysis works to split a water molecule. Initially, the focus is on comprehending the mechanisms behind photocatalysis water splitting. This involves examining many aspects, including light absorption, generation of electron–hole pairs, surface reactions, and the transfer of photogenerated charges. Subsequently, physical and chemical properties, such as energy bands, crystalline structure, specific surface area, and surface reactivity, are discussed. These aspects are fundamental for a comprehensive understanding of the functioning of photocatalysts.

To show how well these materials work in water splitting, examples of well-studied inorganic photocatalysts are given, including TiO_2, bismuth vanadate oxide ($BiVO_4$), and perovskites. It also discusses about ways to make things better, such as doping, changing the surface, optimizing the crystal structure, and engineering defects. These are all meant to make the photocatalysts work better. Finally, current challenges in research and application of these photocatalysts are discussed, along with future perspectives emphasizing the ongoing need for research and development in this field to further advance water splitting technology and its sustainable applications.

6.2 Photocatalysis Mechanisms

Considering the Earth's abundant water resources, the process of water splitting, which has a standard reduction potential of 1.23 V, emerges as an attractive option for artificial photosynthesis. In this context, harnessing solar energy for water splitting stands out as a potential pathway for establishing a sustainable economy based on hydrogen production. Among the various proposed technologies, photocatalysis stands out as a significant advancement within this technological framework, a process in which solar energy can be utilized as a source for redox reactions to produce clean hydrogen [2–4].

For enhanced efficiency in the water splitting process, several fundamental requirements must be accomplished. This entails employing a semiconductor characterized by a narrow band gap to enhance the absorption of solar radiation, valence and conduction bands possessing suitable potentials for driving water oxidation and reduction reactions, and photogenerated charge carriers exhibiting prolonged lifetimes to facilitate efficient conversion into water splitting products [3]. Additionally, sacrificial agents (methanol, ethanol, and glycerol) or co-catalysts are usually added to increase process efficiency. While sacrificial agents operate in the consumption of photogenerated holes, co-catalysts function as sites for capturing photogenerated electrons aiding in the conversion of protons to hydrogen [5, 6]. Once these prerequisites are satisfied, the water splitting reaction follows four main steps as outlined below: the photocatalyst absorbs energy from an external source leading to the excitation of electrons from the VB to the CB, steps 1 and 2; the photoexcited electrons (e^-) and holes (h^+) generated are transported to the surface of the photocatalyst or the cocatalyst, steps 3 and 4; finally, the hydrogen evolution reaction (HER) and the oxygen evolution reaction (OER) proceed on the surface of the photocatalyst or cocatalyst, steps 5 and 6 (Figure 6.2) [7].

The CB minimal edge must, in theory, be more negative than the standard reduction potential of H^+ to H_2 (0 V vs. NHE at pH = 0) for HER to occur. Similarly, OER will happen if the photocatalyst's maximum edge of VB is more positive than water's to O_2 oxidation potential (1.23 V vs. NHE

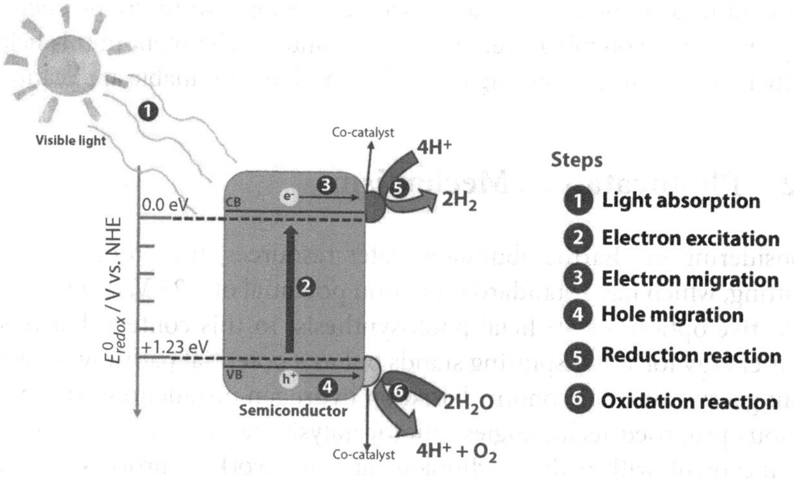

Figure 6.2 Schematic of photocatalytic water splitting reactions in an inorganic semiconductor.

at pH = 0) (Figure 6.2b). Semiconductor materials that exhibit band structures that meet these standards are suitable for conducting HER and OER. Furthermore, photocatalysts with band gap values between 1.23 and 3.26 eV are desirable for these types of reactions [8]. Therefore, it is possible to use the entire spectral range of visible light (400 < λ < 800 nm) since the minimum photon energy thermodynamically required to drive photocatalytic water splitting is 1.23 eV (1,000 nm) [5].

One of the primary obstacles in the hydrogen evolution reaction is the recombination of charge carriers at the photocatalyst's surface. Therefore, control of the reaction steps is essential for the efficiency of the reaction to be maximized. Among these steps, photoexcitation, charge separation, adsorption, surface chemical reactions, and product desorption stand out, all occurring on an extremely small-time scale, from 10^{-15} to 10^{-3} s. In this sense, any problem during charge separation can negatively influence the formation of hydrogen. Another crucial step is the desorption of products from the photocatalyst surface. If this process is not instantaneous, it can result in the recombination of hydrogen and oxygen leading to the formation of water again. This process happens due to the rapid nucleation, growth, detachment, and merging of gaseous bubbles of H_2 and O_2 [5, 6, 8].

Among the numerous essential parameters for greater efficiency during the hydrogen evolution reaction, the structural, morphological, and electronic properties of the photocatalyst stand out. An ideal semiconductor should have smaller particle sizes to facilitate the transport of photogenerated electrons and holes to the surface-active reaction sites minimizing the recombination rate. Furthermore, a high surface area will favor these reactive sites even more leading to greater activity. Specifically, the structure of the photocatalyst is directly related to the synthesis methodology used, among which we can mention the most common: solvothermal, hydrothermal, sol–gel, mechanical mixing, chemical vapor deposition, and impregnation. The electronic characteristics of a semiconductor are related to its band structure, more specifically the band gap. For greater efficiency in H_2 production, it is ideal for the photocatalyst to have a narrow band gap [9–11].

Similarity with the photosynthesis process occurs since chemical energy is produced by the transformation of solar energy, with the generation of protons and the consequent evolution of H_2 through the reaction with electrons that circulate the surface of the photocatalyst. Visible light-absorbing photocatalysts are favored because they make the most use of sunlight [8].

6.3 Properties of Photocatalysts

As previously discussed, the minimum theoretical bandgap required for water splitting is 1.23 eV (~1,000 nm) in the case of one-step photoexcitation. To harvest visible light, photocatalysts with bandgaps smaller than 3.0 eV are typically required. In addition, the photocatalysts need to have band edge potentials that are suitable for water splitting. Their CB must lie at a more negative potential than the H^+/H_2 energy level [0–0.059 pH, V vs. normal hydrogen electrode (NHE)], and the VB has to lie at a more positive potential than the O_2/H_2O energy level (1.23–0.059 V pH, V vs. NHE). (Figure 6.3) Apart from optimal bandgaps and band edge potentials, the overall water splitting efficiency will also be dependent on other processes such as the diffusion of electrons and holes to the photocatalyst surface, and the participation of these photoexcited carriers in hydrogen and oxygen evolution reactions (HER, OER, respectively) [12–14]. The photocatalyst should also present efficient surface reaction kinetics with minimum carrier recombination and be stable over repeated photocatalytic reactions. Photocatalytic performance is typically evaluated based on the amount of evolved hydrogen and oxygen per illumination time [15–17]. However, it is recommended to evaluate the apparent quantum yield (AQY) and solar-to-hydrogen efficiency (STH) instead, since they are independent of the light source and reactor facilitating the comparison between systems from different research groups [12, 14, 18].

After the pioneering work of Fujishima and Honda in 1972 reporting TiO_2 photoanode and a Pt cathode for water splitting photocatalysis, many semiconductors, such as metal oxides, sulfides, oxysulfides, and oxynitrides containing cations with d_0 and d_{10} electronic configuration, have been developed for water splitting under UV light. Since the solar spectrum consists of 42%–43% visible light against only 3%–5% of UV light, the development of visible light-driven photocatalysts is strongly required [19].

So far, only a limited number of semiconductors have been showcased to harvest visible light in a one-step process for overall water splitting. However, these demonstrations are constrained by either limited stability or low quantum yields. To overcome the limitations inherent in single-component systems, two-step photoexcitation Z-scheme systems (named in analogy to the photosynthesis process) provide a strategy to obtain visible light-driven photocatalysts [17, 21]. They comprise two different semiconductors or photocatalysts that facilitate the reduction and oxidation of half-reactions in an overall water splitting. The Z-scheme design allows the employment of photocatalysts with varied bandgap energies. It can enhance the efficiency of photocatalysis by allowing spatial separation of reduction

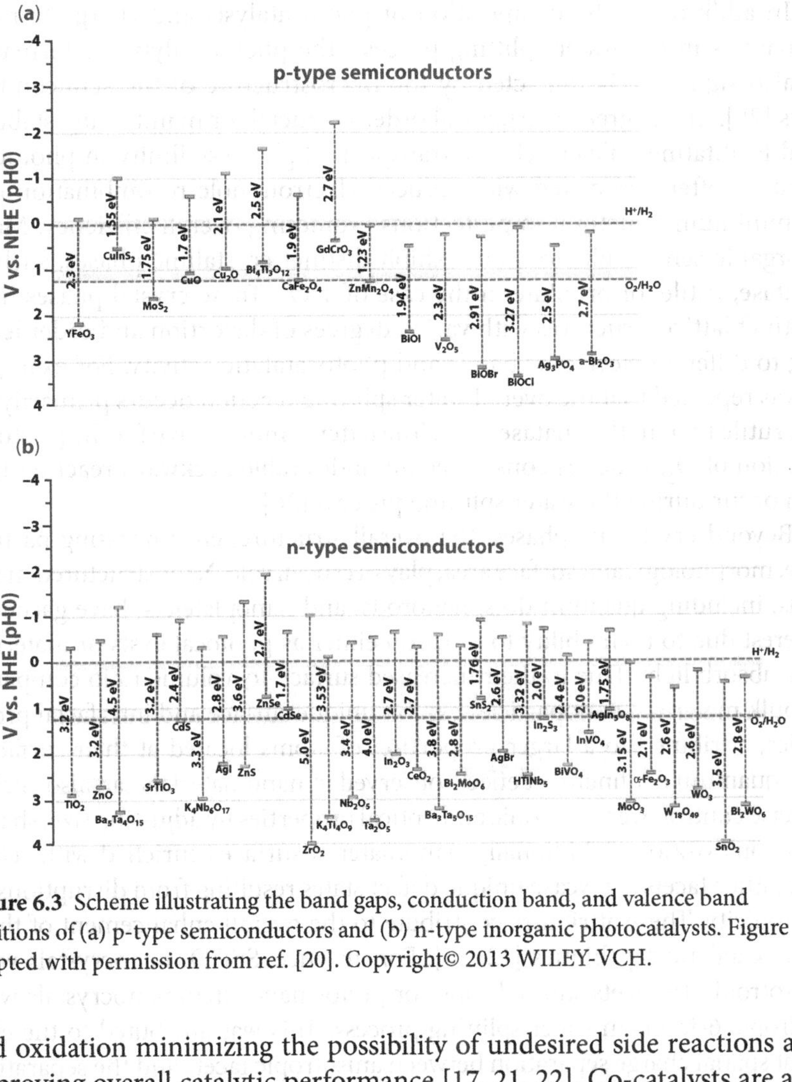

Figure 6.3 Scheme illustrating the band gaps, conduction band, and valence band positions of (a) p-type semiconductors and (b) n-type inorganic photocatalysts. Figure adapted with permission from ref. [20]. Copyright© 2013 WILEY-VCH.

and oxidation minimizing the possibility of undesired side reactions and improving overall catalytic performance [17, 21, 22]. Co-catalysts are also commonly incorporated in overall water splitting systems. Co-catalysts are commonly composed of metal or metal oxide nanoparticles dispersed on the semiconductor surface. Their primary function is to provide sites for redox reactions, thereby catalyzing the reaction by reducing the activation energy. Additionally, co-catalysts facilitate charge migration from the photoexcited semiconductor and contribute to enhancing the system's durability against photodegradation. They efficiently consume photoinduced charge carriers, particularly by shielding the semiconductor surface from oxidation by holes [12, 23, 24].

In addition to the composition of photocatalysts and charge transfer dynamics in the water splitting process, the photocatalytic performance is also significantly impacted by the microstructure of the semiconductors [25]. The degree of structural order is crucial for maintaining stability and facilitating efficient charge transport. High crystallinity in photocatalysts is often associated with reduced electron–hole recombination due to minimum structural imperfections promoting overall efficiency. Many inorganic semiconductors can exhibit distinct crystalline phases, such as anatase, rutile, or brookite in the case of TiO_2. These crystal phases have distinct lattice structures with varied degrees of distortion and order leading to different electronic bands and photocatalytic activity. For example, it was reported that the overall water splitting reaction occurs primarily on the rutile than in the anatase TiO_2. The latter is more active for the photoreduction of O_2, which is considered an undesirable backward reaction that can occur during the water splitting process [26].

Beyond crystalline phases, the overall structure, encompassing particle size, morphology, and surface area, plays a crucial role. Nanostructured materials, including quantum dots, nanorods, and nanoplatelets, have garnered interest due to their ability to function either as photocatalysts or materials that absorb light. Due to their increased surface-to-volume ratio compared to bulk materials, nanoparticles exhibit unique surface and interfacial properties, attributed to a larger proportion of atoms located at their surfaces. The quantum confinement effects observed in nanomaterials can also enable precise control over electronic and optical properties by adjusting size, shape, and composition. Additionally, nanomaterial surfaces enriched with selective lattice facets showcase unique defect states resulting from disruptions in periodicity. These attributes contribute to the overall enhancement of their photocatalytic capabilities [12, 25]. For instance, $SrTiO_3$ nanocrystals with anisotropic 18-facets showed superior performance than nanocrystals with isotropic 6-facets in water splitting process. This was attributed to the efficient spatial charge separation between anisotropic facets and the separation of the reduction and oxidation catalytic sites [27].

In this way, an in-depth comprehension of the physical and chemical properties of inorganic photocatalysts is fundamental for advancing their applications. Precision in tuning composition, bandgaps, optimizing crystallinity, controlling crystalline phases, structures, and morphology while maximizing specific surface area are critical strategies for designing highly efficient photocatalysts. As research in this field progresses, the potential for photocatalysis to address environmental and industrial challenges continues to expand paving the way for sustainable and innovative solutions and environmental and industrial challenges.

6.4 Examples of Inorganic Photocatalysts

The prominence of photocatalysts derived from titanium, tungsten, vanadium, tantalum, and gallium in water splitting applications is not arbitrary. These materials all show the thermodynamic prerequisites: they possess appropriate band positions in relation to water redox potentials and bandgaps that facilitate activation by UV-vis light [28, 29]. In the following discussion, we will delve into the utilization of these materials emphasizing the distinctions among them as well as their respective advantages and drawbacks.

As mentioned in 1972, one of the initial materials discovered with the capability to facilitate water splitting, and subsequently one of the most extensively researched semiconductors, was titanium oxide [30]. The advantages of TiO_2 include chemical inertness, low cost, electrochemical stability, and a bandgap value that makes it less prone to photocorrosion [31]. Nevertheless, its high bandgap value of 3.0 eV also limits its efficiency, failing to exceed 1%, primarily because it allows only absorption of the ultraviolet portion of the light spectrum [32, 33].

A recent approach aimed at overcoming this challenge involves inducing a localized surface plasmonic resonance (LSPR) effect by depositing metals onto the photocatalyst's surface [34]. Nano-sized plasmonic metals, such as silver and copper, can convert incident light energy into a highly concentrated localized electric field and energetic hot electrons. This resonant energy is transmitted from the metal to the adjacent semiconductor through dipole–dipole interactions, thereby augmenting its catalytic activity. An elegant study examined the evolution of hydrogen gas production concerning the deposition of Ag and Cu nanoparticles, both separately and in combination, onto TiO_2 nanorods [34]. When used alone, the photocatalyst yielded the poorest results in terms of average H_2 evolution rate and photocurrent density during illumination. However, when TiO_2 was combined with both metals, it exhibited the most favorable outcomes, with a 66% improvement. This enhancement can be attributed to the migration of hot electrons from silver to copper and the synergistic interplay of different LSPR mechanisms.

Another widely adopted approach to enhance the efficiency of titanium oxide is through doping, which serves to reduce the material's bandgap and enable the absorption of visible light. One example involves doping TiO_2 films with chromium resulting in reductions in the bandgap to values below 2.5 eV [35]. Alternatively, platinum doping can also be employed. In a recent study, the combination of titanium and zinc oxides in a Ti:Zn ratio of 10:1, along with a 0.5 wt% Pt doping, was investigated [36].

The outcomes demonstrated a reduction of 0.2 eV in the bandgap and a substantial increase from 68 μmol h^{-1} g^{-1} for pure TiO_2 to 2,150 μmol h^{-1} g^{-1} when using the doped photocatalyst. Furthermore, platinum nanoparticles, besides their role in bandgap reduction, can serve as electron traps, effectively inhibiting electron–hole recombination.

Another extensively examined photocatalyst based on titanium for water splitting is $SrTiO_3$ [37, 38]. This perovskite-type titanate stands out as a semiconductor with a bandgap closely resembling that of titanium oxide (approximately 3.2 eV) and possesses exceptional optical and electronic properties [39, 40]. Much like TiO_2, it presents non-toxicity, cost effectiveness, and chemical as well as photochemical stability [41, 42].

The most employed strategy to enhance the efficiency of this photocatalyst involves doping it with various transition metals. This doping plays a pivotal role in reducing the bandgap and promoting improved charge separation consequently amplifying the quantity of generated hydrogen gas [43]. Among these transition metals, chromium has been of particular interest. When introduced at a 5% concentration into the $SrTiO_3$ structure, it effectively lowers the bandgap to 2.3 eV [44]. In hydrogen-generation experiments, platinum served as a co-catalyst at a concentration of 0.6 wt%. The findings revealed that the rate of hydrogen production increased with the synthesis temperature likely due to enhanced crystallinity. The highest production rate, reaching 82.6 mmol h^{-1}, was achieved when the material was synthesized at 200°C. Furthermore, the results indicated that the material can be reused for multiple production cycles with minimal loss in the H_2 evolution rate. Even in the third cycle, the rate only dropped to 70.1 mmol h^{-1}, surpassing the performance of all other photocatalysts synthesized at lower temperatures.

Strontium titanate can also undergo doping with rhodium, as demonstrated in a study that assessed the utilization of this photocatalyst alongside a ruthenium co-catalyst (Ru–$SrTiO_3$:Rh) at a concentration of 0.7 wt% to prevent the recombination of photogenerated charges [45]. The reduction in the bandgap was duly validated given that the entire process was conducted using visible light with wavelengths exceeding 400 nm. As the concentration of Fe^{2+} cations, which donate electrons to the photocatalyst, increased in the solution, the quantity of hydrogen gas produced exhibited a gradual augmentation over time, reaching almost 250 μmol after 48 h.

Nevertheless, $SrTiO_3$ is not the sole perovskite utilized in photoelectrochemical water splitting; materials based on tantalum, specifically $NaTaO_3$, are also extensively employed for hydrogen gas production [46–49]. Perovskite structures offer the advantage of considerable flexibility for substitutions allowing for doping and bandgap adjustment [50]. In an effort to enhance performance, a study explored the replacement of 2% sodium in the $NaTaO_3$ structure with lanthanum [51]. This substitution was evaluated using two distinct methods: microwave heating (sol–gel reaction) and furnace heating (solid-state reaction). While both techniques resulted in photocatalysts characterized by high crystallinity, the solid-state reaction yielded nanoparticles with a reduced surface area. Consequently, this led to a lower rate of hydrogen gas production compared to the material synthesized via microwave heating, primarily due to a lower number of active sites.

Bandgap narrowing in $NaTaO_3$ can also be achieved through the process of doping with bismuth. In a recent study, researchers examined the impact of adding varying percentages of Bi on both the bandgap and the rate of hydrogen gas production [52]. The pure material had an initial bandgap of 4.00 eV, which progressively decreased as the percentage of added bismuth increased. It reached its lowest value of 3.53 eV when the maximum percentage of 5% bismuth was introduced. Surprisingly, the highest rate of H_2 production was achieved with the material containing a 3% bismuth doping, while the photocatalysts with 4% and 5% bismuth exhibited a decrease in performance. However, all doped materials presented higher hydrogen evolution than pure $NaTaO_3$. This outcome underscores the delicate nature of the doping process and highlights that an excessive dopant concentration can actually diminish the efficacy of the photocatalyst.

Gallium-based photocatalysts are also utilized for the production of hydrogen through water splitting. One such material is beta-phase gallium oxide, Ga_2O_3, known for its superior stability compared to other phases (α, γ, δ, and ε) [53]. Notably, this material possesses an even wider bandgap than the previously mentioned photocatalysts measuring approximately 4.6 eV [54]. Consequently, methods to enhance its efficiency are of utmost importance. One potential approach involves substituting gallium atoms with nitrogen atoms. Recent research demonstrated that as the nitrogen concentration increased, the bandgap decreased, ultimately reaching a value of 2.34 eV with 5% nitrogen content [55]. In another study, conducted by Sakata *et al.*, (2008), an evaluation was carried out involving the deposition of various metal ions onto the surface of Ga_2O_3 [56]. All cations were introduced at a 2% concentration, and among them, chromium, barium, calcium, and zinc exhibited the highest rates of hydrogen evolution. Notably, zinc demonstrated an enhancement of over sevenfold in comparison to the pure material.

Zinc also demonstrated the ability to enhance the performance of another gallium-based photocatalyst known as GaN [57–60]. This nitride possesses a smaller bandgap in comparison to the oxide variant, with a value of approximately 3.4 eV. One of the notable advantages of GaN is its capacity to inhibit the reverse reaction of water formation. When GaN is combined with 13% ZnO, the resulting materials exhibit a bandgap close to 2.5 eV [61]. This narrower bandgap enables these materials to be activated by visible light, expanding their applicability beyond just UV light. However, for this photocatalyst to effectively produce hydrogen, the use of a co-catalyst, such as Ru_2O_3 or Rh_2O_3, becomes necessary.

The previously mentioned photocatalysts have the capability to drive overall water splitting encompassing both the reduction reaction resulting in hydrogen gas production and the water oxidation reaction leading to oxygen gas production. However, these materials typically exhibit a much higher hydrogen production rate than oxygen production. Moreover, some photocatalysts cannot generate oxygen [62]. Consequently, they can be employed in Z-schemes in conjunction with photocatalysts proficient in water oxidation to facilitate comprehensive water splitting [63].

An illustrative example is $BiVO_4$, an oxide renowned for its robust oxygen generation capacity [64]. However, owing to its low conduction band, it is incapable of producing hydrogen [63]. One of the notable advantages of $BiVO_4$ is its 2.4-eV bandgap, enabling it to absorb visible light without necessitating structural modifications. Nevertheless, its application remains constrained due to the high recombination rate of photogenerated charges [65]. To address this issue and maintain effective charge separation, a co-catalyst, such as NiFeOOH, can be employed [66]. NiFeOOH also can generate charges through light absorption. By incorporating modified $BiVO_4$ with NiFeOOH and Co-Pi, the photocurrent of the composite is increased, consequently enhancing the rate of oxygen generation. Furthermore, due to the presence of NiFeOOH, the photocatalyst was able to produce hydrogen at an accelerated rate. Additionally, it is noteworthy that the monoclinic phase of $BiVO_4$ exhibits greater water oxidation activity compared to the tetragonal phase, and elevating the calcination temperature, up to 450°C, further enhances the photocatalyst's activity [67].

In the context of the Z-scheme for water splitting, $BiVO_4$ is frequently paired with $SrTiO_3$ forming a synergistic combination for the simultaneous production of O_2 and H_2, respectively. For instance, a system comprising Pt/$SrTiO_3$:Rh and $BiVO_4$ has been employed allowing for the utilization of visible light, with peak efficiency occurring at a wavelength of 440 nm [68]. In another significant study, both catalysts were subjected to doping to enhance the overall process efficiency. In this scenario, $SrTiO_3$ was co-doped with lanthanum (La) and rhodium (Rh), while $BiVO_4$ was doped with molybdenum (Mo) [69]. This modified system was also activated by visible light and exhibited superior efficiency compared to the previous one, reaching its optimal performance at a wavelength of 419 nm.

Nonetheless, the predominant approach in Z-schemes involves employing tungsten oxide for oxygen production. Much like bismuth vanadate, tungsten oxide features a bandgap of 2.6 eV making it amenable to visible light utilization [70]. However, it also presents a high recombination rate of photogenerated charges. As a result, it is common practice to combine it with platinum as a co-catalyst [71, 72]. In recent research, the impact of introducing various metal dopants into this photocatalyst was explored [73]. On average, these dopants led to a reduction in the bandgap of approximately 0.25 eV. Among the tested metals, iron, copper, and cobalt exhibited the highest photocurrent density values, with cobalt nearly doubling the activity of WO_3. This boost in performance was attributed to improved charge transfer resistance and enhanced carrier density. Another viable doping option is niobium, which, despite causing a slight increase in the tungsten oxide bandgap, also resulted in elevated photocurrent density values [74].

Tungsten oxide is also employed in Z-schemes alongside the $SrTiO_3$ perovskite for overall water splitting applications enabling the simultaneous generation of oxygen and hydrogen gases. For instance, one approach involves employing an electrode composed of rhodium-doped strontium titanate ($Ru/SrTiO_3$:Rh) in conjunction with a modified form of tungsten oxide (Fe–H–Cs–WO_3) [75]. The outcomes demonstrated that this particular photocatalyst yielded a significantly higher rate of O_2 evolution surpassing unmodified WO_3 by eight times and outperforming $BiVO_4$ by four times. Tungsten oxide has also proven effective in oxygen gas production under visible light conditions, especially when combined with platinum at a concentration of 0.5 wt% as a co-catalyst and coupled with $SrTiO_3$ co-doped with chromium (Cr) and tantalum (Ta) [76].

6.5 Enhancement Strategies

In the relentless pursuit of optimizing the effectiveness and efficiency of inorganic photocatalysts employed in photo-driven water splitting, various strategies are being meticulously investigated. This section provides a comprehensive analysis of these strategies outlining their implications and contributions to the enhancement of the performance of these materials.

In the pursuit of advancing visible light-driven photocatalytic applications, researchers initially focused on extending the utility of photocatalysts beyond traditional UV irradiation, particularly exemplified in the case of TiO_2 for water splitting [28]. Early investigations extended to narrow bandgap semiconductors, like CdS and WO_3, aiming to harness visible light for enhanced photocatalysis [77]. However, challenges, such as the pronounced photocorrosion observed in CdS [78, 79] and the unsatisfactory conduction band of WO_3 [80] for hydrogen production, posed substantial barriers to their efficient performance in visible light-driven water splitting [63, 81–83].

Despite the numerous pivotal considerations, the identification of a cost-effective material system capable of concurrently satisfying all the aforementioned technical requisites for practical water splitting remains an elusive pursuit. Notably, materials characterized by lower bandgaps demonstrate heightened responsiveness to the visible spectrum, but concurrently exhibit an increased susceptibility to photo-corrosion. Conversely, materials characterized by wider bandgaps demonstrate improved resilience during practical applications but respond to a smaller portion of the incident light energy [83].

The focus is on building nanostructures with precisely defined microstructural characteristics such as size, shape, surface area, and pore distribution. These features greatly influence the photocatalytic efficiency of semiconductors. The rational design and synthesis of semiconductors with customized structures are crucial for improving photocatalytic activity. Nanostructured materials have the ability to greatly impact the rates of surface reactions and the paths by which electrons are transported leading to enhanced kinetics of charge transfer. Recent investigations into various semiconductors highlight that optimizing electrode morphology can substantially enhance water splitting photocurrent density (Figure 6.4) [84].

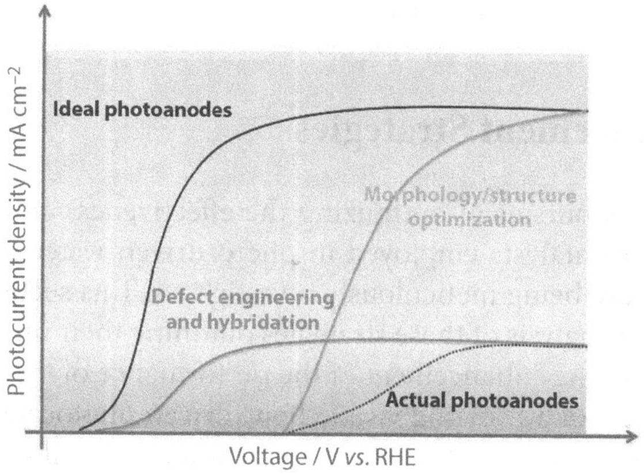

Figure 6.4 Schematic representation of photocurrent density versus voltage curves illustrating an ideal photoanode (black line), currently available models (dashed black line), and two distinct modifications: morphology optimization (orange line) and defect engineering/hybridization (blue line). Reprinted from ref. [85]. Copyright© 2022 Elsevier.

In response to these challenges, significant efforts have been dedicated to addressing the aforementioned aspects, as delineated in the subsequent subtopics.

6.5.1 Semiconductor Heterostructures and Hybridization Strategies

Attaining practical photoelectrochemical (PEC) reduction necessitates semiconductor photocathodes endowed with specific attributes. First, they must exhibit ample reducing capability implying that the conduction band level should be sufficiently cathodically positioned to drive the reduction reaction. Second, exceptional light-absorption properties are essential to harness the entire solar irradiation spectrum effectively. Last, pronounced charge separation is crucial to maximize carrier utilization efficiency. However, meeting all these criteria poses a challenge for single-component semiconductors. Consequently, semiconductor heterostructures, encompassing multiple components integrated into distinctive architectures, have been conceived and implemented to address these multifaceted necessities [86].

Significant emphasis has been placed on the creation of heterojunction nanohybrids through the coupling of TiO_2 nanowire arrays (TiO_2–NAs) with narrow-bandgap semiconductors, including CdS, CdSe, CdTe, and Cu_2O [87, 88]. This approach has garnered notable attention due to its merit in enhancing charge separation efficiency and integrating the distinctive advantages of each constituent. Furthermore, conventional type-II heterojunction nanocompounds may exhibit varying photoresponse ranges, a characteristic that proves advantageous in capturing a broader portion of the solar spectrum [89].

Integration of diverse semiconductors proves effective in broadening the photoelectrode's response to a wider solar spectrum optimizing the utilization of photonic energy [90]. Utilizing several semiconductor-sensitized photoelectrodes promotes efficient charge transfer at interfaces between different sensitizers, thereby enhancing the process of charge separation [91].

Semiconductor heterostructures employing a Z-scheme design [92], such as the Au-mediated Cu_2O–Au–TiO_2 Z-scheme heterostructure, exemplify an effective strategy for enhancing photocatalytic activity [86]. This configuration facilitates swift coupling of electrons that have been stimulated by light and holes that have migrated enhancing the potential for the photoelectrochemical water reduction reaction at the gold interface.

To improve the performance of photoelectrodes in PEC systems, a suggested approach is to combine semiconductor light absorbers with suitable electrocatalysts. This strategy aims to enhance the efficiency of charge transfer (ηCT) and, consequently, the overall performance of the photoelectrodes in generating hydrogen (HER) or oxygen (OER). Significantly, photoelectrodes that have been altered with catalysts have exhibited heightened density of photocurrent at a specific potential, along with a noticeable change toward anodic potential for the initiation of hydrogen HER, or a distinct shift toward cathodic potential for OER, in comparison to unaltered semiconductors. The addition of electrocatalysts to the surface of the semiconductor induces substantial changes in the PEC characteristics of the photoelectrodes [89].

The interaction between the semiconductor and the catalyst can result in different types of junctions, such as the Schottky junction for metallic catalysts, semiconductor–semiconductor junction for compound catalysts, or an adaptive junction for ion-permeable catalysts [93]. When catalysts are applied sparsely with limited coverage and small size, they might maintain optical transparency, hence having negligible impact on the absorption of light by the semiconductor. Nevertheless, catalyst layers or nanoparticles that have extensive coverage can cause unwanted light absorption obstructing the absorption of some incident light by semiconductors. Furthermore, catalysts have the potential to enhance both charge separation inside the semiconductor bulk and charge recombination on the semiconductor surface in addition to facilitating charge transfer [94]. Under certain circumstances, the electrocatalyst can have a direct impact on the separation or recombination of charges, without directly catalyzing the redox reactions [95].

6.5.2 Doping of Inorganic Photocatalysts

Doping, as a prominent strategic approach, emerges as a formidable strategic approach offering a sophisticated means to tailor the intrinsic properties of these materials. This comprehensive discussion navigates the intricacies of doping elucidating its role in modifying photocatalysts and, subsequently, impacting their catalytic efficiency.

The study particularly delves into two distinct categories as follows: one consisting of metal cations with d^0 electron configurations, such as tungsten (W^{6+}), titanium (Ti^{4+}), and niobium (Nb^{5+}), and the other with d^{10} configurations, including tin (Sn^{4+}), gallium (Ga^{3+}), and indium (In^{3+}). A meticulous examination of these dopant types unveils their discernible effects on the photocatalytic efficiency of the material.

Moreover, the incorporation of various elements via semiconductor doping can lead to point defects, which can be observed as the substitution of ions/atoms within the original lattice or the occupation of interstitial sites. Essentially, the ionic radius of substitutional dopants should closely match that of the atoms they replace to minimize structural changes and successfully prevent the formation of secondary phases [94]. Notably, experimental evidence supports the incorporation of heteroatoms, such as Eu [96], Bi [97], and Mn [98], into Cu_2O, for example, as a viable approach to optimizing band gaps and enhancing photocatalytic performance [85].

The co-doping strategy represents a notable approach to enhancing the photocatalytic performance of photocatalysts with a wide bandgap [99]. The primary hurdle associated with TiO_2-based photocatalysts stems from the wide bandgap limiting their functionality to UV light irradiation. To address this limitation, doping metallic or nonmetallic elements into TiO_2 has been established as an effective means to narrow the bandgap [100]. However, the introduction of a single dopant element often results in reduced carrier mobility and the creation of new recombination centers [100]. To address this challenge, researchers have developed the co-doping strategy, which involves incorporating two types of atoms to adjust the bandgap of TiO_2 [101]. This method can be classified into three primary types: charge-compensated n–p co-doping, non-compensated n–p co-doping, and the coupling of dopants mediated by double holes to regulate the bandgap. Numerous efforts have been dedicated to exploring these routes to modulate the bandgap of TiO_2 through co-doping its bulk [101–103]. Furthermore, a comprehensive analysis has been conducted revealing that the calculated band edge positions of TiO_2 support photocatalytic water splitting, although the semiconductor's bandgap value is elevated. For instance, the incorporation of a stable charge-compensated donor–acceptor pair into TiO_2 has been explored resulting in a narrowed bandgap (2.31 eV) [100]. The formation of an intermediate band (IB) through co-doping significantly contributes to narrowing the bandgap. The fully occupied and delocalized IB state, positioned just below the Fermi level, effectively mitigates the formation of electron–hole recombination centers. Additionally, the calculated positions of the conduction band minimum (CBM) and valence band maximum (VBM) of the doped TiO_2 are found to be conductive to overall photocatalytic water splitting. The calculated light absorption spectrum further highlights the robust capability of the co-doping system to absorb visible light [99].

6.5.3 Surface Modifications and Dimension of Materials

Surface modification of inorganic photocatalysts is an essential strategy to optimize catalyst–substrate interaction. The microstructural characteristics, including size, shape, surface area, and pore arrangement, significantly influence the photocatalytic efficiency of semiconductors. Therefore, the deliberate design and synthesis of semiconductors with precisely engineered structures are imperative for enhancing photocatalytic activity [85].

Nanostructures exert a significant impact on the speeds of surface reactions and the paths of electron transport in materials leading to a remarkable improvement in the kinetics of charge transfer [81, 92]. Upon reducing the semiconductor dimensions below its Bohr radius, a pronounced impact on the bandgap energy ensues governed by the quantum confinement effect [104]. This effect becomes a determining factor in the magnitude of the bandgap energy, illustrating the critical influence of particle size on the semiconductor's electronic properties. Understanding and exploiting this effect is essential for the design and optimization of semiconductor nanomaterials, as it opens possibilities for precise adjustments in electronic properties paving the way for innovative applications in photonic and catalytic devices.

The example of $BiVO_4$ highlights the significance of nanosizing, as the close proximity of the CB bottom to the H_2 evolution potential hinders sufficient driving force for H^+ reduction. However, the quantum confinement effect induces a negative shift in the CB edge allowing the efficient utilization of quantum-sized $BiVO_4$ as a promising photoanode for water splitting without requiring co-catalysts or defective structures [105]. In other instances, reducing the particle size of MoS_2 nanocrystals resulted in the emergence of a metallic zone around the particles leading to a notable decrease in excitonic binding energies. Moreover, the fabrication of nanoscale materials promotes the dissociation of charge carriers leading to a more directed absorption of light [106].

The relentless pursuit of novel inorganic photocatalysts and the strategic manipulation of existing elements have driven significant advancements in material science research and the exploration of new perspectives in the water splitting process. Alongside the discovery and synthesis of new materials, it is imperative to meticulously assess their applicability and ensuing properties. In this regard, the exploration of one-dimensional (1D), two-dimensional (2D), and three-dimensional (3D) photocatalysts (as depicted in Figure 6.5) stands out as areas of research and development of great interest [107–109]. These dimensions not only describe the geometry of materials but also encapsulate their unique physical and chemical traits, which exert distinct influences on the efficacy and nature of photocatalytic reactions. The investigation of these photocatalysts, with different dimensions, furnishes deeper insights into how atomic and molecular structures dictate their properties and efficacy in photocatalytic processes.

The efficacy of photocatalysts derived from 0D, 1D, and 2D nanomaterials exhibits significant variation concerning hydrogen production via

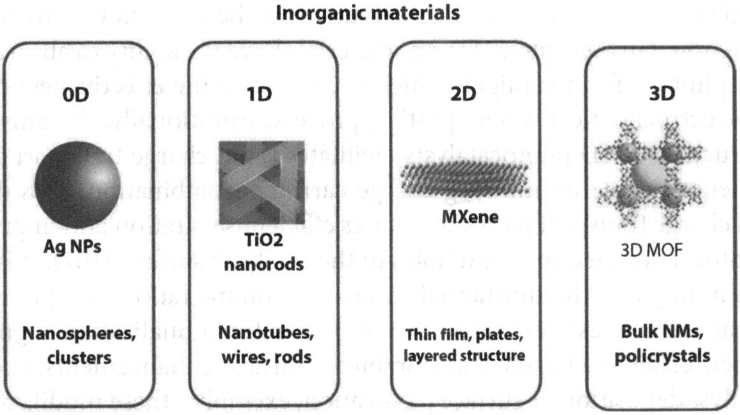

Figure 6.5 Classification of inorganic materials based on dimensionality.

water splitting processes. 0D nanomaterials, such as quantum dots and nanoparticles, have garnered considerable attention owing to their unique optical, electronic, and surface properties [107]. These attributes, intricately linked to particle size, enable 0D nanomaterials to effectively absorb sunlight and catalyze the decomposition of water into hydrogen and oxygen. Furthermore, 0D nanomaterials have a high surface-to-volume ratio increasing the contact area available for the reaction and facilitating the transfer of photoinduced charges, thus contributing to efficiency in hydrogen production [107, 110]. In addition, quantum confinement effects in QDs and nanoparticles result in spatial confinement of charge carriers leading to efficient charge separation.

However, the efficiency of these materials in yielding hydrogen is somewhat constrained by the slow rate of hydrogen evolution when irradiated with visible light. As mentioned before, the performance of quantum dots is intricately linked to their size. Recent studies suggest that an optimal size of approximately 2.8 nm is considered ideal for cadmium selenide (CdSe). At this size, quantum dots exhibit a reduced likelihood of charge carrier recombination compared to smaller counterparts and a higher conduction band edge compared to larger quantum dots. Consequently, quantum dots of this size demonstrated quantum efficiencies up to 20 times greater compared to the other QDs of CdSe of different diameters [111].

On the other hand, 1D nanomaterials, such as nanowires, nanotubes, and nanorods, exhibit notably superior efficiency in producing hydrogen through the water splitting process compared to 0D nanomaterials. These materials possess a high length-to-diameter ratio, which increases the efficiency of light absorption and scattering facilitating the transfer of photoinduced charges and contributing to the efficiency of hydrogen production. Furthermore, 1D nanomaterials have the ability to absorb and retain photons from sunlight, which is crucial for the effectiveness of the photoelectrochemical water splitting process. Additionally, the anisotropic structure of 1D photocatalysts facilitates rapid charge transport along their length while minimizing charge carrier recombination. This directional charge transport pathway ensures efficient separation and migration of photoexcited electrons and holes to the catalyst's surface [107]. It is also worth noting that the substantial surface-to-volume ratio of 1D photocatalysts allows for easy surface modification and functionalization augmenting their catalytic efficacy and durability. Surface enhancements, such as cocatalyst deposition or surface passivation, exemplify these modifications enhancing the kinetics of charge transfer reactions and alleviating surface recombination losses [112].

Studies have indicated that 1D nanomaterials exhibit a charge transfer rate more than 200 times greater than that observed for nanoparticles [107]. Nevertheless, significant advancements are required in current methodologies for the production and utilization of 1D nanomaterials as photocatalysts or co-catalysts to enable their practical application in large-scale hydrogen production. Titanium dioxide stands out as a commonly employed material for fabricating nanotubes intended for hydrogen production via water splitting. Various techniques, including the solvothermal method, hydrothermal method, or template-assisted technique, can be employed for their synthesis, each with its own set of advantages and disadvantages. However, a significant challenge associated with utilizing this material is its band gap value restricting its utilization to only 3%–5% of the solar spectrum [113].

In contrast to 0D and 1D nanomaterials, 2D nanomaterials present notable advantages. Their characteristics include a high specific surface area, which maximizes the exposure of active sites for catalytic reactions, remarkable mechanical properties maintaining structural integrity during catalytic reactions, high activity, and photochemical stability enabling them to efficiently absorb sunlight and catalyze the water decomposition reaction into hydrogen and oxygen. Furthermore, 2D nanomaterials excel in sunlight absorption due to their ultrathin nature and through photon reflection and scattering, thereby enhancing light absorption efficiency [107].

Forming heterostructures with 2D nanomaterials enables the fine-tuning of energy levels directly impacting the potential for hydrogen and oxygen generation. By adjusting these energy levels, the efficiency of hydrogen evolution can be enhanced ensuring that the minimum of the conduction band and the maximum of the valence band are located close to the potentials for hydrogen reduction and oxygen oxidation, respectively. Recent studies investigating $BiVO_4$ thin films have revealed an optimal thickness for hydrogen production using solar light. While increasing the number of layers decreases the band gap, excessive thickness leads to the formation of trap states due to increased imperfections in the film. This impedes the charge transfer process highlighting the importance of finding a balance in film thickness for optimal performance [114].

Concerning 3D materials, the primary properties related to the water splitting process include a three-dimensional porous architecture that provides a large surface area enabling the presence of numerous active sites for redox reactions. These materials feature a porous framework that facilitates efficient molecule transport and facile adjustment of active catalytic sites via surface chemistry modifications [108, 109]. Furthermore, the structural integrity and stability conferred by the three-dimensional architecture of these photocatalysts are indispensable for sustained operation under challenging reaction environments. Moreover, this structural design facilitates the integration of functional elements, including cocatalysts or light-absorbing materials aimed at augmenting catalytic performance. These elements can be seamlessly incorporated within the porous framework or surface deposited to optimize charge separation dynamics [115, 116].

In the study conducted by Luna *et al.* (2020), the synthesis of titania (TiO_2) aerogels incorporating metallic nanoparticles (Au, Pd, and PdAu) was examined for potential application as photocatalysts in hydrogen (H_2) production (see Figure 6.6) [109]. Aerogels, three-dimensional structures, preserve the intrinsic properties of the initial nanoparticles presenting significant potential for the development of effective photocatalysts. The findings revealed that the aerogels possessed high porosity, with an average pore size of approximately 34 nm and a substantial surface area of approximately 450 m^2/g. Moreover, the aerogels exhibited a notable 3.5-fold enhancement in hydrogen generation compared to corresponding powders (see Figure 6.6a). The investigation also delved into the charge carrier dynamics within the aerogels, indicating a significant influence of metallic nanoparticles on anatase's charge carrier dynamics. Furthermore, the aerogels displayed markedly superior photocatalytic activity compared to their powder counterparts underscoring the role of porosity in optimizing mass transport and, consequently, the effectiveness of H_2 production. Additionally, the aerogels demonstrated long-term stability exhibiting a low deactivation rate after 11 h of reaction [109].

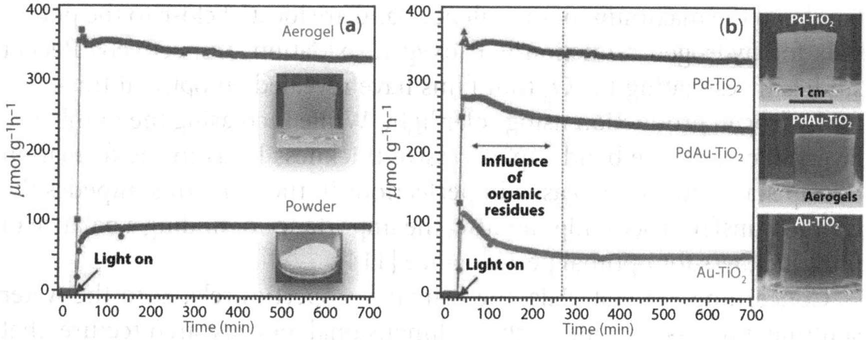

Figure 6.6 Hydrogen evolution utilizing Pd–TiO_2 aerogels and their corresponding powders (a). Hydrogen evolution profiles of TiO_2 aerogels loaded with metals obtained after 11 h of irradiation accompanied by an image depicting the aerogel monoliths featuring Pd, Pd–Au, and Au metal nanoparticles (b). Figure adapted with permission from ref. [109]. Copyright© 2020 Elsevier.

6.6 Summary and Outlook

The research and application of inorganic photocatalysts for use in the water splitting process currently face significant challenges ranging from efficiency in visible light to defect control and stability. Overcoming these obstacles is crucial for advancing the utilization of this technology on a large scale for H_2 production rendering it more viable both technically and economically. Strategies, such as defect engineering, surface modification, and the development of nanostructured photocatalysts, are being explored to optimize the performance of photocatalysts. However, advancing and improving new inorganic semiconductors and associated technology are crucial to driving sustainability and innovation in this field.

Currently, the efficiency of photocatalysts is a central concern. The goal is to achieve a quantum efficiency of 30% with radiation approximately 550–600 nm utilizing solar radiation as the energy source. Although progress has been made in recent decades, such as the utilization of inorganic photocatalysts, it is still necessary to develop photocatalysts with a broader absorption band to fully exploit solar radiation.

To achieve future research objectives in hydrogen production through the water splitting process, it is essential to develop a stable photocatalytic system with significant quantum efficiency. Various modification techniques and chemical additives have been explored to enhance the photocatalytic activity of inorganic semiconductors under visible light, as outlined in the chapter. The addition of metal nanoparticles, metal ion doping, and dye sensitization are among the promising approaches that can increase the efficiency of water splitting for hydrogen production. It is essential to combine various techniques to ensure efficient hydrogen production such as integrating noble metal-loaded dye sensitization with redox mediators. Additionally, the development of specific simulation models for photocatalytic hydrogen production is necessary to deepen our understanding of the process and to design more efficient systems.

In conclusion, continued research and development of inorganic photocatalysts are essential to advance water splitting technology and sustainable hydrogen production. Progress in this field will not only boost the hydrogen economy but also contribute significantly to the transition to renewable energy sources and climate change mitigation. Achieving optimal efficiencies is critical for advancing and ensuring the viability of this technology on an industrial scale.

Acknowledgments

The authors are grateful to the National Council for Scientific and Technological Development (CNPq) (grants #308823/2022 and #406860/2022-0) and FAPESP (grant 2023/17686-4 #) for financial support. The authors are also grateful to the CAPES (Coordination for the Improvement of Higher Education Personnel—Finance Code 001).

AI disclosure statement: The author acknowledges the use of AI-based tools for minor language editing or grammatical correction. However, all intellectual contributions, analyses, and conclusions presented in this chapter are entirely the author's own.

References

1. Dittmer, A., Izsák, R., Neese, F., Maganas, D., Accurate Band Gap Predictions of Semiconductors in the Framework of the Similarity Transformed Equation of Motion Coupled Cluster Theory. *Inorg. Chem.*, 58, 14, 9303–9315, 2019.
2. Guo, Q., Zhou, C., Ma, Z., Yang, X., Fundamentals of TiO_2 Photocatalysis: Concepts, Mechanisms, and Challenges. *Adv. Mater.*, 31, 50, 2019.
3. Sivula, K., Le Formal, F., Grätzel, M., Solar Water Splitting: Progress Using Hematite (α-Fe_2O_3) Photoelectrodes. *ChemSusChem*, 4, 4, 432–449, 2011.
4. Nagtode, V.S., Cardoza, C., Yasin, H.K.A., Mali, S.N., Tambe, S.M., Roy, P., Singh, K., Goel, A., Amin, P.D., Thorat, B.R., Cruz, J.N., Pratap, A.P.,

Green Surfactants (Biosurfactants): A Petroleum-Free Substitute for Sustainability—Comparison, Applications, Market, and Future Prospects. *ACS Omega*, 8, 13, 11674–11699, 2023.

5. Maeda, K. and Domen, K., Development of Novel Photocatalyst and Cocatalyst Materials for Water Splitting under Visible Light. *Bull. Chem. Soc. Jpn.*, 89, 6, 627–648, 2016.
6. Fajrina, N. and Tahir, M., A critical review in strategies to improve photocatalytic water splitting towards hydrogen production. *Int. J. Hydrogen Energy*, 44, 2, 540–577, 2019.
7. Kawawaki, T., Kawachi, M., Yazaki, D., Akinaga, Y., Hirayama, D., Negishi, Y., Development and Functionalization of Visible-Light-Driven Water-Splitting Photocatalysts. *Nanomaterials*, 12, 3, 344, 2022.
8. Sahani, S., Malika Tripathi, K., Il Lee, T., Dubal, D.P., Wong, C.-P., Chandra Sharma, Y., Young Kim, T., Recent advances in photocatalytic carbon-based materials for enhanced water splitting under visible-light irradiation. *Energy Convers. Manag.*, 252, 115133, 2022.
9. Tee, S.Y., Win, K.Y., Teo, W.S., Koh, L., Liu, S., Teng, C.P., Han, M., Recent Progress in Energy-Driven Water Splitting. *Adv. Sci.*, 4, 5, 1–24, 2017.
10. Zuo, F., Wang, L., Feng, P., Self-doped Ti_3^+@TiO_2 visible light photocatalyst: Influence of synthetic parameters on the H_2 production activity. *Int. J. Hydrogen Energy*, 39, 2, 711–717, 2014.
11. Police, A.K.R., Basavaraju, S., Valluri, D.K., Muthukonda V., S., Machiraju, S., Lee, J.S., $CaFe_2O_4$ sensitized hierarchical TiO_2 photo composite for hydrogen production under solar light irradiation. *Chem. Eng. J.*, 247, 152–160, 2014.
12. Chen, S., Takata, T., Domen, K., Particulate photocatalysts for overall water splitting. *Nat. Rev. Mater.*, 2, 10, 17050, 2017.
13. Kranz, C. and Wächtler, M., Characterizing photocatalysts for water splitting: from atoms to bulk and from slow to ultrafast processes. *Chem. Soc. Rev.*, 50, 2, 1407–1437, 2021.
14. Hisatomi, T., Takanabe, K., Domen, K., Photocatalytic Water-Splitting Reaction from Catalytic and Kinetic Perspectives. *Catal. Lett.*, 145, 1, 95–108, 2015.
15. Yan, X., Xia, M., Liu, H., Zhang, B., Chang, C., Wang, L., Yang, G., An electron-hole rich dual-site nickel catalyst for efficient photocatalytic overall water splitting. *Nat. Commun.*, 14, 1, 1741, 2023.
16. Chong, W.-K., Ng, B.-J., Lee, Y.J., Tan, L.-L., Putri, L.K., Low, J., Mohamed, A.R., Chai, S.-P., Self-activated superhydrophilic green ZnIn2S4 realizing solar-driven overall water splitting: close-to-unity stability for a full daytime. *Nat. Commun.*, 14, 1, 7676, 2023.
17. Lin, L., Ma, Y., Vequizo, J.J.M., Nakabayashi, M., Gu, C., Tao, X., Yoshida, H., Pihosh, Y., Nishina, Y., Yamakata, A., Shibata, N., Hisatomi, T., Takata, T., Domen, K., Efficient and stable visible-light-driven Z-scheme overall water splitting using an oxysulfide H_2 evolution photocatalyst. *Nat. Commun.*, 15, 1, 397, 2024.

18. Qin, L., Wang, G., Tan, Y., Plasmonic Pt nanoparticles-TiO_2 hierarchical nano-architecture as a visible light photocatalyst for water splitting. *Sci. Rep.*, 8, 1, 16198, 2018.
19. Zhang, L., Zhang, J., Yu, H., Yu, J., Emerging S-Scheme Photocatalyst. *Adv. Mater.*, 34, 11, 1–14, 2022.
20. Marschall, R., Semiconductor Composites: Strategies for Enhancing Charge Carrier Separation to Improve Photocatalytic Activity. *Adv. Funct. Mater.*, 24, 17, 2421–2440, 2014.
21. Nishioka, S., Osterloh, F.E., Wang, X., Mallouk, T.E., Maeda, K., Photocatalytic water splitting. *Nat. Rev. Methods Primers*, 3, 1, 42, 2023.
22. Abdul Nasir, J., Munir, A., Ahmad, N., Haq, T., Khan, Z., Rehman, Z., Photocatalytic Z-Scheme Overall Water Splitting: Recent Advances in Theory and Experiments. *Adv. Mater.*, 33, 52, 1–41, 2021.
23. Yang, J., Wang, D., Han, H., Li, C., Roles of Cocatalysts in Photocatalysis and Photoelectrocatalysis. *Acc. Chem. Res.*, 46, 8, 1900–1909, 2013.
24. Bai, S., Yin, W., Wang, L., Li, Z., Xiong, Y., Surface and interface design in cocatalysts for photocatalytic water splitting and CO_2 reduction. *RSC Adv.*, 6, 62, 57446–57463, 2016.
25. Gupta, N.M., Factors affecting the efficiency of a water splitting photocatalyst: A perspective. *Renew. Sustain. Energy Rev.*, 71, 585–601, 2017.
26. Maeda, K., Direct splitting of pure water into hydrogen and oxygen using rutile titania powder as a photocatalyst. *Chem. Commun.*, 49, 75, 8404, 2013.
27. Mu, L., Zhao, Y., Li, A., Wang, S., Wang, Z., Yang, J., Wang, Y., Liu, T., Chen, R., Zhu, J., Fan, F., Li, R., Li, C., Enhancing charge separation on high symmetry $SrTiO_3$ exposed with anisotropic facets for photocatalytic water splitting. *Energy Environ. Sci.*, 9, 7, 2463–2469, 2016.
28. Marschall, R., 50 Years of Materials Research for Photocatalytic Water Splitting. *Eur. J. Inorg. Chem.*, 2021, 25, 2435–2441, 2021.
29. Takanabe, K., Photocatalytic Water Splitting: Quantitative Approaches toward Photocatalyst by Design. *ACS Catal.*, 7, 11, 8006–8022, 2017.
30. Fujishima, A. and Honda, K., Electrochemical Photolysis of Water at a Semiconductor Electrode. *Nature*, 238, 5358, 37–38, 1972.
31. Dharma, H.N.C., Jaafar, J., Widiastuti, N., Matsuyama, H., Rajabsadeh, S., Othman, M.H.D., Rahman, M.A., Jafri, N.N.M., Suhaimin, N.S., Nasir, A.M., Alias, N.H., A Review of Titanium Dioxide (TiO_2)-Based Photocatalyst for Oilfield-Produced Water Treatment. *Membr. (Basel)*, 12, 3, 345, 2022.
32. Nogueira, A.E., Ribeiro, L.S., Gorup, L.F., Silva, G.T.S.T., Silva, F.F.B., Ribeiro, C., Camargo, E.R., New Approach of the Oxidant Peroxo Method (OPM) Route to Obtain $Ti(OH)_4$ Nanoparticles with High Photocatalytic Activity under Visible Radiation. *Int. J. Photoenergy*, 2018, 1–10, 2018.
33. Sheppard, L.R. and Wuhrer, R., TiO_2-Based homojunction photo-electrode for solar-driven water splitting. *Int. J. Hydrogen Energy*, 45, 16, 9386–9396, 2020.

34. Sang, L., Yu, Z., Wang, C., Zhao, Y., AgCu/TiO_2 hot-electron device for solar water splitting: the mechanisms of LSPR and time-domain characterization. *Int. J. Hydrogen Energy*, 48, 33, 12215–12226, 2023.
35. Radecka, M., Study of the TiO2–Cr_2O_3 system for photoelectrolytic decomposition of water. *Solid State Ion*, 157, 1–4, 379–386, 2003.
36. Xie, M.-Y., Su, K.-Y., Peng, X.-Y., Wu, R.-J., Chavali, M., Chang, W.-C., Hydrogen production by photocatalytic water-splitting on Pt-doped TiO_2–ZnO under visible light. *J. Taiwan Inst. Chem. Eng.*, 70, 161–167, 2017.
37. Qureshi, M., Garcia-Esparza, A.T., Jeantelot, G., Ould-Chikh, S., Aguilar-Tapia, A., Hazemann, J.-L., Basset, J.-M., Loffreda, D., Le Bahers, T., Takanabe, K., Catalytic consequences of ultrafine Pt clusters supported on $SrTiO_3$ for photocatalytic overall water splitting. *J. Catal.*, 376, 180–190, 2019.
38. Ikeda, S., Hirao, K., Ishino, S., Matsumura, M., Ohtani, B., Preparation of platinized strontium titanate covered with hollow silica and its activity for overall water splitting in a novel phase-boundary photocatalytic system. *Catal. Today*, 117, 1-3, 343–349, 2006.
39. Phoon, B.L., Lai, C.W., Juan, J.C., Show, P.-L., Pan, G.-T., Recent developments of strontium titanate for photocatalytic water splitting application. *Int. J. Hydrogen Energy*, 44, 28, 14316–14340, 2019.
40. Luo, C., Zhao, J., Li, Y., Zhao, W., Zeng, Y., Wang, C., Photocatalytic CO_2 reduction over $SrTiO_3$: Correlation between surface structure and activity. *Appl. Surf. Sci.*, 447, 627–635, 2018.
41. Grabowska, E., Selected perovskite oxides: Characterization, preparation and photocatalytic properties—A review. *Appl. Catal., B*, 186, 97–126, 2016.
42. Kanhere, P. and Chen, Z., A Review on Visible Light Active Perovskite-Based Photocatalysts. *Molecules*, 19, 12, 19995–20022, 2014.
43. Patial, S., Hasija, V., Raizada, P., Singh, P., Khan Singh, A.A.P., Asiri, A.M., Tunable photocatalytic activity of $SrTiO_3$ for water splitting: Strategies and future scenario. *J. Environ. Chem. Eng.*, 8, 3, 103791, 2020.
44. Yu, H., Ouyang, S., Yan, S., Li, Z., Yu, T., Zou, Z., Sol–gel hydrothermal synthesis of visible-light-driven Cr-doped $SrTiO_3$ for efficient hydrogen production. *J. Mater. Chem.*, 21, 30, 11347, 2011.
45. Suzuki, H., Tomita, O., Higashi, M., Abe, R., Z-scheme Water Splitting into H_2 and O_2 Using Tungstic Acid as an Oxygen-evolving Photocatalyst under Visible Light Irradiation. *Chem. Lett.*, 44, 8, 1134–1136, 2015.
46. Wang, M., Ma, Y., Fo, Y., Lyu, Y., Zhou, X., Theoretical insights into the origin of highly efficient photocatalyst NiO/$NaTaO_3$ for overall water splitting. *Int. J. Hydrogen Energy*, 45, 38, 19357–19369, 2020.
47. Sudrajat, H., Ichikuni, N., Onishi, H., Visible light responsive La and Fe co-doped NaTaO3 photocatalysts: Local structure around dopants. *Chem. Phys.*, 531, 110648, 2020.
48. Liu, Y.-L., Yang, C.-L., Wang, M.-S., Ma, X.-G., Yi, Y.-G., Te-doped perovskite $NaTaO_3$ as a promising photocatalytic material for hydrogen production

from water splitting driven by visible light. *Mater. Res. Bull.*, 107, 125–131, 2018.
49. Kato, H., Asakura, K., Kudo, A., Highly Efficient Water Splitting into H_2 and O_2 over Lanthanum-Doped $NaTaO_3$ Photocatalysts with High Crystallinity and Surface Nanostructure. *J. Am. Chem. Soc.*, 125, 10, 3082–3089, 2003.
50. Sebastian, M.T., Cation-Deficient Perovskites, in: *Dielectric Materials for Wireless Communication*, pp. 335–360, Elsevier, Amsterdam, 2008.
51. Yan, S.C., Wang, Z.Q., Li, Z.S., Zou, Z.G., Photocatalytic activities for water splitting of La-doped-$NaTaO_3$ fabricated by microwave synthesis. *Solid State Ion*, 180, 32–35, 1539–1542, 2009.
52. Alves, G.A.S., Centurion, H.A., Sambrano, J.R., Ferrer, M.M., Gonçalves, R.V., Band Gap Narrowing of Bi-Doped $NaTaO_3$ for Photocatalytic Hydrogen Evolution under Simulated Sunlight: A Pseudocubic Phase Induced by Doping. *ACS Appl. Energy Mater.*, 4, 1, 671–679, 2021.
53. Dong, L., Jia, R., Li, C., Xin, B., Zhang, Y., Ab initio study of N-doped β-Ga_2O_3 with intrinsic defects: the structural, electronic and optical properties. *J. Alloys Compd.*, 712, 379–385, 2017.
54. Xue, H., He, Q., Jian, G., Long, S., Pang, T., Liu, M., An Overview of the Ultrawide Bandgap Ga_2O_3 Semiconductor-Based Schottky Barrier Diode for Power Electronics Application. *Nanoscale Res. Lett.*, 13, 1, 290, 2018.
55. Zhao, Z.-C., Yang, C.-L., Meng, Q.-T., Wang, M.-S., Ma, X.-G., Photocatalytic hydrogen production from water splitting with N-doped β-Ga_2O_3 and visible light. *Spectrochim Acta A Mol. Biomol. Spectrosc.*, 211, 71–78, 2019.
56. Sakata, Y., Matsuda, Y., Yanagida, T., Hirata, K., Imamura, H., Teramura, K., Effect of Metal Ion Addition in a Ni Supported Ga_2O_3 Photocatalyst on the Photocatalytic Overall Splitting of H_2O. *Catal. Lett.*, 125, 1–2, 22–26, 2008.
57. Maeda, K., Lu, D., Teramura, K., Domen, K., Direct deposition of nanoparticulate rhodium–chromium mixed-oxides on a semiconductor powder by band-gap irradiation. *J. Mater. Chem.*, 18, 30, 3539, 2008.
58. Hisatomi, T., Maeda, K., Takanabe, K., Kubota, J., Domen, K., Aspects of the Water Splitting Mechanism on $(Ga_{1-x}Zn_x)(N_{1-x}O_x)$ Photocatalyst Modified with $Rh_{2-y}Cr_yO_3$ Cocatalyst. *J. Phys. Chem. C*, 113, 51, 21458–21466, 2009.
59. Maeda, K., Teramura, K., Lu, D., Saito, N., Inoue, Y., Domen, K., Noble-Metal/Cr_2O_3 Core/Shell Nanoparticles as a Cocatalyst for Photocatalytic Overall Water Splitting. *Angew. Chem. Int. Ed.*, 45, 46, 7806–7809, 2006.
60. Sarfraz, M.H., Zubair, M., Aslam, B., Ashraf, A., Siddique, M.H., Hayat, S., Cruz, J.N., Muzammil, S., Khurshid, M., Sarfraz, M.F., Hashem, A., Dawoud, T.M., Avila-Quezada, G.D., Abd_Allah, E.F., Comparative analysis of phyto-fabricated chitosan, copper oxide, and chitosan-based CuO nanoparticles: antibacterial potential against Acinetobacter baumannii isolates and anticancer activity against $HepG_2$ cell lines. *Front. Microbiol.*, 14, 1188743, 2023.
61. Maeda, K., Takata, T., Hara, M., Saito, N., Inoue, Y., Kobayashi, H., Domen, K., GaN:ZnO Solid Solution as a Photocatalyst for Visible-Light-Driven Overall Water Splitting. *J. Am. Chem. Soc.*, 127, 23, 8286–8287, 2005.

62. Lee, J.S., Photocatalytic Water Splitting Under Visible Light with Particulate Semiconductor Catalysts. *Catal. Surv. Asia*, 9, 4, 217–227, 2005.
63. Kudo, A. and Miseki, Y., Heterogeneous photocatalyst materials for water splitting. *Chem. Soc. Rev.*, 38, 1, 253–278, 2009.
64. Wang, L., Zhang, Y., Li, W., Wang, L., Recent advances in elaborate interface regulation of $BiVO_4$ photoanode for photoelectrochemical water splitting. *Mater. Rep.: Energy*, 3, 100232, 2023.
65. Jeon, T.H., Choi, W., Park, H., Cobalt–phosphate complexes catalyze the photoelectrochemical water oxidation of $BiVO_4$ electrodes. *PCCP*, 13, 48, 21392–21401, 2011.
66. Fang, G., Liu, Z., Han, C., Enhancing the PEC water splitting performance of $BiVO_4$ co-modifying with NiFeOOH and Co-Pi double layer cocatalysts. *Appl. Surf. Sci.*, 515, 146095, 2020.
67. Kudo, A., Omori, K., Kato, H., A Novel Aqueous Process for Preparation of Crystal Form-Controlled and Highly Crystalline $BiVO_4$ Powder from Layered Vanadates at Room Temperature and Its Photocatalytic and Photophysical Properties. *J. Am. Chem. Soc.*, 121, 49, 11459–11467, 1999.
68. Kato, H., Hori, M., Konta, R., Shimodaira, Y., Kudo, A., Construction of Z-scheme Type Heterogeneous Photocatalysis Systems for Water Splitting into H_2 and O_2 under Visible Light Irradiation. *Chem. Lett.*, 33, 10, 1348–1349, 2004.
69. Wang, Q., Hisatomi, T., Jia, Q., Tokudome, H., Zhong, M., Wang, C., Pan, Z., Takata, T., Nakabayashi, M., Shibata, N., Li, Y., Sharp, I.D., Kudo, A., Yamada, T., Domen, K., Scalable water splitting on particulate photocatalyst sheets with a solar-to-hydrogen energy conversion efficiency exceeding 1%. *Nat. Mater.*, 15, 6, 611–615, 2016.
70. Costa, M.B., de Araújo, M.A., Tinoco, M.V., de, L., Brito, JF., Mascaro, L.H., Current trending and beyond for solar-driven water splitting reaction on WO_3 photoanodes. *J. Energy Chem.*, 73, 88–113, 2022.
71. Abe, R., Takata, T., Sugihara, H., Domen, K., Photocatalytic overall water splitting under visible light by TaON and WO_3 with an IO_3^-/I^- shuttle redox mediator. *Chem. Commun.*, 30, 3829, 2005.
72. Higashi, M., Abe, R., Teramura, K., Takata, T., Ohtani, B., Domen, K., Two step water splitting into H_2 and O_2 under visible light by $ATaO_2N$ (A = Ca C.OMMAS.R.X.X.X, Ba) and WO_3 with IO^{3-}/I^- shuttle redox mediator. *Chem. Phys. Lett.*, 452, 1–3, 120–123, 2008.
73. Kalanur, S.S., Noh, Y.-G., Seo, H., Engineering band edge properties of WO_3 with respect to photoelectrochemical water splitting potentials via a generalized doping protocol of first-row transition metal ions. *Appl. Surf. Sci.*, 509, 145253, 2020.
74. Kalanur, S.S., Yoo, I.-H., Cho, I.S., Seo, H., Niobium incorporated WO_3 nanotriangles: Band edge insights and improved photoelectrochemical water splitting activity. *Ceram. Int.*, 45, 7, 8157–8165, 2019.

75. Miseki, Y., Fujiyoshi, S., Gunji, T., Sayama, K., Photocatalytic Z-Scheme Water Splitting for Independent H_2/O_2 Production via a Stepwise Operation Employing a Vanadate Redox Mediator under Visible Light. *J. Phys. Chem. C*, 121, 18, 9691–9697, 2017.
76. Abe, R., Sayama, K., Sugihara, H., Development of New Photocatalytic Water Splitting into H_2 and O_2 using Two Different Semiconductor Photocatalysts and a Shuttle Redox Mediator IO_3^-/I^-. *J. Phys. Chem. B*, 109, 33, 16052–16061, 2005.
77. Kudo, A., Development of photocatalyst materials for water splitting. *Int. J. Hydrogen Energy*, 31, 2, 197–202, 2006.
78. Wei, L., Guo, Z., Jia, X., Probing Photocorrosion Mechanism of CdS Films and Enhancing Photoelectrocatalytic Activity via Cocatalyst. *Catal. Lett.*, 151, 1, 56–66, 2021.
79. Ning, X. and Lu, G., Photocorrosion inhibition of CdS-based catalysts for photocatalytic overall water splitting. *Nanoscale*, 12, 3, 1213–1223, 2020.
80. Kalanur, S.S., Yoo, I.H., Eom, K., Seo, H., Enhancement of photoelectrochemical water splitting response of WO_3 by Means of Bi doping. *J. Catal.*, 357, 127–137, 2018.
81. Osterloh, F.E., Inorganic nanostructures for photoelectrochemical and photocatalytic water splitting. *Chem. Soc. Rev.*, 42, 6, 2294–2320, 2013.
82. Chen, X., Shen, S., Guo, L., Mao, S.S., Semiconductor-based photocatalytic hydrogen generation. *Chem. Rev.*, 110, 11, 6503–6570, 2010.
83. Jafari, T., Moharreri, E., Amin, A.S., Miao, R., Song, W., Suib, S.L., Photocatalytic Water Splitting—The Untamed Dream: A Review of Recent Advances. *Molecules*, 21, 7, 900, 2016.
84. Zhou, L., Ge, X., Ren, C., Chen, G., *In-situ* high temperature XRD and TEM study of the thermal stability and sintering behavior of octacalcium phosphate. *J. Alloys Compd.*, 778, 72–76, 2019.
85. Lei, W., Yu, Y., Zhang, H., Jia, Q., Zhang, S., Defect engineering of nanostructures: Insights into photoelectrochemical water splitting. *Mater. Today*, 52, 133–160, 2022.
86. Li, J.M., Tsao, C.W., Fang, M.J., Chen, C.C., Liu, C.W., Hsu, Y.J., TiO_2-Au-Cu_2O Photocathodes: Au-Mediated Z-Scheme Charge Transfer for Efficient Solar-Driven Photoelectrochemical Reduction. *ACS Appl. Nano Mater.*, 1, 12, 6843–6853, 2018.
87. Fang, M.J., Tsao, C.W., Hsu, Y.J., Semiconductor nanoheterostructures for photoconversion applications. *J. Phys. D Appl. Phys.*, 53, 14, 143001, 2020.
88. Chiu, Y.H., Lai, T.H., Kuo, M.Y., Hsieh, P.Y., Hsu, Y.J., Photoelectrochemical cells for solar hydrogen production: Challenges and opportunities. *APL Mater.*, 7, 8, 80901, 2019.
89. Shao, Z., Zhang, Y., Yang, X., Zhong, M., Au-Mediated Charge Transfer Process of Ternary $Cu_2O/Au/TiO_2$-NAs Nanoheterostructures for Improved Photoelectrochemical Performance. *ACS Omega*, 5, 13, 7503–7518, 2020.

90. Tiwari, S., Yadav, P., Ganguli, A.K., Enhancing the activity and stability of Cu_2O nanorods via coupling with a NaNbO3/SnS2 heterostructure for photoelectrochemical water-splitting. *New J. Chem.*, 47, 13, 6294–6304, 2023.
91. Son, H., Lee, J.H., Uthirakumar, P., Van Dao, D., Soon, A., Lee, I.H., Facile synthesis of flexible and scalable Cu/Cu_2O/CuO nanoleaves photoelectrodes with oxidation-induced self-initiated charge-transporting platform for photoelectrochemical water splitting enhancement. *J. Alloys Compd.*, 942, 169094, 2023.
92. Stolarczyk, J.K., Bhattacharyya, S., Polavarapu, L., Feldmann, J., Challenges and Prospects in Solar Water Splitting and CO_2 Reduction with Inorganic and Hybrid Nanostructures. *ACS Catal.*, 8, 4, 3602–3635, 2018.
93. Thorne, J.E., Jang, J.W., Liu, E.Y., Wang, D., Understanding the origin of photoelectrode performance enhancement by probing surface kinetics. *Chem. Sci.*, 7, 5, 3347–3354, 2016.
94. Mouli Thalluri, S., Bai, L., Lv, C., Huang, Z., Hu, X., Liu, L., Thalluri, S.M., Liu, L., Bai, L., Hu, X., Lv, C., Huang, Z., Strategies for Semiconductor/Electrocatalyst Coupling toward Solar-Driven Water Splitting. *Adv. Sci.*, 7, 6, 1902102, 2020.
95. Rossi, R.C. and Lewis, N.S., Investigation of the Size-Scaling Behavior of Spatially Nonuniform Barrier Height Contacts to Semiconductor Surfaces Using Ordered Nanometer-Scale Nickel Arrays on Silicon Electrodes. *J. Phys. Chem. B*, 105, 49, 12303–12318, 2001.
96. Shyamal, S., Hajra, P., Mandal, H., Bera, A., Sariket, D., Satpati, A.K., Malashchonak, M.V., Mazanik, A.V., Korolik, O.V., Kulak, A.I., Skorb, E.V., Maity, A., Streltsov, E.A., Bhattacharya, C., Eu modified Cu_2O thin films: Significant enhancement in efficiency of photoelectrochemical processes through suppression of charge carrier recombination. *Chem. Eng. J.*, 335, 676–684, 2018.
97. Brandt, I.S., Tumelero, M.A., Martins, C.A., Plá Cid, C.C., Faccio, R., Pasa, A.A., Defects controlling electrical and optical properties of electrodeposited Bi doped Cu_2O. *J. Appl. Phys.*, 123, 16, 2018.
98. Ganesan, K.P., Anandhan, N., Gopu, G., Amaliroselin, A., Marimuthu, T., Paneerselvam, R., An enhancement of ferromagnetic, structural, morphological, and optical properties of Mn-doped Cu_2O thin films by an electrodeposition technique. *J. Mater. Sci.: Mater. Electron.*, 30, 21, 19524–19535, 2019.
99. Fu, C.-F., Wu, X., Yang, J., Fu, C.-F., Wu, X., Yang, J., Material Design for Photocatalytic Water Splitting from a Theoretical Perspective. *Adv. Mater.*, 30, 48, 1802106, 2018.
100. Tang, J., Durrant, J.R., Klug, D.R., Mechanism of photocatalytic water splitting in TiO_2. Reaction of water with photoholes, importance of charge carrier dynamics, and evidence for four-hole chemistry. *J. Am. Chem. Soc.*, 130, 42, 13885–13891, 2008.

101. Yin, W.J., Wei, S.H., Al-Jassim, M.M., Yan, Y., Double-hole-mediated coupling of dopants and its impact on band gap engineering in TiO_2. *Phys. Rev. Lett.*, 106, 6, 066801, 2011.
102. Gai, Y., Li, J., Li, S.S., Xia, J.B., Wei, S.H., Design of narrow-gap TiO_2: A passivated codoping approach for enhanced photoelectrochemical activity. *Phys. Rev. Lett.*, 102, 3, 036402, 2009.
103. Zhu, W., Qiu, X., Iancu, V., Chen, X.Q., Pan, H., Wang, W., Dimitrijevic, N.M., Rajh, T., Meyer, H.M., Paranthaman, M.P., Stocks, G.M., Weitering, H.H., Gu, B., Eres, G., Zhang, Z., Band gap narrowing of titanium oxide semiconductors by noncompensated anion-Cation codoping for enhanced visible-Light photoactivity. *Phys. Rev. Lett.*, 103, 22, 226401, 2009.
104. Wang, H., Jia, L., Bogdanoff, P., Fiechter, S., Möhwald, H., Shchukin, D., Size-related native defect engineering in high intensity ultrasonication of nanoparticles for photoelectrochemical water splitting. *Energy Environ. Sci.*, 6, 3, 799–804, 2013.
105. Sun, S., Wang, W., Li, D., Zhang, L., Jiang, D., Solar light driven pure water splitting on quantum sized $BiVO_4$ without any cocatalyst. *ACS Catal.*, 4, 10, 3498–3503, 2014.
106. Lin, L., Miao, N., Huang, J., Zhang, S., Zhu, Y., Horsell, D.D., Ghosez, P., Sun, Z., Allwood, D.A., A photocatalyst of sulphur depleted monolayered molybdenum sulfide nanocrystals for dye degradation and hydrogen evolution reaction. *Nano Energy*, 38, 544–552, 2017.
107. Daulbayev, C., Sultanov, F., Bakbolat, B., Daulbayev, O., 0D, 1D and 2D nanomaterials for visible photoelectrochemical water splitting. A Review. *Int. J. Hydrogen Energy*, 45, 58, 33325–33342, 2020.
108. Pratomo, U., Pratama, R.A., Irkham, I., Sulaeman, A.P., Mulyana, J.Y., Primadona, I., 3D-ZnO Superstructure Decorated with Carbon-Based Material for Efficient Photoelectrochemical Water-Splitting under Visible-Light Irradiation. *Nanomaterials*, 13, 8, 1380, 2023.
109. Luna, A.L., Matter, F., Schreck, M., Wohlwend, J., Tervoort, E., Colbeau-Justin, C., Niederberger, M., Monolithic metal-containing TiO_2 aerogels assembled from crystalline pre-formed nanoparticles as efficient photocatalysts for H_2 generation. *Appl. Catal., B*, 267, 118660, 2020.
110. Kandi, D., Martha, S., Parida, K.M., Quantum dots as enhancer in photocatalytic hydrogen evolution: A review. *Int. J. Hydrogen Energy*, 42, 15, 9467–9481, 2017.
111. Grigioni, I., Bernareggi, M., Sinibaldi, G., Dozzi, M.V., Selli, E., Size-dependent performance of CdSe quantum dots in the photocatalytic evolution of hydrogen under visible light irradiation. *Appl. Catal. A Gen.*, 518, 176–180, 2016.
112. Darrudi, M., Tavakol, H., Momeni, M.M., Electrochemical co-deposition of cobalt and graphene, produced from recycled polypropylene, on TiO_2 nanotube as a new catalyst for photoelectrochemical water splitting. *Int. J. Hydrogen Energy*, 48, 9, 3495–3510, 2023.

113. Ge, M., Li, Q., Cao, C., Huang, J., Li, S., Zhang, S., Chen, Z., Zhang, K., Al-Deyab, S.S., Lai, Y., One-dimensional TiO_2 Nanotube Photocatalysts for Solar Water Splitting. *Adv. Sci.*, 4, 1, 1–31, 2017.
114. Hernández, S., Thalluri, S.M., Sacco, A., Bensaid, S., Saracco, G., Russo, N., Photo-catalytic activity of $BiVO_4$ thin-film electrodes for solar-driven water splitting. *Appl. Catal. A Gen.*, 504, 266–271, 2015.
115. Al-Hamamre, Z., Karimzadeh, Z., Ji, S., Choi, H., Maleki, H., Aerogels-Inspired based Photo and Electrocatalyst for Water Splitting to Produce Hydrogen. *Appl. Mater. Today*, 29, 101670, 2022.
116. Song, L., Xue, T., Shen, Z., Yang, S., Sun, D.T., Yang, J., Hong, Y., Su, Y., Wang, H., Peng, L., Li, J., Metal-organic aerogel derived hierarchical porous metal-carbon nanocomposites as efficient bifunctional electrocatalysts for overall water splitting. *J. Colloid Interface Sci.*, 621, 398–405, 2022.

7

Functional Materials for Water Splitting

Figen Balo[1]* and Lutfu S. Sua[2]

[1]*Department of METE, Engineering Faculty, Firat University, Elazig, Turkiye*
[2]*Deptartment of Management and Marketing, Southern University and A&M College, Baton Rouge, LA, USA*

Abstract
Water, being plentiful in the world, is one of the most proper resources of H_2 power. Through the process of water splitting, hydrogen can be produced directly from the decomposition of water into its constituents. Electrical (electricity), light (electromagnetic irradiation), or thermal (heat) energy resources can all provide the energy needed to break the H–O–H bonds. Sustainable hydrogen production is the essential prerequisite for an envisioned hydrogen economy. Due to the practical challenges faced by overall sustainable power resources, water splitting has been identified as the most suitable sustainable power resource for hydrogen production.

This article first surveys current functional water splitting materials and discusses future requirements in the water splitting area. Then, considering the functional materials used for water separation in recent years, it was aimed to determine the most efficient functional material with the help of analytical hierarchy process. This study will assist in the selection of the materials' optimal group for H_2 production with water splitting for newcomers to the field, industry professionals, and specialists.

Keywords: Water splitting, H_2 production, renewable energy, AHP, MCDM

7.1 Introduction

Fossil fuels provide approximately 90% of the world's energy for transport and industry [1]. High energy consumption, rapidly growing demand, and the global depletion of energy resources are driving researchers to explore renewable energy sources to achieve sustainable national development.

**Corresponding author*: figenbalo@gmail.com

Inamuddin, Tariq Altalhi, Mohammad Luqman and Jorddy Neves Cruz (eds.) Water Splitting: Production of Hydrogen, (171–210) © 2025 Scrivener Publishing LLC

Current needs are being met by exploiting a wide range of available resources. However, a major problem of the modern era is that these resources are also seriously polluting the environment. As a result, there is an urgent need to develop maintainable, ecologically friendly renewable energy sources. Because there is so much water on Earth, it is one of the best sources of H_2 power. Researchers have developed many alternative energy generation techniques, including geothermal, wind, and hydroelectric power. However, the efficiency and usability of these alternative methods are limited by their many practical limitations. Wind energy is incredibly unpredictable making power system management a challenge [2, 3]. Hydropower, another advanced energy source with nearly a century of practical experience, has certain practical drawbacks due to the high cost of dam construction [4]. By reducing external influences, optimizing operations, and developing more affordable and effective technological solutions, this approach has room for further improvement [5, 6]. Due to its high construction and maintenance costs, geothermal energy is a continuous renewable energy resource that harnesses the thermal energy of the Earth's interior. It has negative ecological impacts, including polluted air emissions, air pollution, and land subsidence [7, 8]. Solar energy is another renewable energy source and is considered the most likely replacement for fossil fuels. According to estimates, as little as 0.01% of sun power transformation is enough to solve the global energy crisis. Sun power also helps us produce sustainable, clean, and environmentally friendly gasoline. Although they have limited efficiency and energy storage capability, sun cells can be one viable resource for sun power transformation [9, 10].

Future sustainable energy sources that can be stored include hydrogen [11–13]. The primary benefits of utilizing H_2 as an energy source are that it does not require fossil fuels, is location independent, takes up less space, produces harmless products, and is safe, affordable, clean, and environmentally friendly [14, 15]. In particular, hydrogen can be utilized in place of oil as the energy source for internal engine combustion in fuel cells, where it produces electric power. High energy density, natural availability, and the absence of toxic byproducts after burning are only a few of the many advantages of hydrogen energy sources [16–18]. The so-called "hydrogen economy" is proposed in this scenario. Consequently, there is a growing focus on hydrogen energy and the vehicles that run on it. Not only can it perform similarly to petroleum-sourced ones but it can also be featured without causing any environmental harm. Numerous major automakers worldwide, including Ford, Toyota, BMW, and Hyundai, have made consistent attempts to make hydrogen-powered vehicles a reality in the near future [19–21]. As a result, the development of new techniques

Table 7.1 Water splitting literature.

Topic	References
Sustainable energy sources	[1–15]
Hydrogen generation	[16–28]
Water splitting	[29–37]
Photoelectrochemical water splitting	[38–74]
Z-scheme water splitting	[75–104]

to produce hydrogen, which is considered to be the most suitable renewable energy source, has recently received a great deal of attention [22]. H_2 generation industries are already present to generate H_2, with novel industries in different stages of improvement. The novel industries generally concentrate on doing the generation reaction less or zero emission. The technology utilizes colors to aid in distinguishing industrial groups of H_2 differentiating from clean (pink, white, turquoise, blue, and green) to carbon-intensive (brown/black and gray) H_2. Table 7.1 provides a classification of water splitting literature discussed in this section.

Turquoise H_2 can be generated by natural gas pyrolysis. Unlike blue or gray H_2, this reaction emits solid (not gaseous) C, which can be either stored permanently or utilized as raw material for different industrial reactions (without emitting it in the environment down the value chain as CO_2). Thus, direct C emission is prevented. Nonetheless, this industry would also require addressing the upstream CH_4 emissions and has not proven to be measurable yet.

Gray H_2 depends on the natural gas supporting (by CH_4 gas-heated reforming, CH_4 auto-thermal, or CH_4 steam reformation) the industry. C emissions correlate with steam CH_4 reformers (9 kg of CO_2/kg of H_2), and upstream CH_4 emissions emerging from natural gas resource making gray H_2 an emission-intense reaction.

Green H_2 is produced through electrolyzing sustainable power resources, like wind and solar energy. It is one of the minimum C compact methodologies of producing H_2 and composes directly no emissions. It is easily measurable and estimated to become too cost competitive with increasing distribution, like a lot that has been seen in the improvement of sustainable energies over the recent 10 years.

Utilizing nuclear energy, pink H_2 is produced for water electrolysis. In addition, this approach releases no C. There are a few problems concerning nuclear energy social acceptance, baseload electric supply prioritizing, and scaling up.

Brown or black H_2 defines the most harmful methodology utilized in coal gasification; pollutants emitting 20 kg of CO_2/kg of H_2 in the reaction with H_2 resources that can be accessed through underfloor shafts are mentioned to as white H_2. When crosschecked with global needs, the contributions are unimportant. When combined with C storage and capture, blue H_2 increases gray H_2. Blue H_2 can contribute to rapidly increase the demand for clean H_2 through the use of the groundworks now onsite for gray H_2. Even with a longtime C-capture ratio of approximately 0.95, upstream CH_4 and residual emissions present challenges for this industry to achieve C neutrality [23].

The most up-and-coming of all H_2 generation techniques is water splitting, which is receiving a lot of attention from researchers. The least harmful method of producing hydrogen is water electrolysis, which is followed by coal gasification, steam reforming from hydrocarbons, and partial oxidation of oil [24]. Combining water electrolysis technology with renewable clean energy sources (wind, solar, etc.) has generated significant interest in the production of hydrogen [25, 26]. Hydrogen has been considered one of the most promising energies because of its pleasant and energetically dense properties [27]. Hydrogen is also widely employed in the refining of petroleum, the manufacture of ammonia, and as a possible fuel source for fuel cells. One essential industrial raw material is hydrogen. Direct emissions of pollutants or greenhouse gases that might otherwise contaminate our planet are not possible while using hydrogen energy [28]. Maximum efficiency of the water splitting reaction is currently achieved when photovoltaic cells are connected to water electrolyzers, but the high cost of this technology prevents it from being scaled up globally to meet the demand. An alternative method of producing hydrogen from solar energy is water splitting at semiconductor surfaces. The results of the first use of an electrically induced semiconductor substance for water splitting were published by Honda and Fujishima [29]. Photoelectrochemical water splitting was achieved by absorbing light through a titanium dioxide photoelectrode [30]. Heterogeneous photocatalysis makes it possible to split water using photocatalysis [31]. In heterogeneous photocatalysis, light is applied to semiconductor materials. Carriers of charge are produced within the semiconductor adsorber material through excitation of electrons from the band of valence to the transmission band utilizing adsorbed light power, when the incident light energy is greater than the irradiated semiconductor's band hole. The radiated photons can be thought of as the initiators of the entire process, since the photogenerated carriers of charge must pass through the adsorber and reach the superficies for the surface reaction that follows charge separation. To participate in redox processes with accessible electron acceptors and donors, the excited carriers of charge in the semiconductor diffuse to the semiconductor surface. For example, photoexcited electrons in water can decrease protons to generate H_2. Water can be oxidized to oxygen as a counter-oxidation; so, photocatalytic water splitting is the entire process. All the same, the term "water splitting" is still being misused, as was recently critically examined elsewhere [32]. However, it is incorrect to refer to the hole oxidation reaction as "water splitting" in many experiments that use sacrificial alcohols. The same holds true for the application of electron scavengers; nevertheless, it appears that the erroneous use of "water splitting" and "hydrogen creation" occurs more frequently. The phrase "water splitting" should only be applied when the reactant and solvent are both pure water, without the use of any other reagents such as alcohols. Thermodynamically, one type of photosynthetic process is photocatalytic water splitting [33]. In Gibbs free power, given that the entire water splitting process has a significant positive shift, the process is uphill overall. Common catalysts are generally required for water splitting to occur when there is a charge accumulation of two or four charges, respectively. Therefore, the molecular bonds between hydrogen and oxygen are where solar light is truly stored. Furthermore, since the reverse reaction to water is the thermodynamically more preferred reaction, it must be suppressed. This reaction can occur on the same co-catalysts that catalyze the photosynthetic activity. The ability of photoproduced carriers of charge to recombine on their route to the superficies poses the primary challenge to this photocatalytic action. Recombination is the process by which photoexcited electrons scatter across granule borders and among one semiconductor material's deficiencies returning to the valence band and dissipating their whole potency power from the adsorption of light. Surface defects such as open atom positions serve as recombination sites in addition to bulk defects. As a result, before the photoproduced carriers of charge reach the process areas at the superficies, they may already be lost as intermediates in the heterogeneous photocatalysis reaction. To increase photocatalytic efficiency, the most crucial task and difficulty is to restrict charge carrier recombination [34–36]. For solar water splitting, a number of novel semiconductor materials in visible light have been found recently [37].

According to the materials for a single absorber, the earliest application of titanium dioxide as a photoelectrode was in photoelectrochemical water splitting. Afterward, the powdered form was also studied for photocatalytic water splitting. But without any modifications, such as the development of a heterojunction or the decorating of noble metals, titanium dioxide alone is unable to split water or produce hydrogen photocatalytically. Under UV light, titanium dioxide embellished with noble metals, such as sodium hydroxide or alkali carbonates, has been recorded to facilitate photocatalytic water splitting. This process reduces the back process caused by the carbonate surface covering [38, 39]. Particularly in Japan, a large number of novel metal oxide photocatalysts that are effective at splitting water have been reported and found. Materials having metal cations with d10 (Sb^{5+}, Sn^{4+}, Ge^{4+}, In^{3+}, Ga^{3+}) or d0 (W^{6+}, Ta^{5+}, Zr^{4+}, Nb^{5+}, Ti^{4+}) electron configuration are typically divided into two classes. Different synthesis strategies were used to prepare binary, quaternary, and ternary semiconductor oxides. Solid-state processes were mostly used to fuse diverse metal oxides together to form more complicated compounds. For photocatalytic water splitting, review papers provide excellent overviews of all these materials. Sodium tantalum trioxide is one of the most studied oxide semiconductors by d0 design [40–42]. In literature today, its extraordinarily high activity source is still used [43–45]. Several dopants, such as Sr, have been studied in relation to sodium tantalum trioxide [46, 47]. Strontium titanate is another early metal oxide that has lately gained importance [48]. When strontium titanate was doped with various metal cations by Maeda, it acquired the ability to absorb visible light. Since then, strontium titanate doped with rhodium has become one of the most popular photocatalysts for hydrogen production and, when paired with different O_2 change photocatalysts in z-framework mechanisms, one of the most popular catalysts for hydrogen evolution [49]. For water splitting, the most acknowledged powder-sourced oxide photocatalyst was created by combining Mo-doped bismuth vanadate with La/rhodium-doped strontium titanate. This Z-scheme, which took the form of an immobilized photocatalyst sheet and used Au as an electron mediator, was able to convert solar irradiation into H_2 energy source with a sun-to-H_2 performance of 0.011 [50]. It has been reported that aliovalent doping for defect engineering can eliminate the intrinsic flaws of strontium titanate by using La or Ga doping as one adsorber material for photocatalytic water splitting [51]. It was discovered that strontium titanate's quantum efficiency for water splitting may be greatly increased by Al doping reaching up to 30% at 360 nm. Eventually, at 365 nm, this performance was increased to 56% quantum efficiency [52]. Subsequent research has now been published on this increase, including studies on the quantity of aluminum, long-term stability, and substitute dopants [53–55]. Gallium oxide, when ornamented with nickel oxide, but especially when doped with zinc, is an extremely active oxide with a d10 structure for photocatalytic water splitting in ultraviolet light [56, 57]. But unlike the majority of the aforementioned cases, which only make up a minor portion of solar light irradiation, only ultraviolet light may be used to generate charge carriers; in the case of gallium oxide, ultraviolet-carbon light is required. Titanium dioxide was a key ingredient in material research when scientists were looking for materials that could absorb visible light. Non-metal doping of well-known and stable oxides, such as titanium dioxide, with anions, such as sulfur or nitrogen, was a particularly popular tactic [58]. After treating titanium dioxide with urea, Kisch discovered that the material's visible light activity resulted from surface sensitization caused by melamine condensation products, such as melon, a carbon nitride substance that was initially described by Liebig and Berzelius [59, 60]. Since Antonietti *et al.*'s detailed description of the optical characteristics, sheeted carbon nitride photocatalysts' photocatalytic activity, and electronic structure, numerous research groups worldwide have studied the substance in an effort to maximize its catalytic, electronic, and optical properties [61–64]. The comparatively low band gap and high conduction band and valence band positions of carbon nitride, in comparison to many oxide semiconductors, are noteworthy and intriguing characteristics [65]. In the past 30 years, many more apparent light-adsorbing materials for photocatalysis have been recorded. However, many of these materials (such as bismuth vanadate, tungsten trioxide, and ferric oxide) have band positions that are unsuitable for water splitting, are photocorrosion prone (such as cadmium sulfide and gallium arsenide), or do not simultaneously produce oxygen and hydrogen (such as barium tantalum oxynitride). The tunable absorption of oxynitrides down to 650 nm has drawn a lot of attention to this family of materials [66]. Still, very few materials made of oxynitride can split water. Maeda and his co-workers displayed the activity of water splitting of tantalum oxynitride under apparent light radiation [67]. Throughout the oxynitride synthesis, zirconium dioxide appears to stop Ta^{5+} cations at the tantalum oxynitride superficies from being reduced. When surface flaws are avoided in a reducing atmosphere during nitridation, stoichiometric water splitting occurs following co-catalyst decorating. While a great deal of sunlight is absorbed by many oxynitrides, relatively few of them are able to make use of the absorbed light. It was found that an oxynitride might use up to 600 nm of its absorbed photons. Following Rh–Cr oxide decorating, the complex oxynitride demonstrated water splitting activity and further shielding with amorphous thin titanium n-oxide and silicon n-dioxide protective shifts [68]. A more complicated oxynitride of water splitting was recorded by examining the impact of lattice composition [69]. Similar TiOx coatings were also applied to $CaTaO_2N$ for photocatalytic water splitting of greater than 500 nm in visible light [70]. There are no reports yet of photocatalytic water splitting with different earth-alkali analogs, whose gaps of band decrease from Ca over Ba to Sr. On the other hand, increased charge carrier separation has recently been demonstrated for solutions of perovskite oxynitride solids such as $CaTaO_2N$ and $LaTaON_2$ [71]. The incorporation of sulfur to generate oxysulfides shifts the edges of valence band to higher negative weights in comparison to the first-ever oxides, in a manner similar to the inclusion of nitrogen into oxides. Nevertheless, a large number of oxysulfides fail to show sufficient overpotential for the half-reaction of water oxidation [72]. It was reported [73] that the initial oxysulfide to achieve photocatalytic total water splitting had done so. For all water splitting, numerous adsorber materials with gaps of band in the apparent light area only exhibit activity for a semi process or have band locations that are inappropriate [74].

Consequently, Z-schemes combining two of them can allow overall water splitting [75]. In Z-schemes, the charge carrier flow is inverted as opposed to heterojunctions made of two semiconductors [76]. Furthermore, it was demonstrated recently by Sun and his co-workers that heterojunctions, or Z-schemes, can be intentionally created [77, 78]. In Z-scheme water splitting systems, strontium titanate, tungsten, and rhodium trioksit are perhaps the most typically utilized materials for the evolution of hydrogen and oxygen, respectively. Bismuth vanadate is becoming a very popular oxygen side absorber. Water splitting was reported when it was combined with strontium titanate/rhodium [79]. When coupled with Ta_3N_5, La doping into strontium titanate/rhodium was found to increase the rate of hydrogen generation [80]. As mediators, redox couples in solution are typically employed, to be decreased through electrons photoproduced from the photocatalysts that evolves O_2 and to be oxidized through photoproduced gaps from the catalyst that evolves H_2. For Z-scheme water splitting, solid redox mediators, such as reduced metals [81], carbon [82], and graphene oxide [83, 84], became more common. The photocatalyst layer by Domen and colleagues is a notable example of a Z-scheme. It involves using gold metal immobilized on molybdenum-doped strontium titanate: rhodium/La and bismuth vanadate (bismuth vanadate: molybdenum) [85]. This method has the ability to split water with a power performance of 1.1%, which is the highest performance recorded for this process with oxide powders to date. An STH of 1.0% might be attained by using carbon as the foundation layer. Recently, the system war was further optimized with the use of a zirconium oxide covering and a gold colloid [86]. Another family of materials that has been discussed that absorb visible light and can split

Table 7.2 Water splitting effectiveness of different functional materials under diverse process circumstances.

	Ref.	Mass loading ($mg\ cm^{-2}$)	Cell voltage (V) ($10\ mA\ cm^{-2}$)	Substrate	Catalyst
Benchmark catalysts under acidic condition	[105]	0.226	1.88	RDE	Cu-MOF (8% GO)
	[106]	0.012	1.47	RDE	$Cu_{2-x}S$@IrS_y@IrRu
	[107]	~0.35	1.66	GCE	Co_3O_4–RuCo@NC
	[108]	0.357	1.53	GCE	Ru NCs/Co_2P
	[109]	0.2	1.51	GCE	Ir–SA@Fe@NCNT
	[110]	0.01	1.56	GCE	IrNiCo
	[111]	0.0125	1.58	GCE	IrNi
	[112]	0.031	1.62	GCE	Ir nanowire
	[113]	0.09	1.58	GCE	Ir nanoparticle
Electrocatalysts under different reaction conditions	[114]	2.55	1.91	GCE	NFPGNS
	[115]	0.116	1.74	CFP	DRPC
	[116]	1.0	1.65	NF	$Ni_3ZnC_{0.7}$
	[117]	2.0	1.64	NF	NC@Cu–Co–W–C
	[118]	0.7	1.64	NF	Co/b-Mo_2C@N–CNTs
	[119]	2.0	1.66	NF	Ni/Mo_2C–PC
	[120]	1.0	1.73	NF	Mo_2C/CS
	[121]	0.5	1.6	NF	Ni_3FeN/r-GO
	[122]	2.0	1.64	CC	CVN/CC
	[123]	0.08	1.57	NF	FeB_2–NF
	[124]	1.6	1.58	Ni foil	Ni–B–O@Ni_3B
	[125]	0.3	1.64	GCE	Co–P–B
	[126]	1.0	1.602	CC	Ni_xB/f-MWCNT@CC

(*Continued*)

Table 7.2 Water splitting effectiveness of different functional materials under diverse process circumstances. (*Continued*)

	Ref.	Mass loading (mg cm⁻²)	Cell voltage (V) (10 mA cm⁻²)	Substrate	Catalyst
TMP-based electrocatalysts under different reaction conditions	[127]	0.76	1.63	NF	$Ni_2P/Ni(PO_3)_2$
	[128]	2.0	1.6	Ti plate	Co-Pi/CoP/Ti
	[129]	5.68	1.67	CC	NiS/Ni_2P
	[130]	3.5	1.65	NF	NiO@NF
	[131]	0.28	1.7	CFP	CoP/rGO
	[132]	1.0	1.60	CFP	Co-P@PC
	[133]	0.36	1.64	GCE	Co_2P@N, P-PCN/CNTs
	[134]	1.95	1.67	CFP	Co_2P/NPSC
	[135]	2.0	1.64	NF	Ni/Mo_2C-NCNFs
	[136]	2.0	1.6	NF	NiFeP@C/NF
	[137]	0.88	1.45	GCE	Co/CoP
	[138]	0.4	1.74	GCE	$Cu_{0.3}Co_{2.7}P$/NC
	[139]	1.38	1.589	CFP	$Ni_{1.5}Fe_{0.5}P$/CF
	[140]	1.0	1.67	NF	Ni-Fe-P
	[141]	5.17	1.72	CFP	$Cu_{0.08}Co_{0.92}P$ NAs/CP
	[142]	2.0	1.56	NF	CoFeP NFs/NPCNT
	[143]	2.0	1.52	CC	NiCoP/CC
	[144]	1.6	1.58	NF	NiCoP/NF
	[145]	1.6	1.69	CC	IPNTs (Fe-P nanotube)
	[146]	2.0	1.57	CFP	Ni_2P/CP
	[147]	1.0	1.74	Copper foil	Co-P
	[148]	1.2	1.67	NF	Cu_3P@NF
	[149]	0.5	1.64	NF	Porous Ni-P
	[150]	~0.3	1.63	CFP	CP@Ni-P
	[151]	~0.15	1.57	CFP	Ni_xP_y

(*Continued*)

Table 7.2 Water splitting effectiveness of different functional materials under diverse process circumstances. (*Continued*)

	Ref.	Mass loading (mg cm^{-2})	Cell voltage (V) (10 mA cm^{-2})	Substrate	Catalyst
TMD-based electrocatalysts under different reaction conditions	[152]	1	1.547	CFP	$NiSe_2/NiO_x$ nanoparticles
	[153]	2	2.18	Ti mesh	$CoO/CoSe_2$
	[154]	0.4	1.76	NF	$Co_{0.85}Se$@NC
	[155]	2.9	1.63	CFP	$CoSe_2$/CF
	[156]	2.6	1.59	Cobalt foam	Co_3Se_4
	[157]	0.2	1.58	N	CoS–$Co(OH)_2$@aMoS_{2+x}/NF
	[158]	1.86	1.54	F	$MoS_2/Co_9S_8/Ni_3S_2$/Ni
	[159]	9.7	1.56	NF	MoS_2/Ni_3S_2
	[160]	13	1.50	NF	MoS_2–Ni_3S_2 HNRs/NF
	[161]	0.2	1.47	NF	FeS_2/CoS_2
	[162]	3.9	1.50	Copper foam	Cu@CoS_x/CF
	[163]	4.2	1.58	NF	NiFeSP/NF
	[164]	5.86	1.52	NF	$NiP_{0.62}S_{0.38}$ nanosheets
	[165]	1	1.56	NF	V-doped NiS_2
	[166]	4.09	1.53	Graphdiyne	$NiCo_2S_4$ NW/graphdiyne foam
	[167]	4.1	1.6	NF	Hexagram-like CoMnS/NF
	[168]	3	1.625	NF	Ni0.7Fe0.3S2
	[169]	5	1.60	NF	Co_9S_8@NOSC
	[170]	2.7	1.74	Graphite foam	CoS_2/GF
	[171]	0.7 ± 0.2	1.58	GC	NiS(NiS_2) hollow microspheres

(*Continued*)

Table 7.2 Water splitting effectiveness of different functional materials under diverse process circumstances. (*Continued*)

	Ref.	Mass loading (mg cm^{-2})	Cell voltage (V) (10 mA cm^{-2})	Substrate	Catalyst
Transition metal hydroxides and oxide-based electrocatalysts under different reaction conditions	[172]	2.2	~1.73	Cu plate	CuO@Ni/NiFe hydroxides
	[173]	4.9	1.60	N	NiFe-LDH /$NiCo_2O_4$ arrays
	[174]	20	1.57	F	NC/NiMo/$NiMoO_x$
	[175]	3	1.44@20	NF	NiCo-$NiCoO_2$@NC
	[176]	0.5	2.004	NF	Co@Co_3O_4-NC
	[177]	2	1.58		Co@CoO/G
	[178]	3	1.51	CFP	Li-induced $NiFeO_x$
	[179]	2	1.558	NF	CoFeNi-O
	[180]	1.51	1.53	CC	Oxygen vacancy enriched $CoFe_2O_4$
	[181]	~4	1.583	NF	Co6.25Fe18.75Ni75Ox
	[182]	2	1.7@54	NF	CoMnO@CN superlattices
	[183]	0.2	1.74	Co foil	CoOSeP
	[184]	~0.2	1.58	Fe foil	3D Fe-F-O nanoporous film

(*Continued*)

Table 7.2 Water splitting effectiveness of different functional materials under diverse process circumstances. (*Continued*)

	Ref.	Mass loading (mg cm^{-2})	Cell voltage (V) (10 mA cm^{-2})	Substrate	Catalyst
Transition metal hydroxides and oxide-based electrocatalysts under different reaction conditions	[185]	1	2.06	CF	Co-doped ZnO nanowires
	[186]	~5	1.8	NF	SrTi0.1Fe0.85Ni0.05O3-d/CNT
	[187]	3	~1.68	NF	SrNb0.1Co0.7Fe0.2O3-d
	[188]	~1	1.60	NF	{111}Co_3O_4
	[189]	2	1.55@20	NF	CoO_x@CN
	[190]	1	1.74	NF	Co_3O_4/N-doped carbon
	[191]	1.42	1.591	NF	Ni0.75Fe0.125V0.125–LDHs/NF
	[192]	0.2	1.43	NF	Ni_5Co_3Mo–OH
	[193]	2	1.54	NF	O(H)–NiFe(II,III)–LDH
	[194]	~1.9	1.51	NF	NiFe LDH@Ni NTAs
	[195]	2	~1.4	NF	NiFe LDH–NS@DG
	[196]	1.5	1.51	NF	NiCo–HS@G
	[197]	1	1.66	NF	NiCo–LDH/NF
	[198]	0.2	1.69	NF	Mesoporous CoFe LDH
	[199]	~2.9	1.68	NF	$Ni(OH)_2$ nanosheets/NF

(*Continued*)

Table 7.2 Water splitting effectiveness of different functional materials under diverse process circumstances. (*Continued*)

	Ref.	Mass loading (mg cm^{-2})	Cell voltage (V) (10 mA cm^{-2})	Substrate	Catalyst
Metal-based electrocatalysts under different reactionsConditions.	[200]	12.5	1.58	GCE	IrNi nanoclusters
	[201]	~0.82	~1.55	Graphite foam	Ir/GF
	[202]	1.0	1.687	CC	FeCoNi/NG
	[203]	0.025	1.52	GCE	Fe1.0Co1.1Ni1.4-NC
	[204]	12	1.47	NF	NiFeMo IOS
	[205]	2.0	1.59	CC	Ni_2Co-N
	[206]	0.5	~1.62	Ti foil	$Mo_{0.6}Ni_{0.4}$ nanoparticles
	[207]	4.2	1.59	Copper foil	NiMo microsphere films
	[208]	0.68	1.64	Ti mesh	NiMo hollow nanorod array
	[209]	0.306	1.667	RDE	Fe_3Co_7@PCNSs
	[210]	2.5	1.58	AAO	NiFe nanotube
	[211]	2	1.58	NF	$Ni_{0.9}Fe_{0.1}$/N-CNTs
	[212]	0.1	1.75	QCM	$Ni_{0.5}Co_{0.5}$/NC film
	[213]	1	1.563	CC	Ni QD@NC@rGO
	[214]	0.8	1.62	NF	Ni@NC

water is oxynitrides; however, many of them are not effective at splitting water on their own. Numerous Z-schemes have been recorded to be energetic for water splitting when combined [87–91]. Bismuth vanadate and oxynitrides together work remarkably well for water splitting [92]. Bismuth vanadate might also be mixed with oxysulfides or sulfides for water splitting of Z-scheme [93–96]. Oxychlorides and oxysulfides were also used in Z-scheme water splitting. For apparent light all water splitting, bismuth vanadate is combined with additional oxides that absorb visible light such as zinc rhodium oxide and calcium iron oxide [97, 98]. Z-scheme water splitting has made extensive use of graphitic carbon nitride. Recently, combinations containing tungsten trioksit, titanium dioxide, and ferric oxide have been found [99–101]. Recently, a novel trend to improve photocatalysts using the Z-scheme interfacial contact for transfer of charge has emerged: the 2D adsorber materials' combination [102]. Only a very small number of mesostructured Z-schemes for total water splitting have been described; they include bismuth vanadate mixed with graphitic carbon nitride, black phosphorus/bismuth vanadate, and hematite nanosheets with exfoliated graphitic carbon nitride [103, 104]. Table 7.2 presents the water splitting effectiveness of different functional materials under diverse process circumstances.

H_2 generation and water splitting based on the photocatalysts' features, which result from their composition, nature, and different effects is an important issue. For H_2 generation with water splitting, the used materials and a multiple decision analysis are explained to choose the most efficient material in this study.

7.2 MCDM Analysis

The methodology involves a multi-factor decision making hierarchy where both the alternative functional materials are compared subject to a couple of criteria in an attempt to make an optimal selection. Mass loading and cell voltage are considered when comparing and evaluating 100 different materials. Mass loading and cell voltage values of each material are provided in Table 7.1. The first step involves pairwise comparison of the two factors on a 9-point scale indicating the comparative importance of each criterion against another one.

Comparative assessment of mass loading and cell voltage results in the following relative priorities for the selected criteria. The relative priorities form a priority vector *w* such that:

$$p = [p_i] = [p_1\ p_2 \cdots p_n] \quad where\ p_i > 0\ and\ \textstyle\sum_{i=1}^{n} p_i = 1 \tag{7.1}$$

Expert opinions reveal that cell voltage is evaluated to be the most significant factor in selecting the optimum material with a *66%* weight. Since the values of different characteristics used in Table 7.1 have a wide range of scales, these values need to be normalized for comparison purposes. Normalized values are calculated using Equation 7.2:

$$n_{ij} = {x_{ij}} \Big/ {\sum_{j=1} x_j} \qquad for\ \forall i = 1,\ldots,m \tag{7.2}$$

Table 7.3 Normalized priorities.

	[105]	[106]	[107]	[108]	[109]	[110]	[111]	[112]	[113]	[114]	[115]	[116]	[117]	[118]	[119]
Mass loading	0.0008	0.0000	0.0001	0.0001	0.0004	0.0108	0.0005	0.0042	0.0085	0.0030	0.0085	0.0042	0.0021	0.0085	0.0003
cell voltage	0.0088	0.0091	0.0092	0.0094	0.0092	0.0111	0.0101	0.0096	0.0095	0.0095	0.0097	0.0101	0.0093	0.0095	0.0091
	[120]	[121]	[122]	[123]	[124]	[125]	[126]	[127]	[128]	[129]	[130]	[131]	[132]	[133]	[134]
Mass loading	0.0068	0.0013	0.0042	0.0032	0.0085	0.0240	0.0148	0.0012	0.0042	0.0015	0.0082	0.0085	0.0085	0.0037	0.0017
cell voltage	0.0092	0.0095	0.0093	0.0095	0.0093	0.0097	0.0096	0.0099	0.0093	0.0095	0.0097	0.0095	0.0093	0.0084	0.0101
	[135]	[136]	[137]	[138]	[139]	[140]	[141]	[142]	[143]	[144]	[145]	[146]	[147]	[148]	[149]
Mass loading	0.0058	0.0042	0.0219	0.0085	0.0085	0.0068	0.0068	0.0085	0.0042	0.0051	0.0021	0.0013	0.0006	0.0042	0.0085
cell voltage	0.0092	0.0097	0.0100	0.0091	0.0088	0.0092	0.0098	0.0091	0.0101	0.0097	0.0095	0.0095	0.0091	0.0090	0.0127
	[150]	[151]	[152]	[153]	[154]	[155]	[156]	[157]	[158]	[159]	[160]	[161]	[162]	[163]	[164]
Mass loading	0.0017	0.0123	0.0110	0.0008	0.0079	0.0410	0.0550	0.0008	0.0165	0.0178	0.0248	0.0042	0.0173	0.0173	0.0127
cell voltage	0.0102	0.0095	0.0092	0.0092	0.0090	0.0091	0.0087	0.0085	0.0087	0.0092	0.0088	0.0091	0.0089	0.0093	0.0095

(*Continued*)

Table 7.3 Normalized priorities. (*Continued*)

	[165]	[166]	[167]	[168]	[169]	[170]	[171]	[172]	[173]	[174]	[175]	[176]	[177]	[178]	[179]
Mass loading	0.0212	0.0114	0.0030	0.0093	0.0207	0.0846	0.0127	0.0021	0.0085	0.0127	0.0085	0.0064	0.0169	0.0085	0.0008
cell voltage	0.0093	0.0101	0.0092	0.0101	0.0093	0.0091	0.0084	0.0117	0.0092	0.0088	0.0091	0.0089	0.0092	0.0099	0.0101
	[180]	[181]	[182]	[183]	[184]	[185]	[186]	[187]	[188]	[189]	[190]	[191]	[192]	[193]	[194]
Mass loading	0.0008	0.0042	0.0212	0.0127	0.0042	0.0085	0.0042	0.0060	0.0008	0.0085	0.0080	0.0085	0.0063	0.0042	0.0008
cell voltage	0.0092	0.0120	0.0105	0.0098	0.0093	0.0090	0.0101	0.0093	0.0083	0.0090	0.0088	0.0081	0.0088	0.0097	0.0098
	[195]	[186]	[197]	[198]	[199]	[200]	[201]	[202]	[203]	[204]	[205]	[206]	[207]	[208]	[209]
Mass loading	0.0123	0.0529	0.0035	0.0042	0.0001	0.0508	0.0085	0.0021	0.0178	0.0029	0.0013	0.0106	0.0085	0.0004	0.0042
cell voltage	0.0098	0.0092	0.0090	0.0098	0.0088	0.0085	0.0092	0.0094	0.0092	0.0095	0.0097	0.0092	0.0092	0.0102	0.0091
	[210]	[211]	[212]	[213]	[214]										
Mass loading	0.0106	0.0085	0.0004	0.0042	0.0034										
cell voltage	0.0092	0.0092	0.0102	0.0091	0.0094										

where

$$x_{ij} = the\ value\ of\ the\ j^{th}\ material\ for\ the\ i^{th}\ criterion$$

$$n_{ij} = the\ normalized\ value\ of\ the\ j^{th}\ material\ for\ the\ i^{th}\ criterion$$

Table 7.3 presents the normalized values of the 100 materials evaluated in this study.

As a final step, normalized values should be multiplied with the relative weight of each criterion to reflect their contributions toward the overall attractiveness of each material. To calculate the weighted scores of the alternatives, these normalized values are multiplied by the relative priority values of each criterion using Equation 7.3:

$$y_{ij} = n_{ij} * p_i \quad for\ \forall j = 1,\ldots,n \tag{7.3}$$

where

$$y_{ij} = the\ weighted\ value\ of\ the\ material\ for\ the\ i^{th}\ criterion$$

$$p_i = the\ priority\ of\ the\ i^{th}\ criterion$$

The weighted results in Table 7.4 presents the weight scores of each functional material.

Table 7.4 Weighted priorities.

	[105]	[106]	[107]	[108]	[109]	[110]	[111]	[112]	[113]	[114]	[115]	[116]	[117]	[118]	[119]
Mass loading	0.0003	0.0000	0.0005	0.0005	0.0003	0.0000	0.0000	0.0000	0.0001	0.0036	0.0002	0.0014	0.0028	0.0010	0.0028
cell voltage	0.0073	0.0057	0.0064	0.0059	0.0059	0.0060	0.0061	0.0063	0.0061	0.0074	0.0067	0.0064	0.0064	0.0064	0.0064
	0.0076	0.0057	0.0069	0.0064	0.0061	0.0061	0.0061	0.0063	0.0063	0.0110	0.0069	0.0078	0.0092	0.0073	0.0093
	[120]	[121]	[122]	[123]	[124]	[125]	[126]	[127]	[128]	[129]	[130]	[131]	[132]	[133]	[134]
Mass loading	0.0014	0.0007	0.0028	0.0001	0.0023	0.0004	0.0014	0.0011	0.0028	0.0080	0.0049	0.0004	0.0014	0.0005	0.0027
cell voltage	0.0067	0.0062	0.0064	0.0061	0.0061	0.0064	0.0062	0.0063	0.0062	0.0065	0.0064	0.0066	0.0062	0.0064	0.0065
	0.0081	0.0069	0.0092	0.0062	0.0084	0.0068	0.0076	0.0074	0.0090	0.0145	0.0113	0.0070	0.0076	0.0069	0.0092
	[135]	[136]	[137]	[138]	[139]	[140]	[141]	[142]	[143]	[144]	[145]	[146]	[147]	[148]	[149]
Mass loading	0.0028	0.0028	0.0012	0.0006	0.0019	0.0014	0.0073	0.0028	0.0028	0.0023	0.0023	0.0028	0.0014	0.0017	0.0007
cell voltage	0.0064	0.0062	0.0056	0.0067	0.0062	0.0065	0.0067	0.0060	0.0059	0.0061	0.0066	0.0061	0.0067	0.0065	0.0064
	0.0092	0.0090	0.0069	0.0073	0.0081	0.0079	0.0140	0.0089	0.0087	0.0084	0.0088	0.0089	0.0082	0.0082	0.0071

(Continued)

Table 7.4 Weighted priorities. (*Continued*)

	[150]	[151]	[152]	[153]	[154]	[155]	[156]	[157]	[158]	[159]	[160]	[161]	[162]	[163]	[164]
Mass loading	0.0004	0.0002	0.0014	0.0028	0.0006	0.0041	0.0037	0.0003	0.0026	0.0137	0.0183	0.0003	0.0055	0.0059	0.0083
cell voltage	0.0063	0.0061	0.0060	0.0085	0.0068	0.0063	0.0062	0.0061	0.0060	0.0060	0.0058	0.0057	0.0058	0.0061	0.0059
	0.0067	0.0063	0.0074	0.0113	0.0074	0.0104	0.0098	0.0064	0.0086	0.0197	0.0241	0.0060	0.0113	0.0120	0.0142
	[165]	[166]	[167]	[168]	[169]	[170]	[171]	[172]	[173]	[174]*	[175]	[176]	[177]	[178]	[179]
Mass loading	0.0014	0.0058	0.0058	0.0042	0.0071	0.0038	0.0010	0.0031	0.0069	0.0282	0.0042	0.0007	0.0028	0.0042	0.0028
cell voltage	0.0060	0.0059	0.0062	0.0063	0.0062	0.0067	0.0061	0.0067	0.0062	0.0061	0.0056	0.0078	0.0061	0.0059	0.0060
	0.0075	0.0117	0.0120	0.0105	0.0133	0.0106	0.0071	0.0098	0.0131	0.0343*	0.0098	0.0085	0.0089	0.0101	0.0089
	[180]	[181]	[182]	[183]	[184]	[185]	[186]	[187]	[188]	[189]	[190]	[191]	[192]	[193]	[194]
Mass loading	0.0021	0.0056	0.0028	0.0003	0.0003	0.0014	0.0071	0.0042	0.0014	0.0028	0.0014	0.0020	0.0003	0.0028	0.0027
cell voltage	0.0059	0.0061	0.0066	0.0067	0.0061	0.0080	0.0070	0.0065	0.0062	0.0060	0.0067	0.0062	0.0055	0.0060	0.0059
	0.0081	0.0118	0.0094	0.0070	0.0064	0.0094	0.0140	0.0107	0.0076	0.0088	0.0082	0.0082	0.0058	0.0088	0.0085

(*Continued*)

Table 7.4 Weighted priorities. (*Continued*)

	[195]	[186]	[197]	[198]	[199]	[200]	[201]	[202]	[203]	[204]	[205]	[206]	[207]	[208]	[209]
Mass loading	0.0028	0.0021	0.0014	0.0003	0.0041	0.0176	0.0012	0.0014	0.0000	0.0169	0.0028	0.0007	0.0059	0.0010	0.0004
cell voltage	0.0054	0.0059	0.0064	0.0066	0.0065	0.0061	0.0060	0.0065	0.0059	0.0057	0.0062	0.0063	0.0062	0.0064	0.0065
	0.0082	0.0080	0.0078	0.0068	0.0106	0.0238	0.0072	0.0080	0.0059	0.0226	0.0090	0.0070	0.0121	0.0073	0.0069
	[210]	[211]	[212]	[213]	[214]										
Mass loading	0.0035	0.0028	0.0001	0.0014	0.0011										
cell voltage	0.0061	0.0061	0.0068	0.0061	0.0063										
	0.0097	0.0089	0.0069	0.0075	0.0074										

7.3 Conclusions

The chemical process by which water is split into oxygen and hydrogen is called water splitting. A technological advance based on efficient and cost-effective water splitting could form the basis of a green hydrogen economy. While photosynthesis involves some water splitting, it does not produce hydrogen. The hydrogen fuel cell works on the opposite principle of water splitting. One resource that is free, endless, and good for the environment is hydrogen. Building an energy infrastructure with hydrogen as the main carrier is one of the many alternative energy options that could hold great promise for a clean and sustainable energy future.

Since hydrogen exists in water, which is literally everywhere, electrolysis is the obvious way to collect hydrogen. That raises the question of where the electricity is supposed to come from, though. Once more, solar power is a clear, pollution-free source of electricity. Even better would be if hydrogen could be produced directly by sunlight eliminating the requirement for needless intermediate electricity generation.

One of the most promising methods for generating fuel from renewable sources is photocatalytic water splitting, which produces hydrogen and oxygen by subjecting semiconductors to sunlight. In fact, one of the primary areas of interest is Z-scheme approaches, which imitate photosynthesis in plants and are arguably the most natural approach imaginable [215].

The results obtained from the previous section clearly indicate that one of the transition metal hydroxides and oxide-based electrocatalysts with an F substrate and an NC/NiMo/NiMoOx catalyst [174] has the highest weighted total score indicating that they are the optimum choice based on the criteria employed in this study. Although the number of criteria involves mass loading and cell voltage, the MCDM methodology utilized in this study can easily be extended to include additional criteria of interest.

The main contribution of this study is that it is one of the most comprehensive narrative reviews conducted on water splitting including hundreds of studies within the last 20 years.

AI disclosure statement: The author acknowledges the use of AI-based tools for minor language editing or grammatical correction. However, all intellectual contributions, analyses, and conclusions presented in this chapter are entirely the author's own.

References

1. Ferreira, A., *et al.*, Economic overview of the use and production of photovoltaic solar energy in Brazil. *Renew. Sustain. Energy Rev.*, 81, 181–191, 2018.
2. Díaz-González, F., *et al.*, A review of energy storage Technologies for wind power applications. *Renew. Sustain. Energy Rev.*, 16, 4, 2154–2171, 2012.
3. Georgilakis, P.S., Technical challenges associated with the integration of wind power into power systems. *Renew. Sustain. Energy Rev.*, 12, 3, 852–863, 2008.
4. Hamilton, A.L., Characklis, G.W., Reed, P.M., *Managing financial risk tradeoffs for hydropower generation using snowpackbased index contracts*, American Geophysical Union, USA, 2020.
5. Ellabban, O., Abu-Rub, H., Blaabjerg, F., Renewable energy resources: current status, future prospects and their enabling technology. *Renew. Sustain. Energy Rev.*, 39, 748–764, 2014.
6. Edenhofer, O., *et al.*, *Renewable energy sources and climate change mitigation: special report of the intergovernmental panel on climate change*, Cambridge University Press, Cambridge, 2011.
7. Fridleifsson, I.B., Geothermal energy for the benefit of the people. *Renew. Sustain. Energy Rev.*, 5, 3, 299–312, 2001.
8. Barbier, E., Geothermal energy technology and current status: an overview. *Renew. Sustain. Energy Rev.*, 6, 1–2, 3–65, 2002.
9. Dincer, I. and Acar, C., Review and evaluation of hydrogen production methods for better sustainability. *Int. J. Hydrog. Energy*, 40, 34, 11094–11111, 2015.
10. Turner, J.A., Sustainable hydrogen production. *Science*, 305, 5686, 972–974, 2004.
11. Edalati, K., *et al.*, Enhanced photocatalytic hydrogen production on GaN–ZnO oxynitride by introduction of strain-induced nitrogen vacancy complexes. *Acta Mater.*, 185, 149–156, 2020.
12. Chen, Y., *et al.*, Engineering the atomic interface with single platinum atoms for enhanced photocatalytic hydrogen production. *Angew. Chem. Int. Ed.*, 59, 3, 1295–1301, 2020.
13. Chen, R., *et al.*, Integration of lanthanide–transition-metal clusters onto cds surfaces for photocatalytic hydrogen evolution. *Angew. Chem. Int. Ed.*, 57, 51, 16796–16800, 2018.
14. Safari, F. and Dincer, I., A review and comparative evaluation of thermochemical water splitting cycles for hydrogen production. *Energy Convers. Manag.*, 205, 112182, 2020.
15. Liao, C.-H., Huang, C.-W., Wu, J., Hydrogen production from semiconductor-based photocatalysis via water splitting. *Catalysts*, 2, 4, 490–516, 2012.
16. Hosseini, S.E. and Wahid, M.A., Hydrogen production from renewable and sustainable energy resources: Promising green energy carrier for cleandevelopment. *Renew. Sust. Energ. Rev.*, 57, 850–866, 2016.

17. Li, X.R., Yang, X.C., Xue, H.G., Pang, H., Xu, Q., Metal-organic frameworks as a platform for clean energy applications. *Energy Chem.*, 2, 100027, 2020.
18. Barreto, L., Makihira, A., Riahi, K., The hydrogen economy in the 21st century: a sustainable development scenario. *Int. J. Hydrogen Energ.*, 28, 3, 267–284, 2003.
19. Sharma, S. and Ghoshal, S.K., Hydrogen the future transportation fuel: From production to applications. *Renew. Sust. Energ. Rev.*, 43, 1151–1158, 2015.
20. U.S.D.o, A national vision of America's transition to a hydrogen economy – To 2030 and beyond. *Energy*, 2002, February 1, https://doi.org/10.2172/1216144. http://www.eere.energy.gov/hydrogenandfuelcells/pdfs/vision.
21. Zou, X. and Zhang, Y.U., Noble metal-free hydrogen evolution catalysts for water splitting. *Chem. Soc. Rev.*, 44, 15, 5148–5180, 2015.
22. Kim, J.H., *et al.*, Toward practical solar hydrogen production–an artificial photosynthetic leaf-to-farm challenge. *Chem. Soc. Rev.*, 48, 7, 1908–1971, 2019.
23. Green hydrogen: Energizing the path to net zero, in: *Deloitte's outlook on the global clean hydrogen market*, 2023.
24. Yang, J., Yu, C., Fan, X., Liang, S., Li, S., Huang, H., Ling, Z., Hao, C., Qiu, J., Electroactive edge site-enriched nickel–cobalt sulfide into graphene frameworks for high-performance asymmetric supercapacitors. *Energy Environ. Sci.*, 9, 1299, 2016.
25. Khan, W., Singh, A.K., Naseem, S., Husain, S., Shoeb, M., Nadeem, M., Synthesis and magnetic dispersibility of magnetite decorated reduced graphene oxide. *Nano-Struct.*, 16, 180, 2018.
26. El-Refaei, S.M., Russo, P.A., Pinna, N., Recent advances in multimetal and doped transition-metal phosphides for the hydrogen evolution reaction at different pH values. *ACS Appl. Mater. Interfaces*, 13, 22077, 2021.
27. Li, Y., Li, R., Wang, D., Xu, H., Meng, F., Dong, D., Jiang, J., Zhang, J., An, M., Yang, P., A review: target-oriented transition metal phosphide design and synthesis for water splitting. *Int. J. Hydrog. Energy*, 46, 5131, 2021.
28. Dawood, F., Anda, M., Shafiullah, G., Hydrogen production for energy: an overview. *Int. J. Hydrog. Energy*, 45, 3847, 2021.
29. Fujishima, A. and Honda, K., Electrochemical Evidence for the Mechanism of the Primary Stage of Photosynthesis. *Bull. Chem. Soc. Jpn.*, 44, 1148–1150, 1971.
30. Fujishima, A. and Honda, K., Electrochemical Photolysis of Water at a Semiconductor Electrode. *Nature*, 238, 37–38, 1972.
31. Miseki, Y., Kato, H., Kudo, A., Water splitting into H2 and O2 over Ba5Nb4O15 photocatalysts with layered perovskite structure prepared by polymerizable complex method. *Chem. Lett.*, 35, 1052–1053, 2006.
32. Idriss, H., The elusive photocatalytic water splitting reaction using sunlight on suspended nanoparticles: is there a way forward? *Catal. Sci. Technol.*, 10, 304–310, 2020.

33. Osterloh, F.E., Photocatalysis versus Photosynthesis: A Sensitivity Analysis of Devices for Solar Energy Conversion and Chemical Transformations. *ACS Energy Lett.*, 2, 445–453, 2017.
34. Osterloh, F.E., Inorganic Materials as Catalysts for Photochemical Splitting of Water. *Chem. Mater.*, 20, 35–54, 2008.
35. Chen, X., Shen, S., Guo, L., Mao, S.S., Semiconductor-based Photocatalytic Hydrogen Generation. *Chem. Rev.*, 110, 6503–6570, 2010.
36. Ishaq, T., Ehsan, Z., Qayyum, A., Abbas, Y., Irfan, A., Al-Hussain, S.A., Irshad, M.A., Zaki, M.E.A., Recent Strategies to Improve the Photocatalytic Efficiency of TiO2 for Enhanced Water Splitting to Produce Hydrogen. *Catalysts*, 14, 10, 674, 2024.
37. Sivula, K. and Krol, R.., Semiconducting materials for photoelectrochemical energy conversion. *Nat. Rev. Mater.*, 1, 15010, 2016.
38. Kandiel, T.A., Dillert, R., Robben, L., Bahnemann, D.W., Photonic efficiency and mechanism of photocatalytic molecular hydrogen production over platinized titanium dioxide from aqueous methanol solutions. *Catal. Today*, 161, 196–201, 2011.
39. Yamaguti, K. and Sato, S., Photolysis of water over metallized powdered titanium dioxide. *J. Chem. Soc. Faraday Trans. 1*, 81, 1237, 1985.
40. Kudo, A. and Miseki, Y., Heterogeneous Photocatalyst Materials for Water Splitting. *Chem. Soc. Rev.*, 38, 253–278, 2009.
41. Kato, H. and Kudo, A., New tantalate photocatalysts for water decomposition into H2 and O2. *Chem. Phys. Lett.*, 295, 487–492, 1998.
42. Deo, M., Möllmann, A., Haddad, J., Ünlü, F., Kulkarni, A., Liu, M., Tachibana, Y., Stadler, D., Bhardwaj, A., Ludwig, T., *et al.*, Tantalum Oxide as an Efficient Alternative Electron Transporting Layer for Perovskite Solar Cells. *Nanomaterials*, 12, 5, 780, 2022.
43. Sudrajat, H., Kitta, M., Ito, R., Nagai, S., Yoshida, T., Katoh, R., Ohtani, B., Ichikuni, N., Onishi, H., Water-Splitting Activity of La-Doped NaTaO3 Photocatalysts Sensitive to Spatial Distribution of Dopants. *J. Phys. Chem. C*, 124, 15285–15294, 2020.
44. Ivanova, I., Kandiel, T.A., Cho, Y.-J., Choi, W., Bahnemann, D., Mechanisms of Photocatalytic Molecular Hydrogen and Molecular Oxygen Evolution over La-Doped NaTaO3 Particles: Effect of Different Cocatalysts and Their Specific Activity. *ACS Catal.*, 8, 2313–2325, 2018.
45. Kato, H., Asakura, K., Kudo, A., Highly Efficient Water Splitting into H2 and O2 over Lanthanum-Doped NaTaO3 Photocatalysts with High Crystallinity and Surface Nanostructure. *J. Am. Chem. Soc.*, 125, 3082–3089, 2003.
46. An, L., Kitta, M., Iwase, A., Kudo, A., Ichikuni, N., Onishi, H., Photoexcited Electrons Driven by Doping Concentration Gradient: Flux-Prepared NaTaO3 Photocatalysts Doped with Strontium Cations. *ACS Catal.*, 8, 9334–9341, 2018.

47. Tang, Z.K., Di Valentin, C., Zhao, X., Liu, L.M., Selloni, A., Understanding the Influence of Cation Doping on the Surface Chemistry of NaTaO3 from First Principles. *ACS Catal.*, 9, 10528–10535, 2019.
48. Domen, K., Naito, S., Soma, M., Onishi, T., Tamaru, K., Photocatalytic decomposition of water vapour on an NiO–SrTiO3 catalyst. *J. Chem. Soc., Chem. Commun.*, 2021, 543–544, 1980, DOI: 10.1039/C39800000543.
49. Maeda, K.Z., Scheme Water Splitting Using Two Different Semiconductor Photocatalysts. *ACS Catal.*, 3, 1486–1503, 2013.
50. Wang, Q., Hisatomi, T., Jia, Q., Tokudome, H., Zhong, M., Wang, C., Pan, Z., Takata, T., Nakabayashi, M., Shibata, N., Li, Y., Sharp, I.D., Kudo, A., Yamada, T., Domen, K., Scalable water splitting on particulate photocatalyst sheets with a solar-to-hydrogen energy conversion efficiency exceeding 1. *Nat. Mater.*, 15, 611–615, 2016.
51. Takata, T. and Domen, K., Defect Engineering of Photocatalysts by Doping of Aliovalent Metal Cations for Efficient Water Splitting. *J. Phys. Chem. C*, 113, 19386–19388, 2009.
52. Goto, Y., Hisatomi, T., Wang, Q., Higashi, T., Ishikiriyama, K., Maeda, T., Sakata, Y., Okunaka, S., Tokudome, H., Katayama, M., Akiyama, S., Nishiyama, H., Inoue, Y., Takewaki, T., Setoyama, T., Minegishi, T., Takata, T., Yamada, T., Domen, K., A Particulate Photocatalyst Water-Splitting Panel for Large-Scale Solar Hydrogen Generation. *Joule*, 2, 509–520, 2018.
53. Zhao, Z., Goncalves, R.V., Barman, S.K., Willard, E.J., Byle, E., Perry, R., Wu, Z., Huda, M.N., Moulé, A.J., Osterloh, F.E., Electronic structure basis for enhanced overall water splitting photocatalysis with aluminum doped SrTiO3 in natural sunlight. *Energy Environ. Sci.*, 12, 1385–1395, 2019.
54. Lyu, H., Hisatomi, T., Goto, Y., Yoshida, M., Higashi, T., Katayama, M., Takata, T., Minegishi, T., Nishiyama, H., Yamada, T., Sakata, Y., Asakura, K., Domen, K., An Al-doped SrTiO3 photocatalyst maintaining sunlight-driven overall water splitting activity for over 1000 h of constant illumination. *Chem. Sci.*, 10, 3196–3201, 2019.
55. Han, K., Lin, Y.C., Yang, C.M., Jong, R., Mul, G., Mei, B., Promoting Photocatalytic Overall Water Splitting by Controlled Magnesium Incorporation in SrTiO3 Photocatalysts. *ChemSusChem*, 10, 4510–4516, 2017.
56. Yanagida, T., Sakata, Y., Imamura, H., Photocatalytic Decomposition of H2O into H2 and O2 over Ga2O3 Loaded with NiO. *Chem. Lett.*, 33, 726–727, 2004.
57. Sakata, Y., Matsuda, Y., Yanagida, T., Hirata, K., Imamura, H., Teramura, K., Development of alternative photocatalysts to TiO2: Challenges and opportunities. *Catal. Lett.*, 125, 22–26, 2008.
58. Marschall, R. and Wang, L., Semiconductor Composites: Strategies for Enhancing Charge Carrier Separation to Improve Photocatalytic Activity. *Catal. Today*, 225, 111–135, 2014.

59. Mitoraj, D. and Kisch, H., Surface Control and Photocatalytic Activity of Branched Silver Orthophosphate Dendrites. *Angew. Chem.*, 120, 10123–10126, 2008. Angew. Chem. Int. Ed. 2008, 47, 9975–9978.
60. Liebig, J., Ueber Einige Stickstoff-Verbindungen. *Ann. Pharm.*, 10, 1, 1834.
61. Wang, X., Maeda, K., Chen, X., Takanabe, K., Domen, K., Hou, Y., Fu, X., Antonietti, M., Polymer Semiconductors for Artificial Photosynthesis: Hydrogen Evolution by Mesoporous Graphitic Carbon Nitride with Visible Light. *J. Am. Chem. Soc.*, 131, 1680–1681, 2009.
62. Chen, X., Jun, Y.-S., Takanabe, K., Maeda, K., Domen, K., Fu, X., Antonietti, M., Wang, X., Ordered mesoporous SBA-15 type graphitic carbon nitride: A semiconductor host structure for photocatalytic hydrogen evolution with visible light. *Chem. Mater.*, 21, 4093–4095, 2009.
63. Wang, X., Maeda, K., Thomas, A., Takanabe, K., Xin, G., Carlsson, J.M., Domen, K., Antonietti, M., A metal-free polymeric photocatalyst for hydrogen production from water under visible light. *Nat. Mater.*, 8, 76–80, 2009.
64. Zheng, Y., Liu, J., Liang, J., Jaroniec, M., Qiao, S.Z., Graphitic carbon nitride materials: controllable synthesis and applications in fuel cells and photocatalysis. *Energy Environ. Sci.*, 5, 6717, 2012.
65. Su, F., Mathew, S.C., Lipner, G., Fu, X., Antonietti, M., Blechert, S., Wang, X., mpg-C3N4-Catalyzed Selective Oxidation of Alcohols Using O2 and Visible Light. *J. Am. Chem. Soc.*, 132, 16299–16301, 2010.
66. Maeda, K. and Domen, K., New Non-Oxide Photocatalysts Designed for Overall Water Splitting under Visible Light. *J. Phys. Chem. C*, 111, 7851–7861, 2007.
67. Maeda, K., Lu, D., Domen, K., Direct Water Splitting into Hydrogen and Oxygen under Visible Light by using Modified TaON Photocatalysts with d0 Electronic Configuration. *Chem. A Eur. J.*, 19, 4986–4991, 2013.
68. Pan, C., Takata, T., Nakabayashi, M., Matsumoto, T., Shibata, N., Ikuhara, Y., Domen, K., A complex perovskite-type oxynitride: the first photocatalyst for water splitting operable at up to 600 nm. *Angew. Chem.*, 127, 2998–3002, 2015. Angew. Chem. Int. Ed. 2015, 54, 2955–2959.
69. Pan, C., Takata, T., Kumamoto, K., Khine Ma, S.S., Ueda, K., Minegishi, T., Nakabayashi, M., Matsumoto, T., Shibata, N., Ikuhara, Y., Domen, K., Band engineering of perovskite-type transition metal oxynitrides for photocatalytic overall water splitting. *J. Mater. Chem. A*, 4, 4544–4552, 2016.
70. Xu, J., Pan, C., Takata, T., Domen, K., Photocatalytic overall water splitting on the perovskite-type transition metal oxynitride CaTaO2N under visible light irradiation. *Chem. Commun.*, 51, 7191–7194, 2015.
71. Wang, Y., Kang, Y., Zhu, H., Liu, G., Irvine, J.T.S., Xu, X., Perovskite Oxynitride Solid Solutions of LaTaON2-CaTaO2N with Greatly Enhanced Photogenerated Charge Separation for Solar-Driven Overall Water Splitting. *Adv. Sci.*, 8, 2003343, 2021.
72. Miyoshi, A., Maeda, K., Recent Progress in Mixed-Anion Materials for Solar Fuel Production. *Solar RRL*, 5, 6, 2000521, 2021.

73. Wang, Q., Nakabayashi, M., Hisatomi, T., Sun, S., Akiyama, S., Wang, Z., Pan, Z., Xiao, X., Watanabe, T., Yamada, T., Shibata, N., Takata, T., Domen, K., Oxysulfide photocatalyst for visible-light-driven overall water splitting. *Nat. Mater.*, 18, 827–832, 2019.
74. Fujito, H., Kunioku, H., Kato, D., Suzuki, H., Higashi, M., Kageyama, H., Abe, R., Layered Perovskite Oxychloride Bi4NbO8Cl: A Stable Visible Light Responsive Photocatalyst for Water Splitting. *J. Am. Chem. Soc.*, 138, 2082–2085, 2016.
75. Wang, Y., Suzuki, H., Xie, J., Tomita, O., Martin, D.J., Higashi, M., Kong, D., Abe, R., Tang, J., Mimicking Natural Photosynthesis: Solar to Renewable H2 Fuel Synthesis by Z-Scheme Water Splitting Systems. *Chem. Rev.*, 118, 5201–5241, 2018.
76. Marschall, R., Photocatalysis: Semiconductor Composites: Strategies for Enhancing Charge Carrier Separation to Improve Photocatalytic Activity. *Adv. Funct. Mater.*, 24, 2421–2440, 2014.
77. Jiang, W., Zong, X., An, L., Hua, S., Miao, X., Luan, S., Wen, Y., Tao, F.F., Sun, Z., Photocatalyst for High-Performance H2 Production: Ga-Doped Polymeric Carbon Nitride. *ACS Catal.*, 8, 2209–2217, 2018.
78. Jiang, W., Qu, D., An, L., Gao, X., Wen, Y., Wang, X., Sun, Z., Deliberate construction of direct Z-scheme photocatalysts through photodeposition. *J. Mater. Chem. A*, 7, 18348–18356, 2019.
79. Jia, Q., Iwase, A., Kudo, A., BiVO4-Ru/SrTiO3 Rh Composite Z-Scheme Pho-tocatalyst for Solar Water Splitting. *Chem. Sci.*, 5, 1513–1519, 2014.
80. Kageshima, Y., Kawanishi, T., Saeki, D., Teshima, K., Domen, K., Nishikiori, H., Boosted Hydrogen-Evolution Kinetics Over Particulate Lanthanum and Rhodium-Doped Strontium Titanate Photocatalysts Modified with Phosphonate Groups. *Angew. Chem. Int. Ed. Engl.*, 60, 7, 3654–3660, 2021 Feb 15.
81. Chen, P., Li, Y., Xiao, C., Chen, L., Guo, J.-K., Shen, S., Au, C.-T., Yin, S.-F., Preparation of Helical BiVO4/Ag/C3N4 for Selective Oxidation of C–H Bond under Visible Light Irradiation. *ACS Sustain. Chem. Eng.*, 7, 17500–17506, 2019.
82. Wang, Q., Hisatomi, T., Suzuki, Y., Pan, Z., Seo, J., Katayama, M., Minegishi, T., Nishiyama, H., Takata, T., Seki, K., Kudo, A., Yamada, T., Domen, K., Particulate Photocatalyst Sheets Based on Carbon Conductor Layer for Efficient Z-Scheme Pure-Water Splitting at Ambient Pressure. *J. Am. Chem. Soc.*, 139, 1675–1683, 2017.
83. Iwase, A., Yoshino, S., Takayama, T., Ng, Y.H., Amal, R., Kudo, A., Water Splitting and CO2 Reduction under Visible Light Irradiation Using Z-Scheme Systems Consisting of Metal Sulfides, CoOx-Loaded BiVO4, and a Reduced Graphene Oxide Electron Mediator. *J. Am. Chem. Soc.*, 138, 10260–10264, 2016.

84. Iwase, A., Ng, Y.H., Ishiguro, Y., Kudo, A., Amal, R., Reduced Graphene Oxide as a Solid-State Electron Mediator in Z-Scheme Photocatalytic Water Splitting under Visible Light. *J. Am. Chem. Soc.*, 133, 11054–11057, 2011.
85. Wang, Q., Hisatomi, T., Jia, Q., *et al.*, Scalable water splitting on particulate photocatalyst sheets with a solar-to-hydrogen energy conversion efficiency exceeding 1%. *Nature Mater.*, 15, 611–615, 2016.
86. Okunaka, S., Kameshige, H., Ikeda, T., Tokudome, H., Hisatomi, T., Yamada, T., Domen, K., Scheme Water Splitting under Near-Ambient Pressure using a Zirconium Oxide Coating on Printable Photocatalyst Sheets. *ChemSusChem*, 13, 4906–49, 2020.
87. Abe, R., Takata, T., Sugihara, H., Domen, K., Photocatalytic overall water splitting under visible light by TaON and WO3 with an IO3–/I– shuttle redox mediator. *Chem. Commun.*, 41, 3829–3831, 2005.
88. Higashi, M., Abe, R., Teramura, K., Takata, T., Ohtani, B., Domen, K., Two step water splitting into H2 and O2 under visible light by ATaO2N (A=Ca C.OMMAS.R.X.X.X, Ba) and WO3 with IO3–/I– shuttle redox mediator. *Chem. Phys. Lett.*, 452, 120–123, 2008.
89. Maeda, K., Lu, D., Domen, K., Solar-Driven Z-scheme Water Splitting Using Modified BaZrO3–BaTaO2N Solid Solutions as Photocatalysts. *ACS Catal.*, 3, 1026–1033, 2013.
90. Chen, S., Qi, Y., Hisatomi, T., Ding, Q., Asai, T., Li, Z., Ma, S.S.K., Zhang, F., Domen, K., Li, C., Efficient Visible-Light-Driven Z-Scheme Overall Water Splitting Using a MgTa2O6–xNy/TaON Heterostructure Photocatalyst for H2 Evolution. *Angew. Chem.*, 127, 8618–8621, 2015. Chem. Int. Ed. 2015, 54, 8498–8501.
91. Cui, J., Qi, Y., Dong, B., Mu, L., Ding, Q., Liu, G., Jia, M., Zhang, F., Li, C., One-pot synthesis of BaMg1/3Ta2/3O3-xNy/Ta3N5 heterostructures as H2-evolving photocatalysts for construction of visible-light-driven Z-scheme overall water splitting. *Appl. Catal., B*, 241, 1–7, 2019.
92. Qi, Y., Zhao, Y., Gao, Y., Li, D., Li, Z., Zhang, F., Li, C., Redox-Based Visible-Light-Driven Z-Scheme Overall Water Splitting with Apparent Quantum Efficiency Exceeding 10%. *Joule*, 2, 2393&ndash, 20182402.
93. Ma, G., Chen, S., Kuang, Y., Akiyama, S., Hisatomi, T., Nakabayashi, M., Shibata, N., Katayama, M., Minegishi, T., Domen, K., A perspective on two pathways of photocatalytic water splitting and their practical application systems. *J. Phys. Chem. Lett.*, 7, 3892–3896, 2016.
94. Nakada, A., Higashi, M., Kimura, T., Suzuki, H., Kato, D., Okajima, H., Yamamoto, T., Saeki, A., Kageyama, H., Abe, R., Band Engineering of Double-Layered Sillen-Aurivillius Perovskite Oxychlorides for Visible-Light-Driven Water Splitting. *Chem. Mater.*, 31, 3419–3429, 2019.
95. Yoshino, S., Iwase, A., Ng, Y.H., Amal, R., Kudo, A., CO2 Reduction Using Water as an Electron Donor over Heterogeneous Photocatalysts Aiming at Artificial Photosynthesis. *ACS Appl. Mater. Interfaces*, 3, 5684–5692, 2020.

96. Sun, S., Hisatomi, T., Wang, Q., Chen, S., Ma, G., Liu, J., Nandy, S., Minegishi, T., Katayama, M., Domen, K., Efficient redox-mediator-free Z-scheme water splitting employing oxysulfide photocatalysts under visible light. *ACS Catal.*, 8, 1690–1696, 2018.
97. Srinivasan, N., Sakai, E., Miyauchi, M., Balanced Excitation between Two Semiconductors in Bulk Heterojunction Z-Scheme System for Overall Water Splitting. *ACS Catal.*, 6, 2197–2200, 2016.
98. Takashima, T., Moriyama, N., Fujishiro, Y., Osaki, J., Takeuchi, S., Ohtani, B., Irie, H., Visible-light-induced water splitting on a hierarchically constructed Z-scheme photocatalyst composed of zinc rhodium oxide and bismuth vanadate. *J. Mater. Chem. A*, 7, 10372–10378, 2019.
99. Li, H., Yu, H., Quan, X., Chen, S., Zhang, Y., Uncovering the Key Role of the Fermi Level of the Electron Mediator in a Z-Scheme Photocatalyst by Detecting the Charge Transfer Process of WO3-metal-gC3N4 (Metal = Cu, Ag, Au). *ACS Appl. Mater. Interfaces*, 8, 2111–2119, 2016.
100. Yan, J., Wu, H., Chen, H., Zhang, Y., Zhang, F., Liu, S.F., g-C3N4 Loading Black Phosphorus Quantum Dot for Efficient and Stable Photocatalytic H2 Generation under Visible Light. *Appl. Catal., B*, 191, 130–137, 2016.
101. Pan, Z., Zhang, G., Wang, X., Polymeric Carbon Nitride/Reduced Graphene Oxide/Fe2 O3: All-Solid-State Z-Scheme System for Photocatalytic Overall Water Splitting. *Angew. Chem.*, 131, 7176–7180, 2019. *Chem. Int. Ed.*, 2019, 58, 7102–7106.
102. Liu, X., Zhang, Q., Ma, D., Advances in 2D/2D Z-Scheme Heterojunctions for Photocatalytic Applications. *Sol. RRL*, 5, 1–32, 2020.
103. She, X., Wu, J., Xu, H., Zhong, J., Wang, Y., Song, Y., Nie, K., Liu, Y., Yang, Y., Rodrigues, M.T.F., Vajtai, R., Lou, J., Du, D., Li, H., Ajayan, P.M., Achieving highly efficient all-polymer solar cells by green-solvent-processing under ambient atmosphere. *Adv. Energy Mater.*, 7, 1700025, 2017.
104. Xie, H., Zhao, Y., Li, H., Xu, Y., Chen, X., 2D BiVO4/g-C3N4 Z-scheme photocatalyst for enhanced overall water splitting. *J. Mater. Sci.*, 54, 10836–10845, 2019.
105. Jahan, M., Liu, Z., Loh, K.P., A graphene oxide and copper-centered metal organic framework composite as a Tri-functional catalyst for HER, OER, and ORR. *Adv. Funct. Mater.*, 23, 43, 5363–5372, 2013.
106. Joo, J., Jin, H., Oh, A., Kim, B., Lee, J., Baik, H., Joo, S.H., Lee, K., An IrRu alloy nanocactus on Cu2-xS@IrSy as a highly efficient bifunctional electrocatalyst toward overall water splitting in acidic electrolytes. *J. Mater. Chem. A*, 6, 16130, 2018.
107. Fan, Z., Jiang, J., Ai, L., Shao, Z., Liu, S., Rational design of ruthenium and cobalt- based composites with rich metal-insulator interfaces for efficient and stable overall water splitting in acidic electrolyte. *ACS Appl. Mater. Inter.*, 11, 51, 47894–47903, 2019.
108. Deng, Y., Yang, L., Wang, Y., Zeng, L., Yu, J., Chen, B., Zhang, X., Zhou, W., Ruthenium nanoclusters anchored on cobalt phosphide hollow microspheres

by green phosphating process for full water splitting in acidic electrolyte. *Chin. Chem. Lett.*, 32, 1, 511–515, 2021.
109. Luo, F., Hu, H., Zhao, X., Yang, Z., Zhang, Q., Xu, J., Kaneko, T., Yoshida, Y., Zhu, C., Cai, W., Robust and stable acidic overall water splitting on Ir single atoms. *Nano Lett.*, 20, 3, 2120–2128, 2020.
110. Feng, J., Lv, F., Zhang, W., Li, P., Wang, K., Yang, C., Wang, B., Yang, Y., Zhou, J., Lin, F., Wang, G.-C., Guo, S., Iridium-based multimetallic porous hollow nanocrystals for efficient overall water splitting catalysis. *Adv. Mater.*, 29, 47, 1703798, 2017.
111. Pi, Y., Shao, Q., Wang, P., Guo, J., Huang, X., General formation of monodisperse IrM (M = Ni Co, Fe) bimetallic nanoclusters as bifunctional electrocatalysts for acidic overall water splitting. *Adv. Funct. Mater.*, 27, 27, 1700886, 2017.
112. Fu, L., Yang, F., Cheng, G., Luo, W., Ultrathin Ir nanowires as high-performance electrocatalysts for efficient water splitting in acidic media. *Nanoscale*, 10, 1892–1897, 2018.
113. Fu, L., Zeng, X., Huang, C.Z., Cai, P., Cheng, G., Luo, W., Ultrasmall Ir nanoparticles for efficient acidic electrochemical water splitting. *Inorg. Chem. Front.*, 5, 1121, 2018.
114. Yue, X., Huang, S., Cai, J., Jin, Y., Shen, P.K., Heteroatoms dual doped porous graphene nanosheets as efficient bifunctional metal-free electrocatalysts for overall water-splitting. *J. Mater. Chem. A*, 5, 17, 7784–7790, 2017.
115. Zhang, Z., Yi, Z., Wang, J., Tian, X., Xu, P., Shi, G., Wang, S., Nitrogen-enriched polydopamine analogue-derived defect-rich porous carbon as a bifunctional metal-free electrocatalyst for highly efficient overall water splitting. *J. Mater. Chem. A*, 5, 32, 17064–17072, 2017.
116. Wang, Y., Wu, W., Rao, Y., Li, Z., Tsubaki, N., Wu, M., Cation modulating electrocatalyst derived from bimetallic metal-organic frameworks for overall water splitting. *J. Mater. Chem. A*, 5, 13, 6170–6177, 2017.
117. Qiao, L., Zhu, A., Zeng, W., Dong, R., Tan, P., Ding, Z., Gao, P., Wang, S., Pan, J., Achieving electronic structure reconfiguration in metallic carbides for robust electrochemical water splitting. *J. Mater. Chem. A*, 8, 5, 2453–2462, 2020.
118. Ouyang, T., Ye, Y.-Q., Wu, C.-Y., Xiao, K., Liu, Z.-Q., Heterostructures composed of N-Doped carbon nanotubes encapsulating cobalt and b-Mo2C nanoparticles as bifunctional electrodes for water splitting. *Angew. Chem. Int. Ed.*, 58, 15, 4923–4928, 2019.
119. Yu, Z.-Y., Duan, Y., Gao, M.-R., Lang, C.-C., Zheng, Y.-R., Yu, S.-H., A one-dimensional porous carbon-supported Ni/Mo2C dual catalyst for efficient water splitting. *Chem. Sci.*, 8, 2, 968–973, 2017.
120. Wang, H., Cao, Y., Sun, C., Zou, G., Huang, J., Kuai, X., Zhao, J., Gao, L., Strongly coupled molybdenum carbide on carbon sheets as a bifunctional electrocatalyst for overall water splitting. *ChemSusChem*, 10, 18, 3540–3546, 2017.

121. Y.u. Gu, S., Ren, J., Jia, Y.A., Chen, C., Komarneni, S., Yang, D., Yao, X., Electronic structure tuning in Ni3FeN/r-GO aerogel toward bifunctional electrocatalyst for overall water splitting. *ACS Nano*, 12, 1, 245–253, 2018.
122. Dutta, S., Indra, A., Feng, Y., Han, H., Song, T., Promoting electrocatalytic overall water splitting with nanohybrid of transition metal nitride-oxynitride. *Appl. Catal. B-Environ.*, 241, 521–527, 2019.
123. Li, H., Wen, P., Q.i. Li, C., Xing, J., Lu, C., Adhikari, S., Jiang, L., Carroll, D.L., Geyer, S.M., Earth-abundant iron diboride (FeB2) nanoparticles as highly active bifunctional electrocatalysts for overall water splitting. *Adv. Energy Mater.*, 7, 17, 1700513, 2017.
124. Yuan, W., Zhao, X., Hao, W., Li, J., Wang, L., Ma, X., Guo, Y., Performance of surface-oxidized Ni3B, Ni2B, and NiB2 electrocatalysts for overall water splitting. *Chem. Electro. Chem*, 6, 3, 764–770, 2019.
125. Chunduri, A., Gupta, S., Bapat, O., Bhide, A., Fernandes, R., Patel, M.K., Bambole, V., Miotello, Patel, N., A unique amorphous cobalt-phosphide-boride bifunctional electrocatalyst for enhanced alkaline water-splitting. *Appl. Catal. B-Environ.*, 259, 118051, 2019.
126. Chen, X., Yu, Z., Wei, L., Zhou, Z., Zhai, S., Chen, J., Wang, Y., Huang, Q., Karahan, H.E., Liao, X., Chen, Y., Ultrathin nickel boride nanosheets anchored on functionalized carbon nanotubes as bifunctional electrocatalysts for overall water splitting. *J. Mater. Chem. A*, 7, 2, 764–774, 2019.
127. Tong, J., Li, W., Bo, L., Li, Y., Li, T., Zhang, Q., Simple preparation of Ni2P/Ni(PO3)2 inlayed in nitrogen-sulfur self-doped ultrathin holey carbon nanosheets with excellent electrocatalytic activities for water splitting. *Electrochim. Acta*, 320, 134579, 2019.
128. Ai, L.H., Niu, Z.G., Jiang, J., Mechanistic insight into oxygen evolution electrocatalysis of surface phosphate modified cobalt phosphide nanorod bundles and their superior performance for overall water splitting. *Electrochim. Acta*, 242, 355–363, 2017.
129. Xiao, X., Huang, D., Fu, Y., Wen, M., Jiang, X., Lv, X., Li, M., Gao, L., Liu, S., Wang, M., Zhao, C., Shen, Y., Engineering NiS/Ni2P heterostructures for efficient electrocatalytic water splitting. *ACS Appl. Mater. Inter.*, 10, 5, 4689–4696, 2018.
130. Zheng, J.L., Zhou, W., Liu, T., Liu, S.J., Wang, C.B., Guo, L., Homologous NiO/Ni2P nanoarrays grown on nickel foams: a well matched electrode pair with high stability in overall water splitting. *Nanoscale*, 9, 4409–4418, 2017.
131. Lv, F., Feng, J., Wang, K., Dou, Z., Zhang, W., Zhou, J., Yang, C., Luo, M., Yang, Y., Li, Y., Gao, P., Guo, S., Iridium-tungsten alloy nanodendrites as pH-universal water splitting electrocatalysts. *ACS Cent. Sci.*, 4, 9, 1244–1252, 2018.
132. Wu, J., Wang, D., Wan, S., Liu, H., Wang, C., Wang, X., An efficient cobalt phosphide electrocatalyst derived from cobalt phosphonate complex for all-pH hydrogen evolution reaction and overall water splitting in alkaline solution. *Small*, 16, 15, 1900550, 2020.

133. Li, X., Fang, Y., Li, F., Tian, M., Long, X., Jin, J., Ma, J., Ultrafine Co2P nanoparticles encapsulated in nitrogen and phosphorus dual-doped porous carbon nanosheet/carbon nanotube hybrids: high-performance bifunctional electrocatalysts for overall water splitting. *J. Mater. Chem. A*, 4, 40, 15501–15510, 2016.
134. Li, Y., Cui, M., Li, T., Shen, Y., Si, Z., Wang, H.-G., Embedding Co2P nanoparticles into co-doped carbon hollow polyhedron as a bifunctional electrocatalyst for efficient overall water splitting. *Int. J. Hydrogen Energ.*, 45, 33, 16540–16549, 2020.
135. Li, M., Zhu, Y., Wang, H., Wang, C., Pinna, N., Lu, X., Ni strongly coupled with Mo2C encapsulated in nitrogen-doped carbon nanofibers as robust bifunctional catalyst for overall water splitting. *Adv. Energy Mater.*, 9, 10, 1803185, 2019.
136. Kang, Q., Li, M., Shi, J., Lu, Q., Gao, F., A universal strategy for carbon-supported transition metal phosphides as high-performance bifunctional electrocatalysts towards efficient overall water splitting. *ACS Appl. Mater. Inter.*, 12, 17, 19447–19456, 2020.
137. Xue, Z.-H., Su, H., Yu, Q.-Y., Zhang, B., Wang, H.-H., Li, X.-H., Chen, J.-S., Janus Co/ CoP nanoparticles as efficient mott-schottky electrocatalysts for overall water splitting in wide pH range. *Adv. Energy Mater.*, 7, 12, 1602355, 2017.
138. Song, J., Zhu, C., Xu, B.Z., Fu, S., Engelhard, M.H., Ye, R., Du, D., Beckman, S.P., Lin, Y., Bimetallic cobalt-based phosphide zeolitic imidazolate framework: CoPx phase-dependent electrical conductivity and hydrogen atom adsorption energy for efficient overall water splitting. *Adv. Energy Mater.*, 7, 2, 1601555, 2017.
139. Huang, H.W., Yu, C., Zhao, C.T., Han, X.T., Yang, J., Liu, Z.B., Li, S.F., Zhang, M.D., Qiu, J.S., Iron-tuned super nickel phosphide microstructures with high activity for electrochemical overall water splitting. *Nano Energy*, 34, 472–480, 2017.
140. Xuan, C.J., Wang, J., Xia, W.W., Peng, Z.K., Wu, Z.X., Lei, W., Xia, K.D., Xin, H.L.L., Wang, D.L., Structured Ni-Fe-P nanocubes derived from a prussian blue analogue as an electrocatalyst for efficient overall water splitting. *ACS Appl. Mater. Inter.*, 9, 26134–26142, 2017.
141. Yan, L., Zhang, B., Zhu, J., Li, Y., Tsiakaras, P., Kang Shen, P., Electronic modulation of cobalt phosphide nanosheet arrays via copper doping for highly efficient neutral-pH overall water splitting. *Appl. Catal. B-Environ.*, 265, 118555, 2020.
142. Li, W., Chen, Y., B.o. Yu, Y., Wang, X., Yang, D., 3D hollow Co-Fe-P nanoframes immobilized on N, P-doped CNT as an efficient electrocatalyst for overall water splitting. *Nanoscale*, 11, 36, 17031–17040, 2019.
143. Zhang, J., Wang, G., Liao, Z.Q., Zhang, P.P., Wang, F.X., Zhuang, X.D., Zschech, E., Feng, X.L., Iridium nanoparticles anchored on 3D graphite

foam as a bifunctional electrocatalyst for excellent overall water splitting in acidic solution. *Nano Energy*, 40, 27–33, 2017.

144. Liang, H., Gandi, A.N., Anjum, D.H., Wang, X., Schwingenschlögl, U., Alshareef, H.N., Plasma-assisted synthesis of NiCoP for efficient overall water splitting. *Nano Lett.*, 16, 12, 7718–7725, 2016.
145. Y.a. Yan, B.Y., Ge, X., Liu, Z., Fisher, A., Wang, X., A flexible electrode based on iron phosphide nanotubes for overall water splitting. *Chem. Eur. J.*, 21, 50, 18062–18067, 2015.
146. Sun, H., Xu, X., Yan, Z., Chen, X., Cheng, F., Weiss, P.S., Chen, J., Porous multishelled Ni2P hollow microspheres as an active electrocatalyst for hydrogen and oxygen evolution. *Chem. Mater.*, 29, 19, 8539–8547, 2017.
147. Jiang, N., You, B., Sheng, M., Sun, Y., Electrodeposited cobalt phosphorous derived films as competent bifunctional catalysts for overall water splitting. *Angew. Chem. Int. Ed.*, 54, 21, 6251–6254, 2015.
148. Han, A., Zhang, H., Yuan, R., Ji, H., Du, P., Crystalline copper phosphide nanosheets as an efficient Janus catalyst for overall water splitting. *ACS Appl. Mater. Inter.*, 9, 3, 2240–2248, 2017.
149. Wang, X., Li, W., Xiong, D., Liu, L., Fast fabrication of self-supported porous nickel phosphide foam for efficient, durable oxygen evolution and overall water splitting. *J. Mater. Chem. A*, 4, 15, 5639–5646, 2016.
150. Wang, X., Li, W., Xiong, D., Petrovykh, D.Y., Liu, L., Bifunctional nickel phosphide nanocatalysts supported on carbon fiber paper for highly efficient and stable overall water splitting. *Adv. Funct. Mater.*, 26, 23, 4067–4077, 2016.
151. Li, J., Li, J., Zhou, X., Xia, Z., Gao, W., Ma, Y., Qu, Y., Highly efficient and robust nickel phosphides as bifunctional electrocatalysts for overall water splitting. *ACS Appl. Mater. Inter.*, 8, 17, 10826–10834, 2016.
152. Li, H., Chen, S., Lin, H., Xu, X., Yang, H., Song, L., Wang, X., Nickel diselenide ultrathin nanowires decorated with amorphous nickel oxide nanoparticles for enhanced water splitting electrocatalysis. *Small*, 13, 37, 1701487, 2017.
153. Li, K., Zhang, J., Wu, R., Yu, Y., Zhang, B., Anchoring CoO domains on CoSe2 nanobelts as bifunctional electrocatalysts for overall water splitting in neutral media. *Adv. Sci.*, 3, 6, 1500426, 2016.
154. Meng, T., Qin, J., Wang, S., Zhao, D., Mao, B., Cao, M., *In situ* coupling of Co0.85Se and N-doped carbon via one-step selenization of metal-organic frameworks as a trifunctional catalyst for overall water splitting and Zn-air batteries. *J. Mater. Chem. A*, 5, 15, 7001–7014, 2017.
155. Sun, C., Dong, Q., Yang, J., Dai, Z., Lin, J., Chen, P., Huang, W., Dong, X., Metal- organic framework derived CoSe2 nanoparticles anchored on carbon fibers as bifunctional electrocatalysts for efficient overall water splitting. *Nano Res.*, 9, 8, 2234–2243, 2016.
156. Li, W., Gao, X., Xiong, D., Wei, F., Song, W.-G., Xu, J., Liu, L., Hydrothermal Synthesis of monolithic Co3Se4 nanowire electrodes for oxygen evolution

and overall water splitting with high efficiency and extraordinary catalytic stability. *Adv. Energy Mater.*, 7, 17, 1602579, 2017.

157. Yoon, T. and Kim, K.S., One-step synthesis of CoS-doped beta-Co (OH)2@ amorphous MoS2+x hybrid catalyst grown on nickel foam for high- performance electrochemical overall water splitting. *Adv. Funct. Mater.*, 26, 7386–7393, 2016.
158. Yang, Y., Yao, H., Yu, Z., Islam, S.M., He, H., Yuan, M., Yue, Y., Xu, K., Hao, W., Sun, G., Li, H., Ma, S., Zapol, P., Kanatzidis, M.G., Hierarchical nano-assembly of MoS2/ Co9S8/Ni3S2/Ni as a highly efficient electrocatalyst for overall water splitting in a wide pH range. *J. Am. Chem. Soc.*, 141, 26, 10417–10430, 2019.
159. Wu, Y., Liu, Y., Li, G.-D., Zou, X., Lian, X., Wang, D., Sun, L., Asefa, T., Zou, X., Efficient electrocatalysis of overall water splitting by ultrasmall NixCo3-xS4 coupled Ni3S2 nanosheet arrays. *Nano Energy*, 35, 161–170, 2017.
160. Zhang, J., Wang, T., Pohl, D., Rellinghaus, B., Dong, R., Liu, S., Zhuang, X., Feng, X., Interface engineering of MoS2/Ni3S2 heterostructures for highly enhanced electrochemical overall water splitting activity. *Angew. Chem. Int. Ed.*, 55, 23, 6702–6707, 2016.
161. Li, Y., Yin, J., An, L., Lu, M., Sun, K., Zhao, Y.Q., Gao, D., Cheng, F., Xi, P., FeS2/CoS2 interface nanosheets as efficient bifunctional electrocatalyst for overall water splitting. *Small*, 14, 1801070, 2018. https://doi.org/10.1002/smll.201801070.
162. Liu, Y., Li, Q., Si, R., Li, G.-D., Li, W., Liu, D.-P., Wang, D., Sun, L., Zhang, Y., Zou, X., Coupling sub-nanometric copper clusters with quasi-Amorphous cobalt sulfide yields efficient and robust electrocatalysts for water splitting reaction. *Adv. Mater.*, 29, 13, 1606200, 2017.
163. Bai, J., Meng, T., Guo, D., Wang, S., Mao, B., Cao, M., Co9S8@MoS2 core-shell heterostructures as trifunctional electrocatalysts for overall water splitting and Zn-air batteries. *ACS Appl. Mater. Inter.*, 10, 2, 1678–1689, 2018.
164. Xin, Y., Kan, X., Gan, L.-Y., Zhang, Z., Heterogeneous bimetallic phosphide/sulfide nanocomposite for efficient solar energy driven overall water splitting. *ACS Nano*, 11, 10, 10303–10312, 2017.
165. Gahleitner, G., Hydrogen from renewable electricity: an international review of power-to-gas pilot plants for stationary applications. *Int. J. Hydrogen Energ.*, 38, 5, 2039–2061, 2013.
166. Xue, Y., Zuo, Z., Li, Y., Liu, H., Li, Y., Graphdiyne-supported NiCo2S4 nanowires: A highly active and stable 3D bifunctional electrode material. *Small*, 13, 31, 1700936, 2017.
167. Holladay, J.D., Hu, J., King, D.L., Wang, Y., An overview of hydrogen production technologies. *Catal. Today*, 139, 4, 244–260, 2009.
168. Ren, B., Li, D., Jin, Q., Cui, H., Wang, C., *In-situ* tailoring cobalt nickel molybdenum oxide components for overall water splitting at high current densities. *Chem. Electro. Chem.*, 6, 2, 413–420, 2019.

169. Yamazaki, H., Shouji, A., Kajita, M., Yagi, M., Electrocatalytic and photocatalytic water oxidation to dioxygen based on metal complexes. *Coordin. Chem. Rev.*, 254, 21-22, 2483–2491, 2010.
170. Huang, S., Meng, Y., He, S., Goswami, A., Wu, Q., Li, J., Tong, S., Asefa, T., Wu, M., N-, O-, and S-tridoped carbon-encapsulated Co9S8 nanomaterials: efficient bifunctional electrocatalysts for overall water splitting. *Adv. Funct. Mater.*, 27, 17, 1606585, 2017.
171. Ming, F., Liang, H., Shi, H., Xu, X., Mei, G., Wang, Z., MOF-derived Co-doped nickel selenide/C electrocatalysts supported on Ni foam for overall water splitting. *J. Mater. Chem. A*, 4, 39, 15148–15155, 2016.
172. Qin, Q., Li, P., Chen, L., Liu, X., Coupling bimetallic oxides/alloys and N-doped carbon nanotubes as tri-functional catalysts for overall water splitting and zinc-air batteries. *ACS Appl. Mater. Inter.*, 10, 46, 39828–39838, 2018.
173. Zhang, Z., Liu, S., Xiao, F., Wang, S., Facile synthesis of heterostructured nickel/nickel oxide wrapped carbon fiber: flexible bifunctional gas-evolving electrode for highly efficient overall water splitting. *ACS Sustain. Chem. Eng.*, 5, 1, 529–536, 2017.
174. Zhu, T., Zhu, L., Wang, J., Ho, G.W., *In situ* chemical etching of tunable 3D Ni3S2 superstructures for bifunctional electrocatalysts for overall water splitting. *J. Mater. Chem. A*, 4, 36, 13916–13922, 2016.
175. Zhu, W., Yue, X., Zhang, W., Yu, S., Zhang, Y., Wang, J., Wang, J., Nickel sulfide microsphere film on Ni foam as an efficient bifunctional electrocatalyst for overall water splitting. *Chem. Commun.*, 52, 7, 1486–1489, 2016.
176. Zhang, P., Liu, Y., Liang, T., Ang, E.H., Zhang, X., Ma, F., Dai, Z., Nitrogen-doped carbon wrapped Co-Mo2C dual Mott-Schottky nanosheets with large porosity for efficient water electrolysis. *Appl. Catal., B*, 284, 119738, 2021.
177. Marschall, R., *50 Years of Materials Research for Photocatalytic Water Splitting*, European chemical societies publishing, Belgium, 2021.
178. Sun, L., Luo, Q., Dai, Z., Ma, F., Material libraries for electrocatalytic overall water splitting. *Coord. Chem. Rev.*, 444, 214049, 2021.
179. Ilanchezhiyan, P., Mohan Kumar, G., Tamilselvan, S., Kang, T.W., Kim, D.Y., Highly efficient overall water splitting performance of gadolinium-indium-zinc ternary oxide nanostructured electrocatalyst. *Int. J. Energ. Res.*, 44, 8, 6819–6827, 2020.
180. Wu, S., Liu, J., Cui, B., Li, Y., Liu, Y., Hu, B., He, L., Wang, M., Zhang, Z., Tian, K., Song, Y., Fluorine-doped nickel cobalt oxide spinel as efficiently bifunctional catalyst for overall water splitting. *Electrochim. Acta*, 299, 231–244, 2019.
181. Wang, Z., Zeng, S., Liu, W., Wang, X., Li, Q., Zhao, Z., Geng, F., Coupling molecularly ultrathin sheets of NiFe-layered double hydroxide on NiCo2O4 nanowire arrays for highly efficient overall water-splitting activity. *ACS Appl. Mater. Inter.*, 9, 2, 1488–1495, 2017.

182. Dvořák, F., Szabová, L., Johánek, V., Farnesi Camellone, M., Stetsovych, V., Vorokhta, M., Tovt, A., Skála, T., Matolínová, I., Tateyama, Y., Mysliveček, J., Fabris, S., Matolín, V., Bulk hydroxylation and effective water splitting by highly reduced cerium oxide: The role of O vacancy coordination. *ACS Catal.*, 8, 5, 4354–4363, 2018.
183. Yu, X., Yu, Z.-Y., Zhang, X.-L., Li, P., Sun, B., Gao, X., Yan, K., Liu, H., Duan, Y., Gao, M.-R., Wang, G., Yu, S.-H., Highly disordered cobalt oxide nanostructure induced by sulfur incorporation for efficient overall water splitting. *Nano Energy*, 71, 104652, 2020.
184. Wang, J.W., Liu, W.J., Zhong, D.C., Lu, T.B., Nickel complexes as molecular catalysts for water splitting and CO2 reduction. *Coordin. Chem. Rev.*, 378, 237–261, 2019.
185. Hua, B., Li, M., Zhang, Y.-Q., Sun, Y.-F., Luo, J.-L., All-in-one perovskite catalyst: smart controls of architecture and composition toward enhanced oxygen/hydrogen evolution reactions. *Adv. Energy Mater.*, 7, 20, 1700666, 2017.
186. Xu, X., Chen, Y., Zhou, W., Zhu, Z., Su, C., Liu, M., Shao, Z., A perovskite electrocatalyst for efficient hydrogen evolution reaction. *Adv. Mater.*, 28, 30, 6442–6448, 2016.
187. Huang, L., Chen, D., Luo, G., Lu, Y.-R., Chen, C., Zou, Y., Dong, C.-L., Li, Y., Wang, S., Zirconium regulation induced bifunctionality in 3D cobalt-iron oxide nanosheets for overall water splitting. *Adv. Mater.*, 31, 28, 1901439, 2019.
188. Gu, L.-F., Chen, J.-J., Zhou, T., Lu, X.-F., Li, G.-R., Engineering cobalt oxide by interfaces and pore architectures for enhanced electrocatalytic performance for overall water splitting. *Nanoscale*, 12, 20, 11201–11208, 2020.
189. Xiao, Y., Zhang, P., Zhang, X., Dai, X., Ma, Y., Wang, Y., Jiang, Y., Liu, M., Wang, Y., Bimetallic thin film NiCo-NiCoO2@NC as a superior bifunctional electrocatalyst for overall water splitting in alkaline media. *J. Mater. Chem. A*, 5, 30, 15901–15912, 2017.
190. Liu, Y., Geng, H., Qin, X., Yang, Y., Zeng, Z., Chen, S., Lin, Y., Xin, H., Song, C., Zhu, X., Li, D., Zhang, J., Song, L., Dai, Z., Kawazoe, Y., Oriented attachment revisited: does a chemical reaction occur? *Matter*, 1, 3, 690–704, 2019.
191. Wang, A.-L., Xu, H., Li, G.-R., NiCoFe layered triple hydroxides with porous structures as high performance electrocatalysts for overall water splitting. *ACS Energy Lett.*, 1, 2, 445–453, 2016.
192. Bao, J., Wang, Z., Xie, J., L.i. Xu, F., Guan, M., Zhao, Y., Huang, Y., Li, H., A ternary cobalt-molybdenum-vanadium layered double hydroxide nanosheet array as an efficient bifunctional electrocatalyst for overall water splitting. *Chem. Commun.*, 55, 24, 3521–3524, 2019.
193. Chen, T., Zhang, R., Chen, G., Huang, J., Chen, W., Wang, X., Chen, D., Li, C., Ostrikov, K., Plasma-doping-enhanced overall water splitting: case study of NiCo hydroxide electrocatalyst. *Catal. Today*, 337, 147–154, 2019.

194. Liu, P.F., Yang, S., Zhang, B., Yang, H.G., Defect-rich ultrathin cobalt-iron layered double hydroxide for electrochemical overall water splitting. *ACS Appl. Mater. Inter.*, 8, 50, 34474–34481, 2016.
195. Meng, X., Han, J., Lu, L., Qiu, G., Wang, Z.L., Sun, C., Fe2+-doped layered double (Ni, Fe) hydroxides as efficient electrocatalysts for water splitting and self- powered electrochemical systems. *Small*, 15, 41, 1902551, 2019.
196. Rao, Y., Wang, Y., Ning, H., Li, P., Wu, M., Hydrotalcite-like Ni(OH)2 nanosheets *in situ* grown on nickel foam for overall water splitting. *ACS Appl. Mater. Inter.*, 8, 49, 33601–33607, 2016.
197. Feng, Y., Wang, X., Huang, J., Dong, P., Ji, J., Li, J., Cao, L., Feng, L., Jin, P., Wang, C., Decorating CoNi layered double hydroxides nanosheet arrays with fullerene quantum dot anchored on Ni foam for efficient electrocatalytic water splitting and urea electrolysis. *Chem. Eng. J.*, 390, 124525, 2020.
198. Swesi, A.T., Masud, J., Nath, M., Enhancing electrocatalytic activity of bifunctional Ni3Se2 for overall water splitting through etching-induced surface nanostructuring. *J. Mater. Res.*, 31, 18, 2888–2896, 2016.
199. Lei, L., Huang, D., Zhou, C., Chen, S., Yan, X., Li, Z., Wang, W., Demystifying the active roles of NiFe-based oxides/(oxy)hydroxides for electrochemical water splitting under alkaline conditions, Coord. *Chem. Rev.*, 408, 213177, 2020.
200. Li, L., Wang, P., Shao, Q., Huang, X., Metallic nanostructures with low dimensionality for electrochemical water splitting. *Chem. Soc. Rev.*, 49, 10, 3072–3106, 2020.
201. Zhang, Y., Fu, J., Zhao, H., Jiang, R., Tian, F., Zhang, R., Tremella-like Ni3S2/MnS with ultrathin nanosheets and abundant oxygen vacancies directly used for high speed overall water splitting. *Appl. Catal.*, 257, 117899, 2019.
202. Khalid, M., Honorato, A.M.B., Tremiliosi Filho, G., Varela, H., Trifunctional catalytic activities of trimetallic FeCoNi alloy nanoparticles embedded in a carbon shell for efficient overall water splitting. *J. Mater. Chem. A*, 8, 18, 9021–9031, 2020.
203. Hsieh, C.-T., Huang, C.-L., Chen, Y.-A., Lu, S.-Y., NiFeMo alloy inverse-opals on Ni foam as outstanding bifunctional catalysts for electrolytic water splitting of ultra-low cell voltages at high current densities. *Appl. Catal. B-Environ.*, 267, 118376, 2020.
204. Ahn, S.H. and Manthiram, A., Direct growth of ternary Ni-Fe-P porous nanorods onto nickel foam as a highly active, robust bi-functional electrocatalyst for overall water splitting. *J. Mater. Chem. A*, 5, 6, 2496–2503, 2017.
205. Balogun, M.-S., Qiu, W., Huang, Y., Yang, H., Xu, R., Zhao, W., Li, G.-R., Ji, H., Tong, Y., Cost-effective alkaline water electrolysis based on nitrogen- and phosphorus- doped self-supportive electrocatalysts. *Adv. Mater.*, 29, 34, 1702095, 2017.
206. Zhou, Y.N., Zhu, Y.R., Chen, X.Y., Dong, B., Li, Q.Z., Chai, Y.M., Carbon-based transition metal sulfides/selenides nanostructures for electrocatalytic water splitting. *J. Alloy. Compd.*, 852, 156810, 2021.

207. Tian, J., Cheng, N., Liu, Q., Sun, X., He, Y., Asiri, A.M., Self-supported NiMo hollow nanorod array: an efficient 3D bifunctional catalytic electrode for overall water splitting. *J. Mater. Chem. A*, 3, 40, 20056–20059, 2015.
208. Fang, M., Gao, W., Dong, G.F., Xia, Z.M., Yip, S., Qin, Y.B., Qu, Y.Q., Ho, J.C., Hierarchical NiMo-based 3D electrocatalysts for highly-efficient hydrogen evolution in alkaline conditions. *Nano Energy*, 27, 247–254, 2016.
209. Liu, H., He, Q., Jiang, H., Lin, Y., Zhang, Y., Habib, M., Chen, S., Song, L., Electronic structure reconfiguration toward pyrite NiS2 via engineered heteroatom defect boosting overall water splitting. *ACS Nano*, 11, 11, 11574–11583, 2017.
210. Li, J.W., Xu, W.M., Luo, J.X., Zhou, D., Zhang, D.W., Wei, L.C., Xu, P.M., Yuan, D.S., Synthesis of 3D hexagram-like cobalt-manganese sulfides nanosheets grown on nickel foam: a bifunctional electrocatalyst for overall water splitting. *Nano Micro Lett.*, 10, 1–10, 2018.
211. Gao, X., Yu, Y., Liang, Q., Pang, Y., Miao, L., Liu, X., Kou, Z., He, J., Pennycook, S.J., Mu, S., Wang, J., Surface nitridation of nickel-cobalt alloy nanocactoids raises the performance of water oxidation and splitting. *Appl. Catal. B-Environ.*, 270, 118889, 2020.
212. Luo, P., Zhang, H., Liu, L., Zhang, Y., Deng, J., Xu, C., Hu, N., Wang, Y., Targeted synthesis of unique nickel sulfide (NiS, NiS2) microarchitectures and the applications for the enhanced water splitting system. *ACS Appl. Mater. Inter.*, 9, 3, 2500–2508, 2017.
213. Pei, Z., Tang, Z., Liu, Z., Huang, Y., Wang, Y., Li, H., Xue, Q., Zhu, M., Tang, D., Zhi, C., Construction of a hierarchical 3D Co/N-carbon electrocatalyst for efficient oxygen reduction and overall water splitting. *J. Mater. Chem. A*, 6, 2, 489–497, 2018.
214. Wang, S.G., Qin, J.W., Meng, T., Cao, M.H., Metal-organic framework-induced construction of actiniae-like carbon nanotube assembly as advanced multifunctional electrocatalysts for overall water splitting and Zn-air batteries. *Nano Energy*, 39, 626–638, 2017.
215. Ravi, P., Noh, J., Photocatalytic Water Splitting: How Far Away Are We from Being Able to Industrially Produce Solar Hydrogen? *Molecules*, 23, 27, 21, 7176, 2022 Oct 23.

8

Nanotechnology in Water Splitting Research

Balaji Parasuraman[1], Nazar Riswana Barveen[2], Hariprasath Rangaraju[1] and Pazhanivel Thangavelu[1]*

[1]Smart Materials Laboratory, Department of Physics, Periyar University, Salem, Tamil Nadu, India
[2]Department of Chemical Engineering, Ming Chi University of Technology, New Taipei City, Taiwan

Abstract
The quest for sustainable energy solutions has driven significant research efforts toward the development of clean and efficient technologies for hydrogen production. Among these, water splitting through nanotechnology has emerged as a promising avenue. This chapter aims to provide a concise overview of recent advancements in nanomaterials and nanotechnological approaches employed in the photocatalytic and photoelectrochemical water splitting processes. Furthermore, the integration of nanotechnology in both photocatalytic and photoelectrochemical water splitting systems and their use of advanced nanocomposite materials, co-catalysts, and efficient light absorption mechanisms and overall system were also explored.

Keywords: Nanotechnology, water splitting, nanomaterials, hydrogen evaluation, oxygen evaluation

8.1 Introduction

The future generation's quality of lifestyle will be determined by the creative ideas that exist today. As modern society develops, fossil fuels are used more and more releasing harmful greenhouse gases such as hydrofluorocarbon, methane, carbon monoxide (CO), and carbon dioxide (CO_2) into the environment [1]. These dangerous gasses seriously alter the global

**Corresponding author:* pazhanit@gmail.com

Inamuddin, Tariq Altalhi, Mohammad Luqman and Jorddy Neves Cruz (eds.) Water Splitting: Production of Hydrogen, (211–232) © 2025 Scrivener Publishing LLC

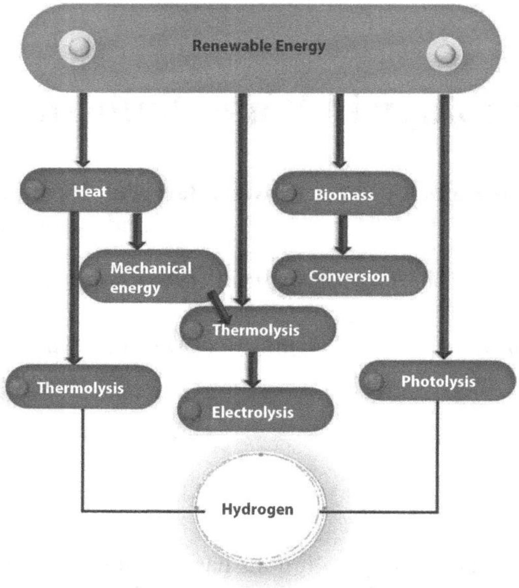

Figure 8.1 The various ways to produce hydrogen.

climate and have a variety of negative consequences on the health of living things. Thus, to preserve a gorgeous pollution-free environment for future generations, renewable and clean alternative development is required in such a severe circumstance. Molecular hydrogen (H_2) has high gravimetric energy density and is an excellent energy carrier due to its environmental friendliness (the combustion product is pure water) [2]. In recent years, hydrogen production (90%) has mainly been done through the steam reforming process, which requires fossil fuels for conversion and emits carbon dioxide as a byproduct. Considering the above information, hydrogen (H_2) appears to be a promising energy source. The scheme in Figure 8.1 shows the several methods that can be utilized for producing hydrogen from renewable energy.

Hybrid nanomaterials are widely used in energy application such as hydrogen production and oxygen production. Here, the process of water splitting, also called water electrolysis, involves using an external source of energy to separate water (H_2O) into hydrogen (H_2) and oxygen (O_2) [3]. It is commonly used to produce hydrogen gas, which is a clean and sustainable energy source. The basic reaction for water splitting is as follows:

$$2\,H_2O\,(l) \rightarrow 2\,H_2\,(g) + O_2\,(g)$$

Thus, reaction occurs in two half-reactions. Equation 8.1 shows that oxidation half-reaction is involved with the anode side, and Equation 8.2 shows that reduction half-reaction is involved with the cathode side [4]. This equation can be described as follows:

$$2\,H_2O\,(l) \rightarrow O_2\,(g) + 4\,H^+\,(aq) + 4\,e^- \qquad (8.1)$$

$$4\,H^+\,(aq) + 4\,e^- \rightarrow 2\,H_2\,(g) \qquad (8.2)$$

The process typically involves the use of an electrolyzer, which is a device that uses an electric current to initiate the HER reaction. Here, the electrolyzer consists of two electrodes (anode and cathode) immersed in an electrolyte (usually water with an electrolyte additive to enhance conductivity) [5]. When an electric current is applied, water molecules are dissociated into hydrogen ions (protons) and hydroxide ions. At the anode, oxygen gas is produced through the oxidation of water, while at the cathode, hydrogen gas is generated through the reduction of protons (Figure 8.2). The overall process requires an external source of electrical energy, which is often obtained from renewable sources, like solar or wind power, to make hydrogen production environmentally friendly.

Hydrogen produced through water splitting is considered a clean and renewable energy carrier, and it can be used in various applications, such as fuel cells for electricity generation and as a fuel for transportation. The challenge lies in making the HER process further effectual and cost effective for widespread use in a sustainable energy economy. Nanotechnology is an emerging and rapidly advancing field that focuses on manipulating and controlling matter at the nanoscale, which refers to dimensions

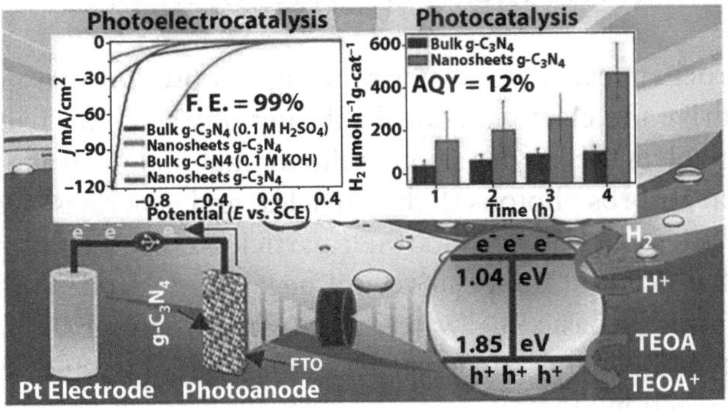

Figure 8.2 Description of water splitting.

typically less than 100 nm. At this scale, the properties of materials can behave differently compared to their bulk counterparts, offering unique opportunities for innovation across various industries. Materials exhibit distinct properties at the bulk scale and unique properties at the nanoscale, where their surface-to-volume ratio increases with size. When used as a catalyst, this surface-to-volume ratio feature increases efficiency over bulk materials by 100%. In the energy industry, we may boost the efficiency of H_2 generation and O_2 generation by employing nanomaterials as a catalyst in the water splitting process [6]. The low noble metal content catalysts, including various transition metals, such as metal oxide [7], metal sulfate [8], tungsten [9], molybdenum [10], chalcogenides [11], carbides [12], and nitrides [13], are extensively researched in an attempt for less costly electrocatalysts. However, most of them suffer from conductivity and stability problems. This has prompted scientists to focus on metal-free electrocatalysts to overcome these challenges in recent years. An overview of nanotechnology's contributions to hydrogen and oxygen economies, as well as sustainable ways to store energy, as a step toward a more sustainable use of energy, is presented in this book chapter.

There are two main types of water splitting methods available here as follows:

1. Photoelectrochemical water splitting
2. Photocatalytic water splitting

While various methods are available for hydrogen production, photoelectrochemical and photocatalytic water splitting offers several advantages making it an attractive and promising approach.

8.2 Photoelectrochemical Water Splitting

Photoelectrochemical (PEC) water splitting is a process that utilizes light energy to drive the electrochemical splitting of water into hydrogen and oxygen. It involves the use of a semiconductor material, such as a photoelectrode, which absorbs photons (light particles) and generates electron–hole pairs [14]. These charge carriers then participate in the electrochemical reactions that occur at the positive and negative electrodes to harvest H_2 gas and O_2 gas. The basic PEC water splitting requires four steps as follows: Light absorption, charge carrier generation, separation, and transport to the surface reaction [15].

Mechanism for PEC Water Splitting

The fundamental concept of PEC water splitting is to convert solar energy to H_2 by providing an outside force to photovoltaic material that is dipped in an electrolyte containing redox pairs. Among them is a semiconductor that is exposed to light and may absorb light. The electricity is then utilized to electrolyze water. Semiconductors are unique in their ability to operate as photocatalysts. As a result, it is essential for triggering the oxidation and reduction reactions that occur in the availability of light, which are shown in Figure 8.3. These photocatalyst electrodes may capture light, which supplies the energy source for the suitable reaction, while the extra voltage necessary for completing the reaction is given by an outside applied electric source. The external source overcomes the sluggish kinetics and provides enough power for the PEC device to initiate the reactions in the appropriate amount and input current. The photoelectrode absorbs photons causing electrons to be excited and generating electron–hole pair through a redox process. The generated holes oxidize the molecules, while the electrons reduce H^+ to H_2. Solar-driven PEC reactions at photoelectrodes need a potential window of 1.23 V to split water into H_2 and O_2. The water splitting reaction requires four electrons for OER and two electrons for HER to oxidize water and decrease protons. The equations for half-cell reactions are as follows:

$$\text{Cathode: } 4H^+ + 4\,e^- \rightarrow 2H_2\,(g) \tag{8.3}$$

$$\text{Anode: } 2H_2O\,(1) \rightarrow O_2\,(g) + 4H^+ + 4\,e^- \tag{8.4}$$

$$\text{Overall: } 2H_2O\,(1) \rightarrow CO_2\,(g) + H_2\,(g) \tag{8.5}$$

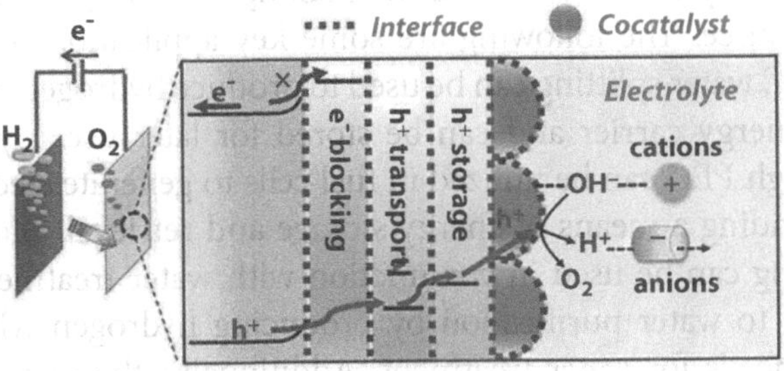

Figure 8.3 Photoelectrochemical water splitting mechanism.

The potential applicable for hydrogen evolution is 0 V_{RHE}, while oxygen evolution is 1.2 V resulting in $\Delta E^0 = 1.2$ V. Here, the pH for the electrolyte influences the reduction and oxidation potential of H_2O. The angle of the graph of pH vs. voltage provides a value of approximately −59 mV/pH. According to the Nernst equations, this means that a unit change in pH decreases the H_2O redox process (redox reaction) potential by approximately 59 mv. Here, the water splitting process in acidic electrolytes is presented in the equation below:

$$\text{Positive Electrode: } 2\ H_2O + 4h^+ \rightarrow O_2 + 4H^+\ \Delta E_{ox}{}^0 = 1.2\ V_{RHE} \tag{8.6}$$

$$\text{Negative Electrode: } 4\ H^+ + 4e^- \rightarrow 2H_2\ \Delta E_{re}{}^0 = 0\ V_{RHE} \tag{8.7}$$

The HER reaction process from the basic electrolytes is presented in the equation below:

$$\text{Positive Electrode: } 2\ OH + 4h^+ \rightarrow O_2 + 2H_2O^+\ \Delta E_{ox}{}^0 = -0.404\ V_{RHE} \tag{8.8}$$

$$\text{Negative Electrode: } 4\ H^+ + 4e^- \rightarrow 2H_2\ \Delta E_{re}{}^0 = -V0.8_{RHE} \tag{8.9}$$

The overall principle of PEC water splitting is to harness the energy from sunlight to initiate the electrochemical reactions required for water splitting [16]. The goal is to produce hydrogen gas, a clean and storable fuel, while releasing oxygen as a byproduct. Researchers are exploring various semiconductor materials and system configurations to enhance the efficiency and stability of PEC water splitting for practical and sustainable energy applications.

Photoelectrochemical (PEC) water splitting has various potential applications, primarily centered around the making of hydrogen as a clean and renewable source. The following are some key applications of PEC water splitting: PEC water splitting can be used to produce hydrogen, which serves as a clean energy carrier and can be stored for later use. Hydrogen produced through PEC can be utilized in fuel cells to generate electricity when needed providing a means of energy storage and retrieval. Moreover, PEC water splitting can be used in conjunction with water treatment processes contributing to water purification by producing hydrogen while addressing energy needs for water treatment. Additionally, the oxygen produced

Table 8.1 Hydrogen evaluation for nanomaterials.

S. no.	Nanomaterials	Hydrogen production proportion (μmol/h)	Reference
1	MoS_2–CdS	5.2	[18]
2	TiO_2 nanotubes	2.31	[19]
3	Mesoporous $SrTiO_2$ nanocrystal	188	[19]
4	CdS QD-sensitized TiO_2/Pt	0.8	[20]
5	CD–CdS	2.55	[21]
6	CNTs/MnO_2–C_3N_4	122	[22]
7	TiO_2–Ni hydroxide/ MWCNT/Cds	12.2	[23]
8	CNTs/$ZnFe_2O_4$	18.1	[24]
9	CQDs/TiO_2	41.26	[25]

during the water splitting process can be used in environmental applications or industrial processes. Despite its potential, PEC water splitting is still in the research and development stage, and practical implementation at a large scale requires addressing challenges related to efficiency, stability, and cost. Ongoing research aims to improve the performance of PEC systems for practical and sustainable energy applications. According to Joy *et al.*, some recent nanomaterials have been used in the photoelectrochemical water splitting process [17], and these previous reports are summarized in Table 8.1.

8.3 Photocatalytic Water Splitting

Photocatalytic water splitting involves using a catalyst activated by light to break water molecules into hydrogen and oxygen. This process typically employs semiconductors as photocatalysts. When light of sufficient energy strikes the catalyst, it excites electrons creating electron–hole pairs. The excited electrons can then participate in water reduction to form hydrogen, while the holes can facilitate water oxidation to generate oxygen [26]. This technology holds promise for renewable energy production, as hydrogen can serve as a clean fuel source when derived from water through such sustainable means.

Mechanism for Photocatalytic Water Splitting

Generally, the photocatalyst mechanism follows this procedure: 2 mol of H_2O is split into 1 mol of O_2 and 2 mol of H_2. The process is described below. The photoexcited electrons participate in the water oxidation reaction process at the surface of the photocatalyst. The water oxidation reaction involves the release of oxygen gas and the generation of protons (H^+) at the photocatalyst surface as shown in Figure 8.4.

$$2H_2O\ (l) \rightarrow O_2\ (g) + 4H^+\ (aq) + 4e^-$$

(or)

$$2H_2O + 4h_{\text{valence band}} \rightarrow O_2 + 4H^+$$

The holes left in the valence band participate in the proton-reduced reaction at the photocatalysts surface. Moreover, the protons (H^+) from the water oxidation reaction are reduced at the photocatalyst surface to from hydrogen gas [27].

$$2H^+ + 2e^-_{\text{conduction band}} \rightarrow 2\ H_2$$

(or)

$$4H^+\ (aq) + 4e^- \rightarrow 2H_2\ (g)$$

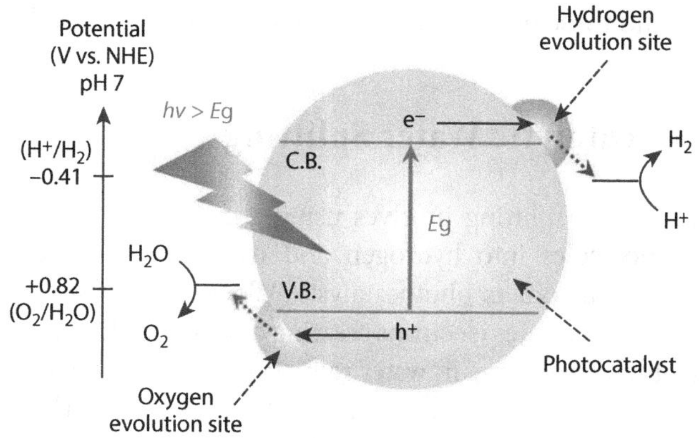

Figure 8.4 The photocatalytic water splitting mechanism.

The overall water reaction can be described below. The combination of water oxidation and reduction reactions results in the overall splitting reaction:

$$2H_2O\ (l) \rightarrow 2H_2\ (g) + O_2\ (g)$$

The photocatalyst generates electron–hole pairs by converting photons with energies greater the 1.2 eV into electron–hole pairs. For photocatalysts to split water, it must have a bandgap large enough to drive the reaction. Losses from material internal resistance and the overall potential of water splitting raise the required bandgap energy to 1.6–2.4 eV. The process of splitting water is strongly endothermic ($\Delta H > O$) [28]. Water splitting happens spontaneously in photosynthesis when the energy of four photons is absorbed and turned into chemical energy via a complicated biochemical process. In particular, the nanomaterials used in photocatalytic water splitting meet the band criteria and are generally enhanced with dopants and or co-catalysts to improve performance. Here, we assume that titanium dioxide (TiO_2), a semiconductor with the appropriate band structure, is commonly used in conjunction with platinum (Pt) as a co-catalyst to boost H_2 generation. As an example, Chi-Hung Liao *et al.* reported that some basic nanomaterials were used in the hydrogen production from previous literature (Table 8.2) [29].

Table 8.2 Hydrogen and oxygen evaluation for nanomaterials.

			Evolution rate (μmol h^{-1})		
S. no.	**Nanomaterials**	**Source of light**	**Hydrogen**	**Oxygen**	**References**
1	Pt/TiO_2	400W Hg lamp	568	287	[30]
2	ZrO_2	400W Hg lamp	142	287	[31]
3	Ru_2O/ZrO_2	400W Hg lamp	11	75	[31]
4	Cu/ZrO_2	400W Hg lamp	14	5	[31]
5	$NiO/Sr_2N_2O_7$	400W Hg lamp	110	6	[32]
6	$NiO/BaTa_2O_6$	400W Hg lamp	629	480	[33]
7	Ru_2O/ZrO_2	400W Hg lamp	11	5	[34]

8.4 Nanomaterials in Hydrogen Evaluation

8.4.1 Noble Metal Electrocatalyst

A noble metal catalyst is considered to be the most efficient electrocatalyst in terms of high HER performance and high efficiency. Noble metals, such as platinum (Pt), iridium (Ir), and ruthenium (Ru), possess high electrocatalytic activity and can improve its catalytic performance when combined with carbon-based materials, alloy formation, or transition metals [35]. The noble metal electrocatalyst exhibits benchmark performance with features such as low overpotential value, low Tafel slope, and high conversion input current. Even though it has remarkable HER performance, it has several challenges when applied as an electrocatalyst in real time such as high-cost, low-abundance, easy to agglomerate, dissolution, low stability, and poor tolerance to poisoning during electrolysis. The capacity to alter the metal composition on the surface of different Pt-based electrocatalysts or doping Pt based materials on other metal components is crucial for increasing electrocatalytic activity with unique features such as tailored increased interfacial active sites, atomic arrangement, and strong electron coupling. One of the best commercial catalysts for HER is 20 wt% Pt deposited on carbon black (Pt/C), which exhibits the lowest overpotential value of ~46 mV (10 mA/cm^2) under neutral and alkaline conditions [36]. Due to its remarkably high HER activity, platinum/carbon is normally employed as a reference material for newly established electrocatalysts. Luo *et al.* proposed a cost-effective strategy to adsorb platinum-based catalysts onto processed vulcan carbon after heat treatment with ethylene glycol solution and Ru precursor, which are meticulously built using an elegant and controllable atomic layer deposition (ALD) approach [37]. When compared to the referred commercial 20 wt% Pt/C, Pt/VC, and Ru/VC catalysts, the Pt/Ru/VC composite catalyst has ideal Pt electronic characteristics, more Pt active sites, and high electron conduction resulting in the highest water splitting activity (Figure 8.5a, b). The P. Kuang group has proposed a novel strategy of tailoring the electronic structure of Pt supported on N-doped hallow carbon sphere (Figure 8.5c), which displays a low overpotential (40 mV), high mass activity (2.07 A mg^{-1}), and excellent TOF (20.18 s^{-1}) (Figure 8.5d, e) [38].

Rather than single noble metal electrocatalysts, bimetallic structures have gained enormous interest in tuning the HER performance. Zhang *et al.* successfully synthesized a high-value Ru–Pt dual metallic dimer structure via the ALD process [39]. The Ru–Pt dimer exhibits superior

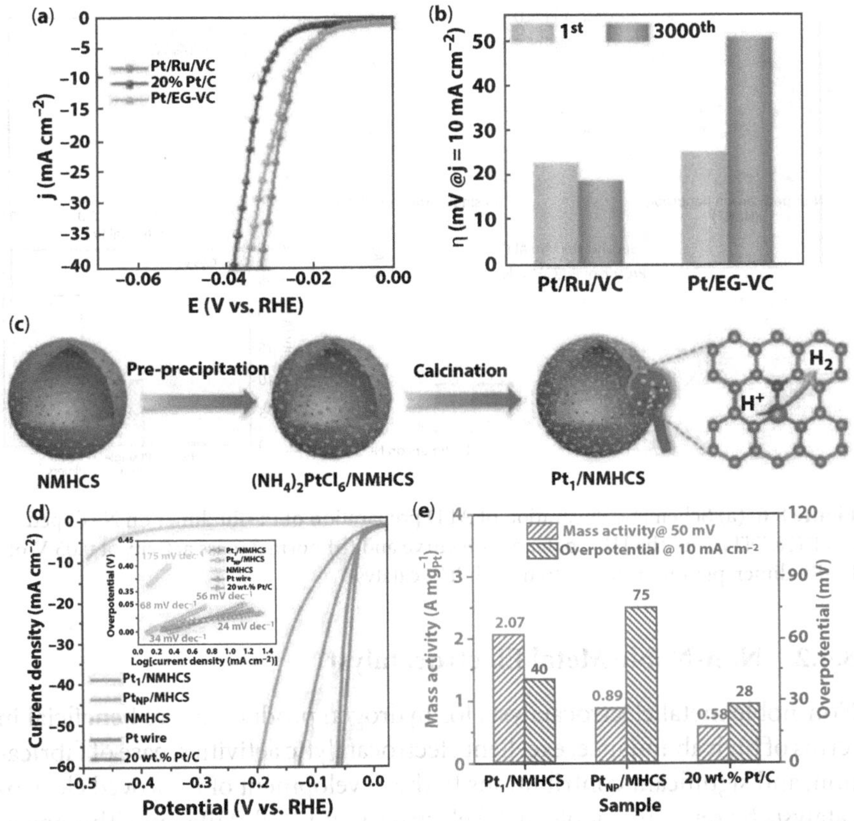

Figure 8.5 (a) LSV curve. (b) Overpotential gained at a current density of 10 mA/cm^2 before and after 3,000 cycles for the catalysts. (c) Schematic diagram for the preparation of Pt1/NMHCS. (d) Polarization curves and Tafel slopes and (e) histogram for mass action and overpotential for Pt1/NMHCS, Pt Nanoparticle/MHCS, and 20% Pt/C.

HER activity and excellent stability compared to the commercial Pt/C catalyst. The enormous HER activity is achieved owing to the effective interaction of Ru and H, which can be modulated through the synergistic effect with Pt. Nanostructured RuIr alloy is loaded onto the N-rich carbon matrix (NrC) through a one-pot pyrolysis process (Figure 8.6). The RuIr/NrC holds the advantages of high porosity, abundantly doped N, excellent synergy, and high HER activity in basic and acidic media. Thus, researchers are still developing strategies to fabricate an extremely effectual electrocatalyst of hydrogen production by overcoming the drawbacks of noble metal catalysts such as high cost and stability.

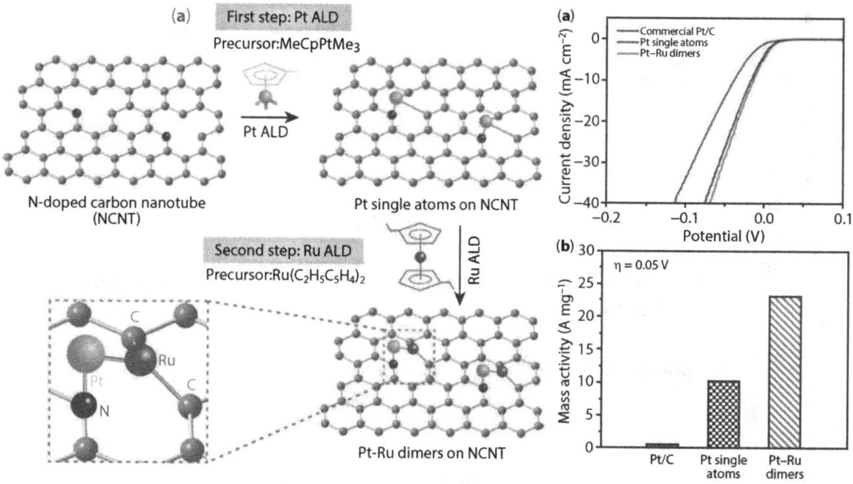

Figure 8.6 (a) Schematic illustration of ALD preparation of Pt–Ru dimer on N_2-doped CNT (NCNTs). (a) The HER polarization curve and (b) normal mass activity at 0.05 V of Pt–Ru dimer, platinum single atom, and Pt/C catalyst.

8.4.2 Non-Noble Metal Electrocatalyst

Non-noble metal electrocatalysts for hydrogen production are beneficial in terms of high abundance, excellent electrocatalytic activities, ease of fabrication, and significant contributions to the development of improved electrocatalysts by replacing noble metal electrocatalyst [40]. Miles and Thompson have used the voltametric technique to examine the catalytic activities of transition metals, like Mo, W, Fe, Ni, and Cu toward HER. Transition-metal-based compounds, such as transition-metal dichalcogenides/sulfides (TMDs/TMSs), transition-metal oxides/hydroxides (TMOs/TMOHs), transition-metal nitrides (TMNs), transition-metal carbides (TMCs), and transition-metal phosphides (TMPs), have been reported as promising HER catalysts having low electrical resistivity and quick charge transfer [41]. Non-noble metal electrocatalysts are promising alternatives to noble metal electrocatalysts. The major limitation of non-noble metal electrocatalysts, which severely restricts them from practical applications, is their susceptibility to corrosion in solid acidic or basic conditions leading to agglomeration into larger particles during catalytic cycling. The evolution of electrocatalysts using the transition metals starts with TMOs and TMOHs because of their adjustable composition, cost friendliness, environmentally friendliness, rapid synthesis, and crystal structure. However, the TMOs hinder the HER performance owing to their poor conductivity, limited active sites, and inappropriate hydrogen adsorption or desorption energies. Similarly, the TMOHs also face issues like unfavorable ΔG_H^* and poor conductivity. Therefore, researchers took the initiative to modify or combine TMOHs with TMOs via oxygen vacancy and surface/interface engineering to regulate more active sites for efficient hydrogen production. Concurrently, their superior stability makes them an ideal candidate for HER electrocatalysts (Figure 8.7) [41].

Generally, M_xS_2 and M_xSe_2 are widely used as TMDs/TMSs in hydrogen production due to their extensive chemical composition and unique crystal structure. Among various TMDs, MS_2 has gained extensive attention in electrochemical research owing to their layer structure, weak interlayer interaction, and strong intralayer interaction. Basically, MS_2 is divided into various phases, such as 1T, 2H, and 3R. The 2H phase, with entirely filled d-orbitals, exhibits semiconducting properties, while the 1T phase, with partially filled d-orbitals, exhibits metallic properties. As a representative of TMDs, MoS_2, according to DFT calculations, reveals an electronic structure similar to Pt, and the ΔG_H^* of MoS_2 (edge sites) is close to 0 indicating

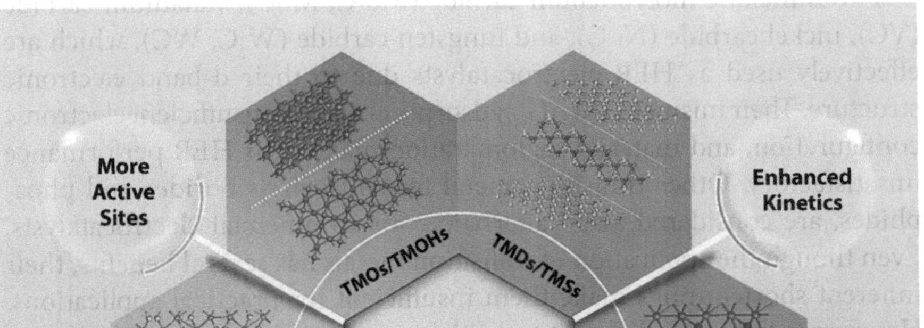

Figure 8.7 Schematic classification of different components in heterostructure catalysts.

its superior potential as a HER electrocatalyst. However, during the synthesis process, the formation of polysulfides are unavoidable, which makes it difficult to understand the catalytic mechanism. Significant improvement in the HER performance of TMDs/TMSs can be achieved by tuning the density of grain boundaries and doping with other metal components.

Generally, the TMNs exhibit high corrosion resistance, high electrical properties, and unique chemical and physical properties, which makes them a suitable candidate for HER. Meanwhile, the introduction of nitrogen atom into the transition metal matrix can tune the d-band properties and electronic structure, and provides the ability to shrink the d-band. Additionally, the small atomic radii of the nitrogen atom can fit into the lattice gap of transition metals, which is beneficial for efficient electron transfer. However, TMN stability is a great concern owing to the formation of strong M=H bonds of the single phase of TMNs making it unsuitable as an electrocatalyst for extreme electrochemical conditions. Thus, the development of TMN heterostructures with other metals can greatly tune their surface structure, chemical stoichiometry, volume, surface area, mass transport, and electronic and catalytic properties.

TMCs include molybdenum carbide (Mo_2C, MoC), vanadium carbide (VC), nickel carbide (Ni_3C), and tungsten carbide (W_2C, WC), which are effectively used as HER electrocatalysts due to their d-band electronic structure. Their mismatched ΔG_{H^*} adsorption energy, insufficient electronic configuration, and material agglomeration make their HER performance unsatisfactory. Other transition metal forms, such as borides and phosphides, are considered as an alternative for commercial electrocatalysts. Even though the application of transition metals has several benefits, their inherent shortcomings make them insufficient for practical applications. However, combining different transition metals with other metal components or carbon matrix to hypothesize a new multicomponent catalyst is expected to improve HER activity by enhancing their strength of interaction with reaction intermediates and catalysts, and improving hydrogen adsorption and HER kinetics.

8.4.3 Carbon-Based Metal-Free Electrocatalyst for HER

In the quest for novel electrocatalytic nanomaterials for HER, metal-free electrocatalysts are emerging as promising candidates due to their unique merits such as cost effectiveness, earth abundance, environmental friendliness, high conductivity, and resistance to a wide pH range [42]. Especially, carbon materials, such as graphene, carbon quantum dots, carbon nanotubes, fullerene, graphite, graphene oxide, and graphdiyne, with various architectures, have proven to be effective electrocatalysts for H_2 production [42]. The fascinating strategy to progress the electrocatalytic process of bare carbon material is to perform heteroatom doping (O, N, S, P, B, or F), which has the ability to profoundly tune the electronic structure of carbon by creating specific defects [43]. Substitutional doping of heteroatoms can be achieved by replacing carbon atoms at different planes such as edge or basal planes. Co-doping (NS, NB, NP, NO) and tri-doping (NBO) of carbon materials also tends to exhibit high catalytic activity in terms of both acid and alkaline conditions [44]. Another strategy to add active sites without requiring substitutional doping is by adsorbing heteroatoms containing molecules or layers onto the surface of carbon nanomaterials. The coupling properties between heteroatoms and lattice defects considerably enhance catalytic activity by providing high conductivity due to electron-rich defective structures and optimizing the Gibbs free energy for hydrogen adsorption (ΔG_{H^*}), which is very close to that of commercial Pt catalysts.

On the other hand, 3D architected carbon materials are preferred for electrocatalysis owing to their effective electron transport, enhanced mass transport, large specific surface area with more redox active sites, mechanical integrity, and efficient diffusion of electrolytes [45]. Other than doping, self-assembly and structural modification are two factors in fabricating a carbon-based metal-free electrocatalyst. Concurrently, biomass, being a perfect carbon-rich precursor, provides environmentally favorable properties as well as great electrical conductivity [46]. Biomass-derived carbon-based materials have proven to be effective electrocatalysts for H_2 production owing to its ability to tune morphology, structure, specific surface area, ion transport channels, electrical conductivity, and active sites through various processes such as pyrolysis, hydrothermal treatment, molten-salt carbonization, and ionothermal carbonization [46]. The interaction between the electrocatalyst and the adsorbed H* during HER should be moderate to enable proton–electron transfer, without hindering the release of H_2 gas. The evolution of carbon-based metal-free electrocatalysts started in 2014, when the research group of Zheng *et al.* found the remarkable HER performance of the g-C_3N_4/N-graphene composite, which is comparable to that of metal-based catalysts (Cu, MoS_2, or Ni) [47]. The synergistic effect raised from the highly conductive N-graphene and the good catalytic activity from the g-C_3N_4, which results in the enhanced hydrogen production. In the same year, Zheng *et al.* demonstrated an N- and P-doped (dual atom) graphene catalyst with a synergistically enhanced HER performance [48]. The electrocatalytic activity for N-doped graphene is observed with an overpotential value of ~420 mV (Figure 8.8), which

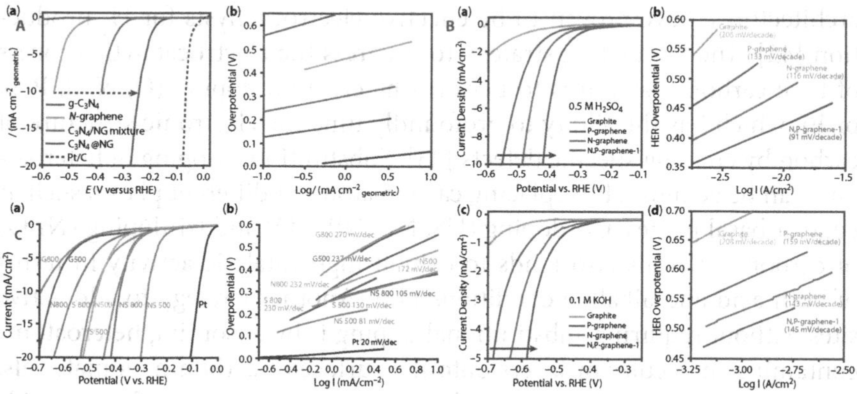

Figure 8.8 (a) The HER polarization curves (a) and Tafel plots (b) for four metal-free electrocatalysts and 20% Pt/C (electrolyte: 0.5 M H_2SO_4, scan rate: 5 mV s^{-1}).
(b) HER polarization curve and corresponding Tafel plots of N- and/or P-doped graphene electrocatalysts in (a, b) 0.5 M H_2SO_4 (pH = 0) and (c, d) 0.1 M KOH (pH = 13).
(c) (a) CV spectra for samples generated for various CVD temperatures and dopants in contrast to undoped graphene and (b) Tafel plot of various graphene samples.

indicates enhanced HER activity compared to individual graphite, N-graphene, and P-graphene. In a parallel study, Y. Ito *et al.* improved the HER performance of nanoporous graphene by N and S doping prepared at two different CVD temperatures (500° and 800°C) [49]. When compared to undoped and single N- or S-doped samples, N and S co-doping significantly improves graphene HER catalysis.

8.5 Summary and Future Perspectives

Hydrogen production from water electrolysis is becoming an advanced hydrogen economy technique. The book chapter summarizes the research progress of electrocatalysts for efficient hydrogen production. The benchmark electrocatalyst for HER is Pt/C. Owing to the cost and shortage of noble metal electrocatalysts, the growth of nanostructured non-noble metal electrocatalysts, carbon-based materials, noble metal/carbon matrix, and non-noble metal/carbon matrix has gained superior attention. The tremendous exertions have commercialized hydrogen production in large scale in neutral, acidic, and alkaline environment. To overcome the sluggish reaction rate of HER, the researchers have tuned their focus toward the fabrication of rapid, efficient, and low-cost electrocatalysts. Various parameters, such as morphology, catalytic mechanism, active site formation, exposure, density, and electronic structure, are mainly interrelated with the modification of electrocatalysts for hydrogen production. The major factors determining HER performance of electrocatalysts are the synergistic effect among the various components in the heterostructure, tuning of adsorption energy for reaction intermediates, and charge transfer at the heterointerface. The state-of-the-art strategies to develop an electrocatalyst for HER are the following: 1) improving the active sites via designing dimensional architecture and porous structuring, 2) enhancing conductivity by combining with noble or non-noble metals with highly conductive materials such as Ni foam, graphene, Cu foam, and carbon materials, and 3) improving intrinsic catalytic properties such as doping, defect engineering, component regulation, and interface engineering. To date, remarkable research progress has been achieved in the designing and development of HER electrocatalysts, but there are still some challenges that require further attention. First, the understanding of catalytic mechanism: Although electrolysis is based on the simplest chemical reaction, a deep understanding of electrolysis through advanced characterization methods, such as microscopy, computational methods, and DFT calculations, is highly recommended. Second, the evaluation parameters of the catalysts should be standardized: The superior HER electrocatalyst is determined by identifying the overpotential values at -10 mA/cm^{-2} and Tafel slope values. However, other factors, such as turnover frequency, mass/specific activity, onset potential, and stability, should be taken into consideration. Third, many of the electrocatalysts employed for HER are surprisingly found to exhibit multifunctional properties such as OER and ORR.

The multifarious properties of electrocatalysts should be explored to decrease costs and enhance the experimental setup for overall water splitting.

Acknowledgment

This work was supported by DST-SERB SURE Project under grant number SUR/2022/005307. The financial support provided by the Department of Science and Technology enabled the successful completion of this book chapter.

AI disclosure statement: The author acknowledges the use of AI-based tools for minor language editing or grammatical correction. However, all intellectual contributions, analyses, and conclusions presented in this chapter are entirely the author's own.

References

1. Perera, F., Pollution from Fossil-Fuel Combustion is the Leading Environmental Threat to Global Pediatric Health and Equity: Solutions Exist. *Int. J. Environ. Res. Public Health*, 15, 16, 2017.
2. Hota, P., Das, A., Maiti, D.K., A short review on generation of green fuel hydrogen through water splitting. *Int. J. Hydrogen Energy*, 48, 523–541, 2023.
3. Herbaut, M., Siaj, M., Claverie, J.P., Nanomaterials-Based Water Splitting: How Far Are We from a Sustainable Solution? *ACS Appl. Nano Mater.*, 4, 907–910, 2021.
4. You, B. and Sun, Y., Innovative Strategies for Electrocatalytic Water Splitting. *Acc. Chem. Res.*, 51, 1571–1580, 2018.
5. Chatenet, M., Pollet, B.G., Dekel, D.R., Water electrolysis: from textbook knowledge to the latest scientific strategies and industrial developments. *Chem. Soc. Rev.*, 51, 4583–4762, 2022.
6. Fiorio, J.L., Gothe, M.L., Kohlrausch, E.C., *et al.*, Nanoengineering of Catalysts for Enhanced Hydrogen Production. *Hydrogen*, 3, 218–254, 2022.
7. Tahir, A., Arshad, F., Haq, T., Roles of Metal Oxide Nanostructure-Based Substrates in Sustainable Electrochemical Water Splitting: Recent Development and Future Perspective. *ACS Appl. Nano Mater.*, 6, 1631–1647, 2021.
8. Wang, M., Zhang, L., He, Y., Zhu, H., Recent advances in transition-metal-sulfide-based bifunctional electrocatalysts for overall water splitting. *J. Mater. Chem. A Mater.*, 9, 5320–5363, 2021.
9. Sohail, U., Pervaiz, E., Ali, M., Role of tungsten carbide (WC) and its hybrids in electrochemical water splitting application- A comprehensive review. *FlatChem.*, *35*, 100404, 2022.
10. Diao, J., Li, X., Wang, S., promoting water splitting on arrayed molybdenum carbide nanosheets with electronic modulation. *J. Mater. Chem. A Mater.*, 9, 21440–21447, 2021.
11. Liu, Y., Guo, Y., Liu, Y., A Mini Review on Transition Metal Chalcogenides for Electrocatalytic Water Splitting: Bridging Material Design and Practical Application. *Energy Fuels*, 37, 2608–2630, 2023.
12. Jian, J. and Sun, J., A Review of Recent Progress on Silicon Carbide for Photoelectrochemical Water Splitting. *Sol. RRL*, *4*, 2000111, 2020.
13. Butchosa, C., Guiglion, P., Zwijnenburg, M.A., Carbon Nitride Photocatalysts for Water Splitting: A Computational Perspective. *J. Phys. Chem. C*, 118, 24833–24842, 2014.

14. Kumar, P. and Kumar, A., Hydrogen Generation via Photoelectrochemical Splitting of Water, in: *Handbook of Ecomaterials*, pp. 1807–1844, Springer International Publishing, Cham, 2019.
15. AlSalka, Y., Schwabe, S., Geweke, J., Electrochemical and Photoelectrochemical Water Splitting: Operando Raman and Fourier Transform Infrared Spectroscopy as Useful Probing Techniques. *Energy Technol.*, 11, 2200788, 2023.
16. Kumar, M., Meena, B., Subramanyam, P., Recent trends in photoelectrochemical water splitting: the role of cocatalysts. *NPG Asia Mater.*, 14, 88, 2022.
17. Joy, J., Mathew, J., George, S.C., Nanomaterials for photoelectrochemical water splitting – review. *Int. J. Hydrogen Energy*, *43*, 4804–4817, 2018.
18. He, G., Zhang, Y., He, Q., MoS_2/CdS Heterostructure for Enhanced Photoelectrochemical Performance under Visible Light. *Catalysts*, 9, 379, 2019.
19. Puangpetch, T., Sreethawong, T., Yoshikawa, S., Chavadej, S., Hydrogen production from photocatalytic water splitting over mesoporous-assembled SrTiO3 nanocrystal-based photocatalysts. *J. Mol. Catal. A Chem.*, 312, 97–106, 2009.
20. Zhang, J., Zhu, W., Liu, X., Stable hydrogen generation from vermiculite sensitized by CdS quantum dot photocatalytic splitting of water under visible-light irradiation. *Dalton Trans.*, 43, 9296, 2014.
21. Zhu, C., Liu, C., Zhou, Y., Carbon dots enhance the stability of CdS for visible-light-driven overall water splitting. *Appl. Catal., B*, 216, 114–121, 2017.
22. Wang, N., Li, J., Wu, L., MnO_2 and carbon nanotube co-modified C_3N_4 composite catalyst for enhanced water splitting activity under visible light irradiation. *Int. J. Hydrogen Energy*, 41, 22743–22750, 2016.
23. Wang, J., Wang, Z., Zhu, Z., Synergetic effect of Ni $(OH)_2$ cocatalyst and CNT for high hydrogen generation on CdS quantum dot sensitized TiO_2 photocatalyst. *Appl. Catal., B*, 204, 577–583, 2017.
24. Dang, H., Qiu, Y., Cheng, Z., Hydrothermal preparation and characterization of nanostructured CNTs/$ZnFe_2O_4$ composites for solar water splitting application. *Ceram. Int.*, 42, 10520–10525, 2016.
25. Hao, X., Jin, Z., Xu, J., Functionalization of TiO_2 with graphene quantum dots for efficient photocatalytic hydrogen evolution. *Superlattices Microstruct.*, 94, 237–244, 2016.
26. Nishioka, S., Osterloh, F.E., Wang, X., Photocatalytic water splitting. *Nat. Rev. Methods Primers*, 3, 42, 2023.
27. Kim, J.H., Hansora, D., Sharma, P., Toward practical solar hydrogen production – an artificial photosynthetic leaf-to-farm challenge. *Chem. Soc. Rev.*, 48, 1908–1971, 2019.
28. Yano, J. and Yachandra, V., Mn_4Ca Cluster in Photosynthesis: Where and How Water is Oxidized to Dioxygen. *Chem. Rev.*, 114, 4175–4205, 2014.

29. Liao, C.-H., Huang, C.-W., Wu, J.C.S., Hydrogen Production from Semiconductor-based Photocatalysis via Water Splitting. *Catalysts*, 2, 490–516, 2012.
30. Sayama, K. and Arakawa, H., Effect of carbonate salt addition on the photocatalytic decomposition of liquid water over Pt–TiO_2 catalyst. *J. Chem. Soc., Faraday Trans.*, 93, 1647–1654, 1997.
31. Yoshino, S., Iwase, A., Yamaguchi, Y., Photocatalytic CO_2 Reduction Using Water as an Electron Donor under Visible Light Irradiation by Z-Scheme and Photoelectrochemical Systems over $(CuGa)_{0.5}$ ZnS_2 in the Presence of Basic Additives. *J. Am. Chem. Soc.*, 144, 2323–2332, 2022.
32. Kudo, A., Kato, H., Nakagawa, S., Water Splitting into H_2 and O_2 on New $Sr_2M_2O_7$ (M = Nb and Ta) Photocatalysts with Layered Perovskite Structures: Factors Affecting the Photocatalytic Activity. *J. Phys. Chem. B*, 104, 571–575, 2017.
33. Fujiwara, T., Sasahara, A., Happo, N., Single-Crystal Model of Highly Efficient Water-Splitting Photocatalysts: A $KTaO_3$ Wafer Doped with Calcium Cations. *Chem. Mater.*, 32, 1439–1447, 2020.
34. Sayama, K. and Arakawa, H., Photocatalytic decomposition of water and photocatalytic reduction of carbon dioxide over zirconia catalyst. *J. Phys. Chem.*, 97, 531–533, 1993.
35. Li, C. and Baek, J.B., Recent Advances in Noble Metal (Pt, Ru, and Ir)-Based Electrocatalysts for Efficient Hydrogen Evolution Reaction. *ACS Omega*, 5, 31–40, 2020.
36. Wang, S., Lu, A., Zhong, C.J., Hydrogen production from water electrolysis: role of catalysts. *Nano. Converg.*, 8, 2021.
37. Luo, W., Gan, J., Huang, Z., *et al.*, Boosting HER Performance of Pt-Based Catalysts Immobilized on Functionalized Vulcan Carbon by Atomic Layer Deposition. *Front. Mater.*, 6, 2019.
38. Kuang, P., Wang, Y., Zhu, B., *et al.*, Pt Single Atoms Supported on N-Doped Mesoporous Hollow Carbon Spheres with Enhanced Electrocatalytic H_2-Evolution Activity. *Adv. Mater.*, 33, 2008599, 2021.
39. Zhang, L., Si, R., Liu, H., Atomic layer deposited Pt-Ru dual-metal dimers and identifying their active sites for hydrogen evolution reaction. *Nat. Commun.*, 10, 4936, 2019.
40. Wu, T., Sun, M.Z., Huang, B.L., Non-noble metal-based bifunctional electrocatalysts for hydrogen production. *Rare Met.*, 41, 2169–2183, 2022.
41. Song, A., Song, S., Duanmu, M., Recent Progress of Non-Noble Metallic Heterostructures for the Electrocatalytic Hydrogen Evolution. *Small Sci.*, 3, 2300036, 2023.
42. Paul, R., Dai, Q., Hu, C., Dai, L., Ten years of carbon-based metal-free electrocatalysts. *Carbon Energy*, 1, 19–31, 2019.
43. Zhang, L., Xiao, J., Wang, H., Shao, M., Carbon-Based Electrocatalysts for Hydrogen and Oxygen Evolution Reactions. *ACS Catal.*, 7, 7855–7865, 2017.

44. Zhai, Q., Pan, Y., Dai, L., Carbon-Based Metal-Free Electrocatalysts: Past, Present, and Future. *Acc. Mater. Res.*, 2, 1239–1250, 2021.
45. Zhou, W., Jia, J., Lu, J., Recent developments of carbon-based electrocatalysts for hydrogen evolution reaction. *Nano Energy*, 28, 29–43, 2016.
46. Wang, Q., Guo, R., Wang, Z., Progress in carbon-based electrocatalyst derived from biomass for the hydrogen evolution reaction. *Fuel*, 293, 120440, 2021.
47. Zheng, Y., Jiao, Y., Zhu, Y., Hydrogen evolution by a metal-free electrocatalyst. *Nat. Commun.*, 5, 3783, 2014.
48. Zheng, Y., Jiao, Y., Li, L.H., Toward design of synergistically active carbon-based catalysts for electrocatalytic hydrogen evolution. *ACS Nano*, 8, 5290–5296, 2014.
49. Ito, Y., Cong, W., Fujita, T., High catalytic activity of nitrogen and sulfur co-doped nanoporous graphene in the hydrogen evolution reaction. *Angew. Chem. Int. Ed.*, 54, 2131–2136, 2015.

9

Hydrogen Utilization: Fuel Cells and Energy Storage

Srijita Basumallick

PG-Department of Chemistry, Asutosh College under Calcutta University, Kolkata, India

Abstract

This chapter provides a unique method of utilizing hydrogen using hydrogen fuel cells. It discusses the basics of fuel cell thermodynamics and the kinetics of fuel cell reactions. It describes in detail different types of hydrogen fuel cells, like acid, alkali, and phosphoric acid, and high-temperature fuel cells. Catalysts used in hydrogen oxidation reactions, abbreviated as HOR, and oxygen reduction reactions, abbreviated as ORR, are discussed in detail. Different types of proton exchange membranes and electrodes are included. Finally, the problems of hydrogen transport are discussed.

Keywords: Hydrogen fuel cell, electrocatalyst for ORR, phosphoric acid fuel cell, proton exchange membrane, porous electrode

Importance of Fuel Cells

A fuel cell is just like a normal electrochemical cell containing a cathode, anode, membrane, and electrolytic solution. To move ahead toward a carbon-free society, we need to implement hydrogen fuel cell technology [1] that has pure water as the only end product. Together with that, hydrogen has a high energy density. When 1 U of hydrogen is burned, it can provide three times more energy than a similar amount of gasoline combustion [2]. As the earth contains three times more water than land, the splitting of water by electrolysis to produce hydrogen makes it a clean renewable

Email: srijita.basumallick@asutoshcollege.in

Inamuddin, Tariq Altalhi, Mohammad Luqman and Jorddy Neves Cruz (eds.) Water Splitting: Production of Hydrogen, (233–244) © 2025 Scrivener Publishing LLC

energy source. When talking about the efficiency of a fuel cell, it does not follow the Carnot principle [3], and 100% efficiency is possible. Usually, for gaseous reactions, $-\Delta G$ (= Wrev – PΔV). That is, ΔG is equal to reversible work reduced by expansion work. In the case of electrochemical reactions, there is no pressure–volume-related work done. Here, W_{rev} may be obtained in a simpler way. It is related to the 4e transfer from the anodic to the cathodic chamber. For example, at the anode, the HOR is $2H_2 = 4H^+ + 4e$, and at the cathode, the ORR is $O_2 + 4H^+ + 4e^- = 2H_2O$ giving rise to the overall reaction $2H_2 + O_2 = 2\ H_2O$ at normal temperature and pressure. Here, the transport of four electrical charges occurs under the total potential difference V, which is the cell potential. If the cell potential is taken to be the equilibrium cell potential, assuming the reaction takes place in a reversible way, the net work done by this process is 4FVe, which is nothing but free energy change, where the number of electron transfers is 4, F denotes the Faraday constant, and Ve is the cell potential when a cell reaction occurs reversibly. There is no other work obtained during the electrochemical cell reaction, such as pressure–volume work, surface work, or work against a gravitational field. So, the free energy of an electrochemical reaction will not be diminished by any amount of other work (unlike the PV-related work done in the case of a gaseous reaction). Thus, it can really reach 100% efficiency provided that the potential at which an electrochemical cell reaction has taken place is its equilibrium potential or reversible potential. Cell potential changes in infinitely small steps near the equilibrium of the reaction giving equilibrium or reversible cell potential. But actually, the efficiency of fuel cells can be given as V/Ve. Of course, this relationship works well when reactants are completely converted to final products, and no other by-products have reduced coulombic efficiency. Unlike heat engines, electrochemical reactions (water formation) are more like cold combustion.

During the movement of pistons in a heat engine, both sides of the piston contain heat energy, and total energy transfer is impossible. In the case of cold combustion, there is no such heat equilibrium, and the only loss of energy is due to entropy change.

Basic Principles for Increasing the Efficiency of Fuel Cells

As we have seen, the efficiency of energy output divided by the maximum or theoretical energy output of a fuel cell is given by V/Ve. There are three impediments that cause the potential to be different from equilibrium potential. These are as follows: 1. Internal resistance of a cell imposed by electrolytes (as well as membrane material). 2. Concentration overpotential [4].

3. Activation overpotential [5]. Thus, for good energy conversion, the electrolyte resistance of the cell must be as small as possible. This requires excess electrolytes. To reduce concentration overpotential, the limiting current density (affected by the slow diffusion of reactants from bulk electrolyte to electrode) needs to be maximized. This can be done by agitation or stirring. Now, activation overpotential depends on exchange current. High exchange current with small polarization is seen on highly active catalyst surfaces. As discussed before, only slow and reversible reactions give the highest efficiency, but they reduce power output. Power P = IV (I = current, V = potential). If I is very low, even if V is equal to Ve, the power output will be low. Thus, at constant voltage, power output is proportional to current, and the maximum is the very fast reaction rate at the electrode. But a higher reaction rate indicates irreversibility, and efficiency decreases with an increase in irreversible behavior. More efficiency needs reversibility; therefore, a slower reaction rate and less power are needed. From the Tafel equation, at very high current output or when I is equal to I_L (limiting current density), the power reduces to zero. There is a maxima in the power vs. current plot near the limiting current [6]. But as the limiting current is reached, the power drops to zero. Apart from that particular region (the $I = I_L$ region), an increase in power reduces efficiency, as shown in Figure 9.1 [7].

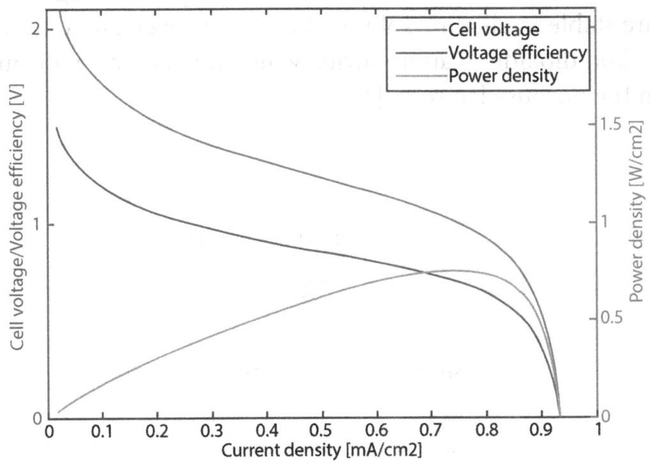

Figure 9.1 Cell voltage (red) and voltage efficiency or overall efficiency (blue) as well as power density (green) vs. current density (from Dynamic modelling and optimal control strategy of energy storage elements in hybrid electrical vehicle "fuel cell and ultra capacitor").

Porous Electrode

Use of any oxidants [8] in fuel cells is allowed, yet the best is O_2 from air. The overall reaction in acid solution is $O_2 + 4H_3O^{+} + 4e = 6H_2O$, and in alkaline solution, $O_2 + 2H_2O + 4e = 4\ OH^{-}$. The reaction rate for these reactions is very low. A catalyst is needed for these reactions. To compensate for the low yield in electrons per second or to increase current density, the surface area of electrodes can be increased by the introduction of porosity [9]. But a plane electrode has two phases, whereas in a porous electrode, there are three phases, as shown in Figure 9.2 below. The thickness of the diffusion layer, or water meniscus, varies at the boundary interface, as shown in Figure 9.2. With a decrease in the thickness of the water, the meniscus concentration gradient increases with an increase in diffusion-limiting current. But at very low thicknesses of the meniscus (as shown in Figure 9.2), reduced products over cathode (which can be an oxidized product on the anode) cannot diffuse back to bulk solution reducing the limiting current at the very low thickness of the electrolyte layer in the extreme geometry shown in the diagram.

Types of Fuel Cells

1. *Alkaline fuel cell*

Alkaline fuel cell [10, 11] functions at relatively low temperatures, and no auxiliary power source is needed. Due to their high pH range, a variety of catalysts are stable in alkaline solutions [12]. However, air containing oxygen has carbon dioxide as an impurity, which forms carbonate and blocks the flow in the porous electrode [13].

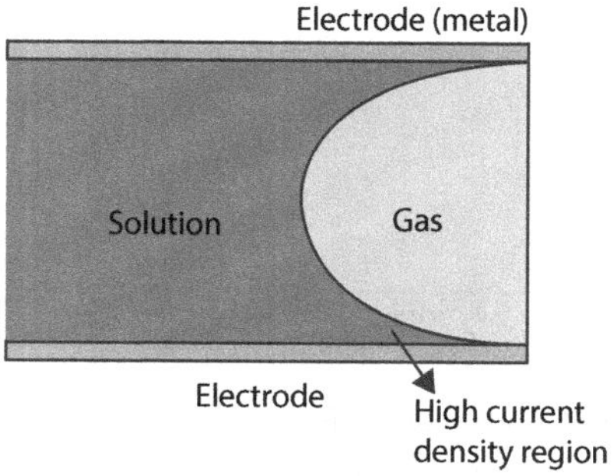

Figure 9.2 Three phases are shown for single pore. Also, at extremely low electrolyte thickness (low contact angle) giving rise to high current density.

$$\textbf{\textit{Anode reaction}}: \quad 4OH^- + 2H_2 = 4H_2O + 4e$$

$$\textbf{\textit{Cathode reaction}}: \quad 4e + 2H_2O + O_2 = 4OH^-$$

$$\textbf{\textit{Fuel cell reaction}}: \quad 2H_2 + O_2 = 2H_2O$$

2. *Phosphoric acid fuel cell*

A combination of 98% H_3PO_4 and 2% water electrolyte is common in phosphoric acid fuel cells [14]. Typical high-temperature phosphoric acid fuel cells are heated to greater than 200°C at atmospheric pressure. At this temperature (>150°C), phosphoric acid is polymerized to form pyrophosporic acid, which has a higher ionic conductivity than pure H_3PO_4 [15]. This cell functions at very high temperatures, approximately 200°C to 205°C. Thus, it does not need any high-pressure equipment. The high temperature in the system produces free steam. Reforming of hydrogen gas [16] is possible simultaneously. Thus, no external source of energy is required. The high temperature increases the exchange current density automatically. But extremely concentrated phosphoric acid can freeze below 40°C and thus needs auxiliary heating. The catalyst used in phosphoric acid fuel cells is exposed to high temperatures as well as high acid strengths. Thus, this precious metal (platinum) degrades easily, but as carbon monoxide increases at high temperatures to 2%, it is beneficial for blocking and reducing the degradation of platinum catalysts. Alloys of platinum transition metals [17] are better than pure platinum. Platinum supported on TiC [18] rather than C performs better as the high conductivity of TiC reduces IR drop through the support at the electrode interface.

$$\textbf{\textit{Anode reaction:}} \quad H_2 = 2H^+ + 2e$$

$$\textbf{\textit{Cathode reaction:}} \quad 2e + 2H^+ + \frac{1}{2}O_2 = H_2O$$

$$\textbf{\textit{Fuel cell reaction}}: \quad H_2 + \frac{1}{2}O_2 = H_2O$$

3. *High-temperature fuel cell*
It functions at more than 650°C and thus uses a solid electrolyte for a solid oxide electrolyte (yttria–zirconia). At higher temperatures, oxygen mobility and conductance in the solid oxide electrolyte (yttria-zirconia) are high. At very high temperatures, exchange current increases greatly, but at high temperatures, corrosion and loss of activity of electrodes and other structural materials are inevitable. Molten carbonate fuel cells working at 600°C to 700°C are a great example.

Coal gasification: $C + H_2O = CO + H_2$

$$CO_3^{2-} + H_2 = CO_2 + H_2O + 2e$$

$$CO + CO_3^{2-} = 2CO_2 + 2e$$

Overall anode reaction : $H_2 + CO + 2CO_3^{2-} = 3CO_2 + H_2O + 4e$

Cathode reaction: $4e + 2CO_2 + O_2 = 2CO_3^{2-}$

Fuel cell reaction : $H_2 + CO + O_2 = CO_2 + H_2O$

Overall power plant reaction: $C + O_2 = CO_2$

4. *Solid polymer electrolyte fuel cell*
Here, a humidified membrane of perfluorosulfonic acid polymer is used as an electrolyte. This polymeric membrane material conducts protons.

Catalyst in the Fuel Cell

Catalysts play an important role in fuel cells. At high temperatures, the proton exchange membrane fuel cell has overall electrode kinetics that increase with an increase in exchange current density. Water that forms in a fuel cell can be split back into hydrogen. The latter step is facilitated at high temperatures. At high temperatures, the tolerance of the electrocatalyst to poisonous gases is increased. On the other hand, at high temperatures, the corrosion of electrodes is much faster. Not only that, but degradation of the catalyst and proton exchange membrane can also occur. A large number of phosphoric acid and phosphoric acid-dissociated ions block the active sites of the platinum catalyst. In a fuel cell, an anodic reaction or hydrogen oxidation reaction is much faster than a cathodic reaction

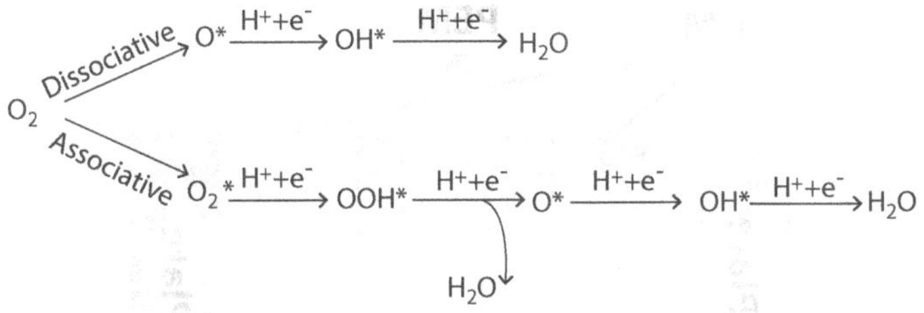

Figure 9.3 Scheme showing 2e and 4e transfer mechanism.

or oxygen reduction reaction. Thus, developing a better oxygen reduction reaction catalyst is a major challenge. Pt is at the top of the volcano graph [19], which acts as the best catalyst in this particular reaction. On top of the catalyst, a third layer may be deposited. This last layer having a zeolite effect, that is, a porous structure that allows size-selective reactant migration through it, might help exclude phosphate ions. Phosphate ions block active catalyst sites [20, 21]. Small organic molecules, like melamine [22], cyanide ions, and sulfhydryl groups, modified on catalyst surfaces can prevent poisoning effects. But at high temperatures, dissociation of these small molecules from the catalytic surface is possible. It is worth mentioning that there are two types of reaction paths during the oxygen reduction reaction. The first is a dissociative path that consists of two electron transfers, and the other is an associative path that consists of four electron transfers [23]. Figure 9.3 shows dissociative path as well as associative path in a scheme. Metals that bind too strongly with oxygen form oxygen radicals, followed by a dissociative path, whereas metals that bind weekly with oxygen undergo an associative path. The d-band of platinum plays an important role in bond formation. If d orbital energy is moved downward with respect to the Fermi level, the overall strength of the bond between the intermediate and platinum is reduced a lot. This might be due to highly unstable anti-bonding electrons. A single-atom catalyst is a new concept that can design energy levels of different d orbitals and Fermi levels to produce bonds with more characteristics of covalent nature compared to ionic nature, subsequently increasing their strength. Tanaka et al. proposed that phosphate anions bind with three oxygen atoms on the platinum 111 [20] surface but with one or two oxygen atoms on the platinum 100 and 110 surfaces [21].

Membrane Function

In PEMFC, there are several layers on top of the electrode, namely, the diffusion medium, then the catalytic layer, followed by the polymer electrolyte

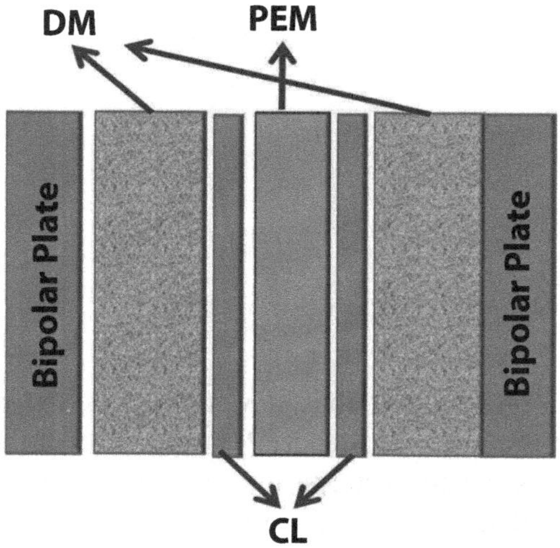

Figure 9.4 Schematic representation of proton-exchange membrane fuel cell (PEMFC). CL, catalyst layer; DM, diffusion medium, PEM, polymer electrolyte membrane.

membrane. Figure 9.4 shows schematic representation of PEMFC. An electrode is typically made up of porous carbon medium. The catalyst layer consists of nanosized catalysts, like platinum, deposited on top of larger carbon particles. Polymer electrolyte membranes can selectively diffuse H^+, but electrons can only flow through the outer circuit. After diffusion, H^+ reaches the cathode section to bind with reduced oxygen to form water vapor. The water vapor thus generated needs to be partially removed. Fluorinated polymers and hydrocarbon membranes require a minute electrolyte medium to conduct proton, where excess water can block available pathways in the diffusion medium as well as the catalyst layer. Apart from that, excess water can condense to create internal pressure and damage the catalytic layer [24]. Condensed water can carry ionic impurities and degrade the quality of the electrolyte. Residual water is in the diffusion medium and catalytic layer, while the fuel cell is at rest. The polymer electrolyte membrane is mainly made up of Nafion. It has sulfonic acid groups that help in the conduction of ions through the hydrophilic environment and hydrophobic backbone. Proton migration inside the polymer membrane follows two mechanisms, namely, the vehicular or diffusion mechanism (proton moves inside water channels inside the polymer) and the Grotthuss proton hopping mechanism (proton jumps from one water drop to the other inside the polymer matrix). Apart from that, there are several water fluxes possible, such as the following: 1. Electro-osmotic drag is when flow occurs due to a potential difference. Proton as well as associated water migrates from anode to cathode under electrostatic attraction force. 2. Diffusion is concentration-driven flow. It follows Fick's simple law of diffusion. As water is generated in the cathode compartment, diffusion takes place in a direction from the cathode to the anode. As discussed earlier, protons migrate from anode to cathode through membrane diffusion and end up reacting with O_2 to produce water in the cathode compartment. During this migration of proton (weakly bound with a water molecule), water also migrates to the cathode leaving a dry anode. On the other hand, cathodes get excess water due to both direct water formation and electro-osmotic drag force. Thus, water can diffuse back from the excess water region or cathode region to the anode region to maintain equilibrium. 3. Hydraulic permeability refers to the diffusion of water from areas of high pressure to low pressure. This occurs due to the capillary effect. The difference in pressure between H_2 and O_2 also creates a pressure difference, but in practice, it is negligible. 4. Thermoosmosis or temperature-driven flow [25] is a kind of diffusion that is difficult to study and depends on the membrane type and its conditioning. It can vary, especially during the start-up and shutdown of the fuel cell. These four effects give rise to net water flux.

Due to the Joule–Thomson effect, hydrogen is highly flammable, necessitating strict safety measures throughout the transportation process. Transporting hydrogen poses several challenges. First, hydrogen molecules are small and can leak easily through conventional materials requiring specialized storage and transport infrastructure to prevent losses.

Several advanced materials have been used for the safe transportation of hydrogen gas such as the following: 1. Metal Hydrides: These materials can absorb and release hydrogen through reversible chemical reactions. Examples include complex metal hydrides like sodium alanate and magnesium hydride. 2. Carbon-Based Materials: Carbon nanotubes, graphene, and activated carbon have high surface areas and can adsorb hydrogen molecules onto their surfaces. 3. Metal–Organic Frameworks (MOFs): MOFs are porous materials made of metal ions or clusters connected by organic linkers. Their high porosity and tunable structures make them promising for hydrogen storage. 4. Nanostructured Materials: Nanoparticles of metals, like palladium or platinum, can absorb hydrogen due to their high surface area and catalytic properties.

Conclusion

This chapter describes lucidly the importance of hydrogen fuel cells, their uniqueness, and the production of hydrogen by simple water splitting reactions. The kinetic and thermodynamic factors of fuel cell reactions are discussed, with an emphasis on the oxygen reduction reaction, which has an important role in the efficient performance of fuel cells. Different types of proton exchange membranes and electrodes, with an emphasis on porous electrodes, have been discussed. Finally, the problems of transporting hydrogen have also been included.

AI disclosure statement: The author acknowledges the use of AI-based tools for minor language editing or grammatical correction. However, all intellectual contributions, analyses, and conclusions presented in this chapter are entirely the author's own.

References

1. Abdelkareem, M.A., Elsaid, K., Wilberforce, T., Kamil, M., Sayed, E.T., Olabi, A., Environmental aspects of fuel cells: A review. *Sci. Total Environ.*, 752, 141803, 2021.
2. Nicoletti, G., Arcuri, N., Nicoletti, G., Bruno, R., A technical and environmental comparison between hydrogen and some fossil fuels. *Energy Convers. Manage.*, 89, 205–213, 2015.
3. Lutz, A.E., Larson, R.S., Keller, J.O., Thermodynamic comparison of fuel cells to the Carnot cycle. *Int. J. Hydrogen Energy*, 27, 1103–1111, 2002.
4. Wang, W., Wei, X., Choi, D., Lu, X., Yang, G., Sun, C., Chapter 1 - Electrochemical cells for medium- and large-scale energy storage: fundamentals, in: *Advances in Batteries for Medium and Large-Scale Energy Storage*, C. Menictas, M. Skyllas-Kazacos, T.M. Lim (Eds.), pp. 3–28, Woodhead Publishing, Waltham, Massachusetts, 2015.
5. Xu, H. and Ni, M., Chapter 4 - Numerical simulation of hybrid systems based on solid oxide fuel cells, in: *Hybrid Energy System Models*, A. Berrada and R. El Mrabet (Eds.), pp. 91–127, Academic Press, Netherlands, 2021.
6. Akroot, A., Ekici, O., Koksal, M., Process modeling of an automotive pem fuel cell system. *Int. J. Green Energy*, 16, 1–11, 2019.
7. Heddad, O., Ziet, L., Gana, C., Dana, W., Chettoueh, K., Dynamic Modeling and Optimal Control Strategy of Energy Storage Elements in Hybrid Electrical Vehicle "Fuel Cell and Ultracapacitor", in: *Artificial Intelligence in Renewable Energetic Systems*, M. Hatti (Ed.), pp. 366–374, Springer International Publishing, Cham, 2018.

8. Esan, O.C., Shi, X., Dai, Y., An, L., Zhao, T.S., Operation of liquid e-fuel cells using air as oxidant. *Appl. Energy*, 311, 118677, 2022.
9. Breiter, M.W., Properties of Porous Electrodes, in: *Electrochemical Processes in Fuel Cells*, M.W. Breiter (Ed.), pp. 238–253, Springer Berlin Heidelberg, Berlin, Heidelberg, 1969.
10. Ferriday, T.B. and Middleton, P.H., Alkaline fuel cell technology - A review. *Int. J. Hydrogen Energy*, 46, 18489–18510, 2021.
11. Hamada, A.T., Orhan, M.F., Kannan, A.M., Alkaline fuel cells: Status and prospects. *Energy Rep.*, 9, 6396–6418, 2023.
12. Firouzjaie, H.A. and Mustain, W.E., Catalytic Advantages, Challenges, and Priorities in Alkaline Membrane Fuel Cells. *ACS Catal.*, 10, 225–234, 2020.
13. Gülzow, E. and Schulze, M., Long-term operation of AFC electrodes with CO2 containing gases. *J. Power Sources*, 127, 243–251, 2004.
14. Sammes, N., Bove, R., Stahl, K., Phosphoric acid fuel cells: Fundamentals and applications. *Curr. Opin. Solid State Mater. Sci.*, 8, 372–378, 2004.
15. Krueger, R., Vilciauskas, L., Melchior, J.-P., Bester, G., Kreuer, K.-D., The Mechanism of Efficient Proton Conduction in Diphosphoric Acid Elucidated via First-Principles Simulation and NMR. *J. Phys. Chem. B*, 119, 15866–75, 2015.
16. Flavio, C., Christian Gonçalves, A., Tatiana Duque, M., Roberto Batista de, L., Antonio Carlos Chaves, R., Leandro de Lima, C., André Mychell Barbieux Silva, S., Monah Marques, M., José William Diniz, C., Geovany Albino de, S., Lucas Fernandes, A., Diericon de Sousa, C., Adão Marcos Ferreira, C., Thiago Soares Silva, R., Pedro Henrique Machado, G., Guilherme Botelho Meireles de, S., Production of Hydrogen and their Use in Proton Exchange Membrane Fuel Cells, in: *Advances In Hydrogen Generation Technologies*, E. Murat (Ed.), IntechOpen, Rijeka, 2018, Ch. 4.
17. Ramaswamy, N., Kumaraguru, S., Jarvis, K., Ferreira, P., Improving Durability of Fuel Cells with Platinum-rich Alloy Cathode Catalysts. *J. Electrochem. Soc.*, 170, 054504, 2023.
18. Saha, S., Rajbongshi, B.M., Ramani, V., Verma, A., Titanium carbide: An emerging electrocatalyst for fuel cell and electrolyser. *Int. J. Hydrogen Energy*, 46, 12801–12821, 2021.
19. Nørskov, J.K., Rossmeisl, J., Logadottir, A., Lindqvist, L., Kitchin, J.R., Bligaard, T., Jónsson, H., Origin of the Overpotential for Oxygen Reduction at a Fuel-Cell Cathode. *J. Phys. Chem. B*, 108, 17886–17892, 2004.
20. He, Q., Yang, X., Chen, W., Mukerjee, S., Koel, B., Chen, S., Influence of phosphate anion adsorption on the kinetics of oxygen electroreduction on low index Pt(hkl) single crystals. *PCCP*, 12, 12544–12555, 2010.
21. Tanaka, A., Adzic, R., B, N., Oxygen reduction on single crystal platinum electrodes in phosphoric acid solutions. *J. Serb. Chem. Soc.*, 64, 695–705, 1999.

22. Chung, Y.-H., Kim, S.J., Chung, D.Y., Park, H.Y., Sung, Y.-E., Yoo, S.J., Jang, J.H., Third-body effects of native surfactants on Pt nanoparticle electrocatalysts in proton exchange fuel cells. *Chem. Commun.*, 51, 2968–2971, 2015.
23. Tao, J., Wang, X., Xu, M., Liu, C., Ge, J., Xing, W., Non-noble metals as activity sites for ORR catalysts in proton exchange membrane fuel cells (PEMFCs). *Ind. Chem. Mater.*, 1, 388–409, 2023.
24. Kumbur, E.C., Sharp, K.V., Mench, M.M., Liquid droplet behavior and instability in a polymer electrolyte fuel cell flow channel. *J. Power Sources*, 161, 333–345, 2006.
25. Zaffou, R., Yi, J., Kunz, H., Fenton, J., Temperature-Driven Water Transport Through Membrane Electrode Assembly of Proton Exchange Membrane Fuel Cells, Electrochemical and Solid State Letters. *Electrochem. Solid-State Lett.*, 9, A418–A422, 2006.

10

Catalyst for Anodic Oxygen Evolution Reaction

Soner Çakar[1,2] and Mahmut Özacar[2,3*]

[1]*Department of Chemistry, Faculty of Science, Zonguldak Bulent Ecevit University, Zonguldak, Türkiye*

[2]*Biomaterials, Energy, Photocatalysis, Enzyme Technology, Nano & Advanced Materials, Additive Manufacturing, Environmental Applications and Sustainability Research & Development Group (BIOENAMS R & D Group), Sakarya University, Sakarya, Türkiye*

[3]*Department of Chemistry, Faculty of Science, Sakarya University, Sakarya, Türkiye*

Abstract

Oxygen evolution reaction (OER) is a crucial reaction in various systems such as rechargeable metal–air batteries, regenerative fuel cells, and water splitting. Especially in recent years, OER has also gained importance due to the increasing interest in clean energy. Catalyst systems used in OER are generally limited in commercial use in these applications due to their high cost and low stability properties. Catalysts used in OER are examined in two classes: noble metal- and non-noble metal-essenced catalysts. In recent years, the OER activities and applications of carbon-essenced, perovskite, and metal organic framework (MOF)-derived structures have attracted attention. In this book chapter, the properties, challenges, and activities of the catalyst types used in OER and the features that need to be realized and considered in the future are investigated.

***Keywords*:** Oxygen evolution reaction, noble metal catalysts, non-noble metal catalysts, electrocatalysts

**Corresponding author*: mozacar@sakarya.edu.tr

Inamuddin, Tariq Altalhi, Mohammad Luqman and Jorddy Neves Cruz (eds.) Water Splitting: Production of Hydrogen, (245–276) © 2025 Scrivener Publishing LLC

10.1 Introduction

In electrocatalytic water splitting process, the overall reaction involves two half-cell reactions as follows: H_2 and O_2 evolution reactions (HER and OER). These reactions, carried out in both acidic and basic environments, are given in (Equations 10.1–10.5) [1]. While HER is carried out effectively in an acidic environment due to high proton content, the formation of OER is difficult (Equations 10.1, 10.2). OER is more dominant in the basic environment due to the high OH^- content (Equations 10.3, 10.4). The total water electrolysis reaction is given in Equation 10.5, and an energy of 1.23 V is required for this reaction to occur. Proton exchange membrane (PEM)-type water electrolyzer systems are designed to realize this reaction. OER, which is slower in an acidic environment and occurs with higher energy, plays a decisive role in systems such as PEM water electrolyzers. In the developed systems, the OER reaction takes place at the anode. The OER catalysts to be developed are expected to have high catalytic activity, low cost, and long time stability [2].

Acidic medium

$$2H^+_{(aq)} + 2e^- \rightarrow H_{2\,(g)} \; E_0 = +0.0 \text{ V vs SHE} \tag{10.1}$$

$$H_2O_{(l)} \rightarrow \frac{1}{2}O_{2\,(g)} + 2H^+_{(aq)} + 2\,e^- \; E_0 = +1.23 \text{ V vs SHE} \tag{10.2}$$

Basic medium

$$2H_2O_{(l)} + 2e^- \rightarrow H_{2\,(g)} + 2OH^- \; E_0 = -0.83 \text{ V vs SHE} \tag{10.3}$$

$$2OH^- \rightarrow \frac{1}{2}O_{2\,(g)} + H_2O_{(l)} + 2\,e^- \; E_0 = +0.40 \text{ V vs SHE} \tag{10.4}$$

Overall water splitting

$$H_2O_{(l)} \rightarrow H_{2\,(g)} + \frac{1}{2}O_{2\,(g)} \; E_0 = +1.23 \text{ V vs SHE} \tag{10.5}$$

Among OER catalysts, Ir- and Ru-essenced catalyst systems are preferred owing to their properties like high catalytic activities and stabilities. However, noble metal-essenced catalysts have disadvantages like high cost, difficulties in the synthesis stage, and limited applicability [3]. Therefore, it is necessary to evolve non-noble metal-essenced (Mn, Co, Ni, etc.) nanomaterials that can be an alternative to noble metal-essenced catalysts. This chapter examines and reviews a general evaluation of the catalysts used in OER and the latest studies. Finally, the future status and challenges of these catalysts are discussed.

10.2 Noble Metal Catalysts for OER

Noble metal catalysts are the most tremendous and fruitful catalysts for OER. Noble metal-essenced catalysts are divided into three main groups as follows: single-atom noble metal catalysts, noble metal oxide catalysts, and others. These catalysts are further detailed and assessed below (*vide infra*).

10.2.1 Noble Metal Single-Atom Catalysts for OER

Single-atom noble metal catalysts were first investigated in the 1960s and are still extensively studied. Noble metals, like Ir, Pd, Pt, Rh, Ru, and Ag, have tremendous activity as potential catalysts in OER. However, their low stability in acidic environments and high costs make their commercialization difficult. Ir and Ru are noteworthy among noble metals in terms of their activities. According to experimental results, the order of reactivity of single-atom noble metal-essenced catalysts is Ag < Pt < Rh < Pd < Ir < Ru. Intensive studies have been performed in the last years, especially on the stability of single-atom noble metal-essenced catalysts in the OER in an acidic medium. For this purpose, studies on controlling the shape and size of noble metals by using different developed synthesis methods are carried out. To both increase their stabilities and provide higher efficiencies, studies on the production of nanostructures and ultrathin layer nanostructures rather than bulk structures are being carried out intensively.

Reier *et al.* [4] synthesized Ru, Ir, and Pt catalysts as bulk and nanoparticles and comparatively examined the effects of these catalysts on electrocatalytic OER. The effect of electrochemical surface characteristics for these noble metals has been comprehensively explored for the first time in this study. When the electrochemical behavior of these catalysts was examined, it was determined that Ru nanoparticles had a lower passivation potential than bulk Ru material, Ir nanoparticles lost their voltammetric metallic feature compared to bulk Ir, and Pt nanoparticles are more efficient than bulk Pt. It was determined that the OER activity of catalysts in nanoparticle structures was higher than those in bulk structures, and Ru showed the highest activity followed by Ir and Pt. Additionally, studies on the stabilities of these noble metal catalysts have been carried out. In another study, Ravichandran *et al.* [5] examined the electrocatalytic OER activities of Pt–Ru–Ir triple metallic electrodes with different molar ratios. In the synthesis of these noble metal catalysts, they used silica template coded KIT-6 and reduced them with $NaBH_4$. They synthesized three different metallic triple catalyst nanostructures in the molar ratios of $Pt_{70}Ru_{25}Ir_5$, $Pt_{70}Ru_{20}Ir_{10}$, and $Pt_{70}Ru_{15}Ir_{15}$ and characterized them via XRD, TEM, and XPS. The lowest overpotential value was found in the sample coded $Pt_{70}Ru_{15}Ir_{15}$, and this sample also has the lowest Tafel slope value. In addition, high mass activity, specific capacity, and extended stability properties were observed for the catalyst with $Pt_{70}Ru_{15}Ir_{15}$ ratios. In this study, the effect of Ir and Ru doping on Pt noble metal was examined, and it was found that the added Ru and Ir particularly increased both stability and OER activity.

Another study on bimetallic nanoparticles [6] examined the synthesis and the effects on OER of Pt–M (M: Ni, Cr, Er) alloys. In this study, which used Zn–MOF-supported carbon structure and specifically targeted loft OER activity and stability in 1 M KOH alkaline solution, it was observed that the sample coded Pt–Ni@PCN920 maintained its stability and signal for 9 h at 0.5 V. It has been shown that the bimetallic strategy, in which metals abundant in the earth's crust are doped to noble metal catalysts, has an effect on catalyst stability, and it has also been shown that it will contribute to this stability in carbon structures containing N. Zhu *et al.* [7], in their study examining the OER activities of noble metal-essenced Pt–Ir–Pd trimetallic nanocages-essenced catalysts, developed a cage-like $Pt_{39}Ir_{10}Pd_{11}$ catalyst system by doping Ir and Pd onto Pt. Figure 10.1 depicts TEM and EDS mapping images of this developed trimetallic Pt–Ir–Pd nanocage structure. It is seen from Figure 10.1 that the nanocage trimetallic catalyst has a cubic-shaped and angular structure with a length of 12.3 nm and a wall thickness of 1 nm. The, $\eta10$ and Tafel slope values of this developed nanocage Pt–Ir–Pd trimetallic catalyst were consecutively found to be 372 mV and 128.7 mV dec^{-1}. Also, the stability of the Pt–Ir–Pd trimetallic catalyst was measured at 5,000 cycles under 1.53 V and compared with commercial Pt/C. Only a 10% OER activity loss was observed for this trimetallic catalyst developed at 5,000 cycles. This nanocage study led to the development a strategy for coating noble metals as a thin layer to increase the stability of Pt electrodes. These studies on Pt catalysts, which are expected to have the highest efficiency among the noble metals, not

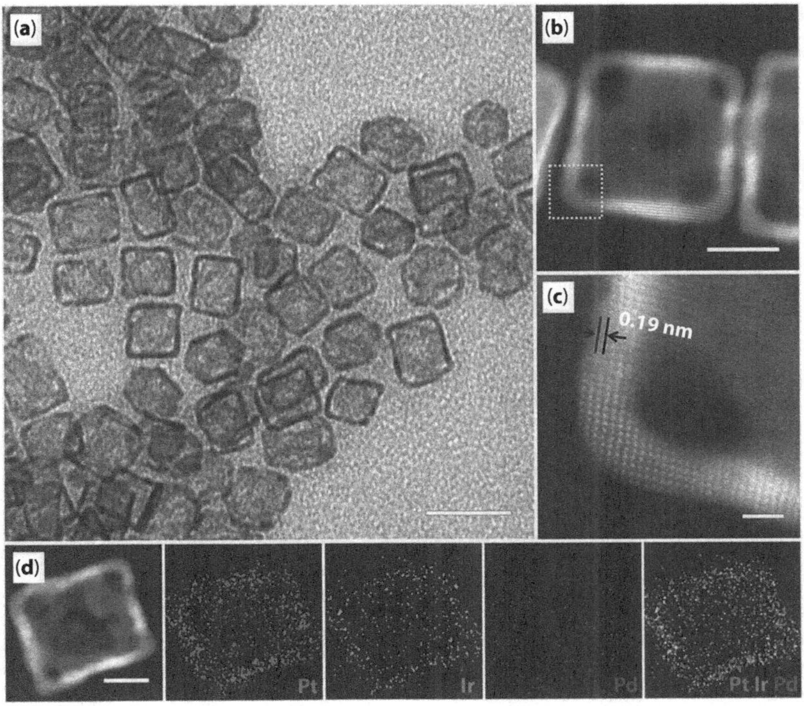

Figure 10.1 (a) TEM image of the Pt–Ir–Pd alloy cubic nanocages (scale bar: 20 nm). (b) HAADF-STEM image of an individual Pt–Ir–Pd nanocage (scale bar: 5 nm). (c) Atomic-resolution HAADF-STEM image obtained from the district marked in (b) (scale bar: 1 nm). (d) EDX mapping of Pt, Ir, and Pd for the Pt–Ir–Pd nanocage (scale bar: 5 nm). Reproduced with permission from ref. [7]. Copyright 2020 by Wiley.

only improved the low OER efficiency and low stability of the Pt catalyst but also accelerated the studies on Ir and Ru, which are more effective and efficient noble metals than the Pt catalyst.

In a study conducted on Ir noble metal [8], the coating of Ir–nanodendrite (Ir–ND) structures on antimony tin oxide (ATO) substrate and its effect on OER were examined. Additionally, the electrocatalytic performances of Ir–ND structures developed in this study were investigated in PEM electrolyzer cells. The $\eta 10$ and Tafel slope values of the developed Ir–ND structures were consequtively determined as 280 mV and 60 mV dec^{-1}. It has been stated that the developed Ir–ND structures have high chemical darkness and are suitable for use in PEM electrolyzer applications. In two different studies on Ir bimetallic catalysts, the OER efficiencies of the Ir–Cu bimetallic alloy structure in acidic [9] and alkaline [10] environments were examined. The $\eta 10$ and Tafel slope values of Ir–Cu bimetallic catalysts coated on glassy carbon electrodes (GCE) were consecutively 280 mV and

Table 10.1 Selected single-atom noble metal-essenced catalysts for OER.

Catalysts	Electrolyte	Electrode/ substrate	Overpotential (10 mA cm^{-2}) [mV vs. RHE]	Tafel slope [mV dec^{-1}]	Reference
Ru Ir Pt	0.1 M $HClO_4$	RDE	-	44 63 145	[4]
$Pt_{70}Ru_{15}Ir_{15}$	0.5 M H_2SO_4	RDE	450	114	[5]
Pt–Ni@ PCN920 Pt–Er@ PCN920 Pt–Cu@ PCN920	1 M KOH	Ni foam	-	59 86 105	[6]
Pt–Ir–Pd	0.1 M $HClO_4$	RDE	408	128.7	[7]
Ir–ND	0.05 M H_2SO_4	ATO	280	60	[8]
Ir–Cu	1 M KOH	GCE	340	38	[9]
$Cu_{1.11}Ir$	0.05 M H_2SO_4	GCE	286	43.8	[10]
$Ir_{44}Pd_{10}$	0.1 M $HClO_4$	RDE	226	53.9	[11]
IrRu	0.1 M $HClO_4$	GCE	194	47.5	[12]
Ir–Co	1 M $HClO_4$	PTE /Ti PTL	-	37	[13]
$NiFeIr_{0.03}$/Ni NW@NSs	1 M KOH	RDE	200	44.6	[14]
Ir NW	0.1 M $HClO_4$	GCE	295	47.0	[15]
Pt NW Ir NW Rh NW RuO_2 NW Ir/C	0.5 M H_2SO_4	GCE	530 260 410 225 300	259 57 106 52 77	[16]
IrMnOF@Ir	0.1 M $HClO_4$	Carbon paper	180	58.3	[17]

PCN920: N-doped porous carbon, ND: nanodentrite, NW: nanowire, NSs: nanosheets, RDE: rotating disc electrode, ATO: antimony tin oxide, GCE: glassy carbon electrode, PTE: porous transport electrodes, PTL: porous transport layer.

60 mV dec^{-1} in an acidic environment, while they were determined as 340 mV and 38 mV dec^{-1} in a basic environment. It is stated that Cu atoms added to the Ir noble metal show smaller Tafel slope values and provide enhanced activity and durability. Especially in Ir structures surrounded by metal abundant in the earth's crust, such as Cu, higher OER activity and stability were observed compared to pristine Ir. Similarly, Zhu *et al.* [11], in their study in which they developed the $Ir_{44}Pd_{10}$ cubic nanocage type catalyst system, determined the η10 and Tafel slope values of the 1.11-nm-thick ultrathin-walled porous structure catalyst as 226 mV and 53.9 mV dec^{-1}. With these values they obtained, they showed that the catalyst they developed had high OER efficiency and specific activity. It is observed that the catalyst developed by doping Pd noble metal instead of Cu metal exhibits more effective OER performance. Another study on bimetallic noble metals [12] examined the effect of ultrafine IrRu nanoparticles on OER. It was determined by TEM results that this nanostructure, which was prepared by converting HCOONa-encapsulated Ir and Ru nanoparticles into ultrafine IrRu catalysts with the effect of heat was approximately 1.6 nm in size. It has been reported that the developed IrRu nanoparticles showed more effective OER activity than commercial IrO_2 and RuO_2 and maintained their stability in the 10-h stability test. Additionally, it has been observed that lower overpotential values are obtained from IrPd, IrCu, and similar bimetallic catalyst systems. In recent studies, the use and properties of noble metal catalyst-essenced OERs on half-cell systems have been examined. As an example of this, recently, Jimenez *et al.* [13] obtained a very low Tafel slope value of 37 mV dec^{-1} of the Ir–Co catalyst with the porous transport electrodes (PTE) prepared using the PVD-essenced catalyst production technique and carried out stability measurements over 20 h. This innovative PTE-essenced half-cell has enabled the measurement of their stability at higher current densities. Some single-atom noble metal catalysts used in OER [4–17] are listed in Table 10.1.

10.2.2 Noble Metal Oxide Catalysts for OER

In terms of comparing Ir and Ru, which are effective and highly productive among noble metals, with IrO_2 and RuO_2, Cherevko *et al.* [18] stated in their study conducted in 2016 that although the OER activity efficiency trend was $IrO_2 < RuO_2 \approx Ir < Ru$, the solubility trend was Ru ≫ Ir > RuO2 ≫ IrO_2. Strategies, such as heteroatom doping, core-shell structures, and surface engineering, have been evolved specifically to boost the stability and OER efficiency of IrO_2 and RuO_2 catalysts. Some of the noble metal oxide-essenced catalysts for OER are shown in Table 10.2.

Table 10.2 Selected noble metal oxide-essenced catalysts for OER.

Catalysts	Electrolyte	Electrode/ substrate	Overpotential (10 mA cm^{-2}) [mV vs RHE]	Tafel slope [mV dec^{-1}]	Reference
RuO_2 IrO_2	0.5 M H_2SO_4	GCE	- -	40 60	[19]
TiN/IrO_2	0.5 M H_2SO_4	GCE	313	65.5	[20]
IrO_2–TiO_2-245	0.1 M $HClO_4$	RDE	-	42	[21]
IrO_2/SnO_2	0.5 M H_2SO_4	PTFE sheets	-	79	[22]
IrO_x NRs/ Sb–SnO_2	0.1 M $HClO_4$	RDE	240	42	[23]
$Ru_{0.9}Ir_{0.1}O_2$	0.5 M H_2SO_4	Gold Slab	-	48	[24]
RuO_2@IrO_2	0.5 M H_2SO_4	Ti electrode	-	86	[25]
RuO_2@IrO_2-20	0.5 M H_2SO_4	RDE	275	62	[26]
IrO_2–WO_3-75	0.5 M H_2SO_4	GCE	105.3	104	[27]
40 wt % IrO_2/ Ti-doped SnO_2	0.5 M H_2SO_4	GCE	264	49	[28]
Ru/RuO_2 clusters	0.5 M H_2SO_4	GCE	219	44.9	[29]
Ru@RuO_2 core-shell	0.5 M H_2SO_4	RDE	320	86	[30]
$Li_{0.52}RuO_2$	0.5 M H_2SO_4	GCE	156	83.3	[31]
Ni–Co codoped RuO_2	0.1 M $HClO_4$	Ti plate	-	32	[32]
$Sn_{0.1}RuO_2$@N-doped carbon	0.5 M H_2SO_4	Carbon paper	178	60.5	[33]
RuO_2/MoO_3	0.1 M $HClO_4$	Ti plate	260	49	[34]
Ce–RuO_2	0.5 M H_2SO_4	GCE	191	59.1	[35]

(*Continued*)

Table 10.2 Selected noble metal oxide-essenced catalysts for OER. (*Continued*)

Catalysts	**Electrolyte**	**Electrode/ substrate**	**Overpotential (10 mA cm^{-2}) [mV vs RHE]**	**Tafel slope [mV dec^{-1}]**	**Reference**
Mn–RuO_2 (300)	0.5 M H_2SO_4	Pt mesh	217	54.46	[36]
$Ru_{0.95}Co_{0.05}O_y$	1.0 M KOH	GCE	217	50.83	[37]
$Mo_{0.15}$–RuO_2	0.5 M H_2SO_4	RDE	147	38	[38]

NRs: nanorods, GCE: glassy carbon electrode, RDE: rotating disc electrode, PTFE: polytetrafluoroethylene.

In the first study examining the OER efficiency of RuO_2 and IrO_2, Mattos Costa *et al.* [19] stated that they showed high catalytic activity in an acidic environment. This study also synthesized different ratios of $Ru_xIr_{1-x}O_2$ catalysts via the sol–gel method and examined their OER efficiency. At the same time, the effectiveness of catalysts developed with accelerated life tests was also examined. The metal doping strategy has been extensively utilized to increase the active areas and OER efficiencies of catalysts, like IrO_2 and RuO_2, and to obtain stable structures. The most commonly used additive metals for this purpose are Ni, Sn, Mo, Co, Cu, W, Ti, Mn, Fe, Ce, etc.

Zhang and coworkers [20] investigated using TiN as a support material for IrO_2 nanoparticles in electrochemical water oxidation to address the sluggish OER in an acidic environment. In the study, a colloidal method was used to prepare a TiN-supported IrO_2 composite catalyst, and it was found that ultrafine IrO_2 nanoclusters scattered on the TiN support provided wider active surface area and revealed more unsaturated surface Ir atoms leading to loftier OER activity. The TiN/IrO_2 catalyst system prepared by loading 31 wt% IrO_2 exhibited a mass-normalized OER activity of 874.0 A g^{-1} (IrO_2), approximately 5.0-fold the unsupported IrO_2 activity at 1.6 V potential. The paper highlights the potential benefits of using TiN as a carrier for IrO_2 catalyst in reducing precious metal usage and improving OER activity. The study also discusses the electrochemical stability and prospects for practical application.

Oakton and coworkers [21] examined the wide surface area, high activity, and stable IrO_2–TiO_2 catalyst. In this study, the high catalytic activity of iridium hydroxo (*OH) species was examined using the operando XAS technique. In 2012, Xu *et al.* [22] used the IrO_2/SnO_2 electrocatalyst for the solid polymer electrolyte (SPE) water electrolysis system and examined the OER effect. It is stated that IrO_2/SnO_2 with high activity value not only improves efficiency but also effectively separates adsorbed hydroxy specimens and creates more active reaction sites of IrO_2. After this study on SnO_2 doping, Shi *et al.* [23] very recently examined the OER activities of Sb-doped tin oxide/IrO_x nanorods. In this study, it is stated that the nanorod structure from the IrO_x/Sb–SnO_2 catalysts prepared in the form of nanoparticles and nanorods is more effective and efficient, the doped Sn and Sb structures were examined with XPS and operando XAS techniques, and doping reduces the oxidation state of Ir and lowers the energy barrier.

In studies [24–26], where different ratios of RuO_2@IrO_2 catalysts were used, the OER efficiency was examined on different substrates. Audicton *et al.* [24] synthesized pure and crystalline ruthenium iridium oxide catalysts with a 10.5-nm crystallite size and a 53.3-m^2 g^{-1} surface area. For $Ru_{0.9}Ir_{0.1}O_2$, the Tafel slope and normalized current values were consecutively obtained as 48 mV dec^{-1} and 1.57 A C^{-1}. They state that the $Ru_{0.9}Ir_{0.1}O_2$ catalyst is more stable in 1,000-h stability measurements than the pure RuO_2 catalyst due to its low Ir contribution. Ma *et al.* [25] and Li *et al.* [26] examined the OER efficiency of core-shell RuO_2@IrO_2 structures and found that *O phase by oxidative desorption is a bit difficult on IrO_2, but they state that although the *O phase is formed more easily in RuO_2, its reactivity decreases due to its stability. Therefore, they report that RuO_2 doped with IrO_2 shows a more effective OER activity, even if it is doped at a low ratio.

There are very recent studies on doping WO_3 [27] and Ti–SnO_2 [28] to increase the OER efficiency of IrO_2. Uzgören *et al.* [27] stated that recycling WC–Co wire drawing dies, and doping them to IrO_2 via electrodeposition technique not only increased catalytic activity but also reduced costs. Similarly, in the study conducted by Li *et al.* [28] to reduce the costs of IrO_2 and increase their catalytic activity, it was stated that Ti-doped SnO_2 caused an increase in conductivity and stability.

In a study in which Ru/RuO_2 cluster structures were created via laser ablation technique to increase the reactive properties of RuO_2 catalysts [29], Wang *et al.* examined OER activities in alkaline and acidic environments. They reported that the electrochemically active surface areas and conductivity features of the developed Ru/RuO_2 clusters structures increased, and the overpotential and Tafel slope values of these clusters at 0.5 M H_2SO_4 were consecutively 219 mV (at 10 mA cm^{-2}) and 44.9 mV dec^{-1}. Figure 10.2 shows the RuO_2 cluster structures developed in this study, the surface of which is decorated with Ru/RuO_2. In a study on Ru@RuO_2 core-shell structures [30], Jiang *et al.* reported that the Ru@RuO_2 nanorods' catalytic activities in the OER were 6.5-fold loftier than the IrO_2 catalyst.

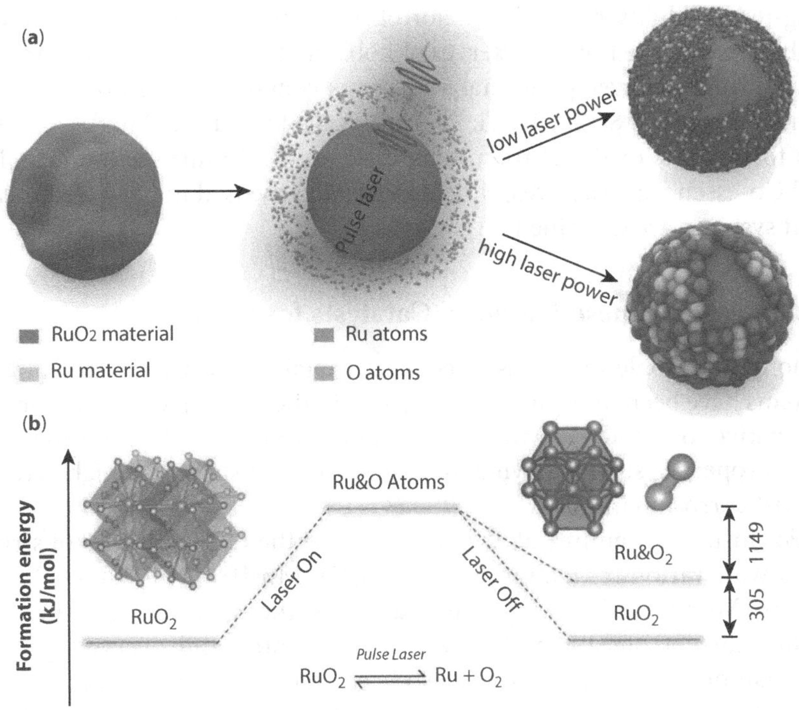

Figure 10.2 Image of the generation of lychee-shaped RuO_2@Ru/RuO_2 structure. (a) Schematic demonstration of the *in situ* surface decoration. Ru/RuO_2 clusters with 5-nm dimensions are constituted at the RuO_2 particle surface following laser ablation. (b) Standard enthalpy variation during the formation of small Ru/RuO_2 spheres by the decomposition of RuO_2. Reproduced with permission from ref. [29]. Copyright 2021 by Elsevier.

Many studies have been performed on doping different metal structures to increase the catalytic OER activity of RuO_2 and improve its stability. For this purpose, the doping of metals, such as Li [31], Ni–Co [32], Sn [33], Mo [34], Ce [35], Mn [36], Co [37], and Mo [38], to RuO_2 has been carried out. In these metal-doping studies, it was seen that the electrochemical surface areas and stability of RuO_2-essenced catalysts were boosted and their OER activities improved.

10.3 Non-Noble Metal Catalysts for OER

Although noble metal-essenced catalysts, such as Ir and Ru, are widely used for OER, their inability to maintain their electrocatalytic activity and long-term stability remains a major obstacle. In addition, because of the high costs and the limited reserves of noble metals worldwide, researchers have turned to developing catalyst systems containing non-noble metals. Elements, such as Si, Sn, Te, Tl, Pb, Bi, Mn, Ti, Fe, Cu, Co, Mo, W, and Ag, can form stable oxides at room temperatures [2]. In this section, Mn, Ni, and Co-essenced structures, which are frequently used in OER electrocatalyst systems, are examined.

10.3.1 Manganese-Essenced Catalysts for OER

Among non-noble metal-essenced electrocatalysts, Mn-essenced catalyst systems have been intensively investigated in the last years as an important alternative to Ir and Ru. Mn has an important place in OER research due to its properties such as having different oxidation states and high activity against corrosion in water.

For instance, Ramirez *et al.* [39] examined the OER activities of structures with various oxidation levels (Mn(II), Mn(III), Mn(IV), etc.), like MnOx, Mn_2O_3, and Mn_3O_4. It is stated that the α-Mn_2O_3 structure has higher current density, and lower onset potential and overvoltage values than other structures. It has also been stated that the OER activity of α-Mn_2O_3 is similar to that of the NiO_x structure. In a similar study, Kölbach *et al.* [40] stated that the α-Mn_2O_3 coating on the fluorine-doped tin oxide (FTO) substrate, which they realized using the electrodeposition technique, showed a high electrochemical active area and good electrical conductivity. At the same time, this developed α-Mn_2O_3 catalyst system was observed to be stable in a basic environment below 2 h. Guo *et al.* [41], in 2014, synthesized Mn_3O_4 nano-octahedrons structures with ethylenediamine-assisted hydrothermal method and characterized their crystal and morphological properties. They stated that the synthesis mechanism of Mn_3O_4 nano-octahedrons structures occurs as "dissolution–recrystallization" and "capping-molecule-assisted growth." It has been reported that the obtained Mn_3O_4 nano-octahedron structures with a 50-nm size have higher current density and more stable properties. In another study examining the effect of morphological properties of synthesized Mn-essenced catalysts on OER efficiency, Gupta *et al.* [42] synthesized catalysts using the hydrothermal method and investigated the Mn^{3+}/Mn^{4+} oxidation states on the surface using XRD and XPS techniques. It is reported that the MnO_2 nanowire structure shows better OER activity due to its features like higher mass activity values and higher surface area. In the study [43] where the OER efficiency of polymorphs in MnO_2 structures was examined, the ORR and OER activities of low-cost MnO_2 electrocatalysts were examined. OER efficiency and stability measurements of α-MnO_2 and β-MnO_2 nanorod-structured electrocatalysts were carried out in H_2 and air atmosphere. It has been stated that these developed low-cost electrocatalysts have particularly high OER efficiency and stability and are also more effective and stable than the mercantile Pt/C electrode.

Because of the low corrosion properties of oxide structures, synthesis, characterization, and OER studies are carried out on phosphide structures that can provide an alternative to oxide structures. Kaneti *et al.* [44] synthesized manganese–cobalt phosphide York-shell sphere structures and examined their OER activities. It is stated that in these developed binary metal phosphide structures, active areas increase thanks to Mn^{3+}/Mn^{4+} and Co^{2+}/Co^{3+} redox conversion, and electron transfer by oxygen adsorption/desorption increases. In another study on binary systems compatible with the manganese element, Browne *et al.* [45] stated that the Mn_3O_4 structure, especially added to RuO_2, showed higher OER activity and lower overpotential value than IrO_2 and RuO_2. This study also explains OERs and mechanisms in RuO_2 and Mn_3O_4 aqueous solutions in detail. In another study, Sumboja *et al.* [46] stated that NiMn-layered double hydroxide (LDH) has the characteristics of an extremely porous structure, wide surface area, lofty degree of crystallinity, broad interlayer spacing, a high OER efficiency, and a stable structure.

Metal doping strategy to Mn-essenced electrocatalysts has recently been studied to increase their stability and electroactivity values. Mosa *et al.* [47] examined the OER performances of thermally stable mesoporous cesium-doped manganese oxide (Cs–MnO_x) structures. In this study, the active oxidation sites of Cs-doped Mn species were examined in detail using density function theory. It is stated that by adding very low amounts of Cs, the active oxidation sites increase, and the OER efficiency increases. Frydendal *et al.* [48], in the study examining the OER activities of Au–MnO_x structures, stated that Au doping dramatically increases the current density and also has a high oxidation state. Sun *et al.* [49] studied the OER activity of the $Ru_{0.75}Mn_{0.25}O_2$ Ru–Mn binary oxide structure. It is stated that the Ru–Mn oxide structure synthesized in this study has a high electrochemically active surface area value in porous structure, and no current decrease was observed in 30-h stability measurements. In a similar study, Kakati *et al.* [50] examined the change in OER efficiency depending on Ir doping to β-MnO_2 and δ-MnO_2 polymorphs in Ir–MnO_2 electrocatalysts. They state that the active alanine increases with Ir–O–Ir and Ir–O–Mn binding and, therefore, increases OER performance. Zand *et al.* [51] investigated the OER activities of SiO_2-doped MnO_2 electrocatalysts. It is reported that Si–O covalent bonding increases electrochemical activity and, therefore, increases the amount and efficiency of O_2 catalyzed on the surface. Luo *et al.* [52] examined the OER activities in alkaline media of synthesized Mn_3O_4@$CoMn_2O_4$–Co_xO_y electrocatalysts using a cation-exchange-essenced synthesis process. With this innovative material,

Table10.3 Selected Mn-essenced catalysts for OER.

Catalysts	**Electrolyte**	**Electrode/ substrate**	**Overpotential (10 mA cm⁻²) [mV vs RHE]**	**Tafel slope [mV dec⁻¹]**	**Reference**
MnO_x Mn_2O_3 Mn_3O_4	0.1 M KOH	FTO	230 170 290	- - -	[39]
α-Mn_2O_3	1.0 M KOH	FTO	340	300	[40]
Mn_3O_4	0.1 M KOH	GCE	-	71.5	[41]
MnO_2 NW MnO_2 NT MnO_2 NRs	0.1 M KOH	Gold Electrode	- - -	28 36 39	[42]
α-MnO_2 β-MnO_2	0.1 M KOH	GCE	508 531	213.5 196.3	[43]
Mn–Co phosphide	1.0 M KOH	GCE	330	59	[44]
Mn/Ru oxide	1.0 M NaOH	Ti electrode	259	48	[45]
NiMn LDH	1.0 M KOH	RDE	350	40	[46]
Cs-MnO_x	0.1 M KOH	RDE	420	-	[47]
Au-MnO_x	1.0 M KOH	GCE	400	-	[48]
$Ru_{0.75}Mn_{0.25}O_2$	1.0 M KOH	GCE	229	52.1	[49]
Ir/β-MnO_2	0.5 M H_2SO_4	Carbon paper	319	123	[50]
SiO_2–MnO_2	0.1 M H_2SO_4	FTO	640	270	[51]
Mn_3O_4@ $CoMn_2O_4$– Co_xO_y	0.1 M KOH	GCE	310	52	[52]

NW: nanowire, NT: nanotube, NRs: nanorods, LDH: layered double hydroxide, FTO: fluorine-doped tin oxide, GCE: glassy carbon electrode.

a much more effective and efficient OER activity has been observed than Pt, Ir-essenced commercial electrocatalysts. Some of the Mn-essenced catalysts for OER are given in Table 10.3.

10.3.2 Nickel-Essenced Catalysts for OER

Like Mn-essenced electrocatalysts, Ni-essenced electrocatalysts have recently been widely used as alternative systems to Ru and Ir. In this context, NiO, NiO, NiOOH, $Ni(OH)_2$, NiS, NiSe, NiP, NiM_xO_y, etc. structures are the most studied Ni-essenced catalysts.

Luan *et al.* [53], in the study examining the OER activities of various morphological nanostructures of 2D α-$Ni(OH)_2$, such as ($Ni(OH)_2$-NB (nano bud), $Ni(OH)_2$-NF (nanoflower) $Ni(OH)_2$-NP (nanopetal), $Ni(OH)_2$-NS (nanosheet), stated that the $Ni(OH)_2$-NS catalyst has a higher active boundary size resulting from its nanolayered structure and, therefore, has a more effective OER efficiency. It is also noted that it shows higher OER activity than commercial IrO_2, and 96% of the inceptive current density is preserved after 10 h. In another study, Wang *et al.* [54] examined the OER activity of Ni–Co oxide hierarchical nanosheet (NCO–HNSs) catalysts. They state that the developed nanosheet layers are rich in Ni^{3+} on the surface and have an extensive surface area. It is also stated that the current density of this developed catalyst system was preserved in 18-h stability measurements.

In studies on the OER activities of NiFe-layered double hydroxide (LDHs) nanosheet catalysts, Wang *et al.* [55] state that the catalyst containing Ni vacancies shows high OER activity. They state that Ni vacancy structures have a high adsorption capacity for OER intermediates. This study indicates that forming vacancies, especially in nanolayered structures, increases OER activity. To tide over the disadvantages of NiFe LDHs nanosheets, like faint conductibility behavior and inability to control active sites, Bai *et al.* [56] doped Sn to these catalysts. It is reported that with Sn doping, active sites are formed in NiFe LDH structures, and conductivity increases. Similarly, Wang *et al.* [57] perused the OER activity of cobalt-doped Ni–Mn LDH nanoplates. In this developed catalyst system, an effective OER efficiency is achieved by creating mixed valence regions of Ni^{2+}/Co^{2+}–Co^{3+}/Mn^{3+}. It is also stated that this developed catalyst system showed high stability in the 9-h stability measurement. In the study [58] on doping-reduced graphene oxide (rGO) to evolve the low conductivity behavior of NiCo–LDH structures, it was stated that a lower onset potential value was observed than commercial IrO_2, and its OER activity was higher.

Zhu *et al.* [59] reported that Ni and Mo binary oxide ($NiMoO_4$) doping affects the surface design, boosts the number of active sites by creating double cation deficiencies, and accordingly improves the OER activity. In the study on the OER activity of binary metal oxide structures by Dinesh *et al.* [60], overpotential and Tafel slope values were obtained consecutively to be 256 mV and 114 mV dec^{-1}, and 20-h stability was observed. In their study on the metal oxide/binary metal oxide system, which can be an alternative to the binary metal oxide strategy, Landon *et al.* [61] carried out advanced characterizations of the $NiO/NiFe_2O_4$ catalyst such as Raman, X-ray absorption near-edge structure (XANES), XAS, and temperature-programmed reaction spectroscopy (CH_3OH-TPRS). In this study, especially the effectiveness and efficiency of Fe in the catalyst system are discussed.

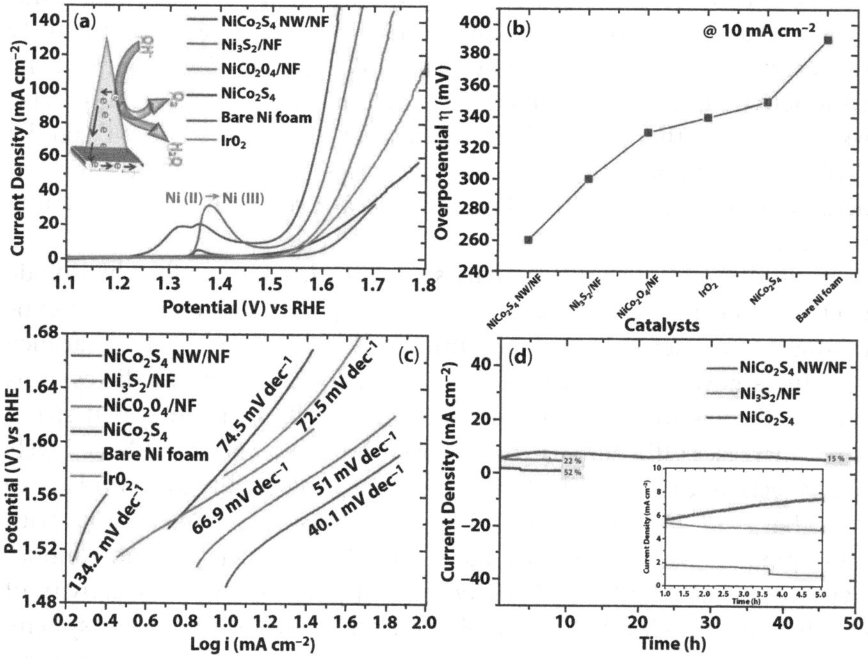

Figure 10.3 (a) OER polarization curves (iR corrected) of $NiCo_2S4$ NW/NF, Ni_3S_2/NF, $NiCo_2O_4$/NF, $NiCo_2S_4$, uncovered Ni foam, and IrO_2 with a scan rate of 10 mV s^{-1}. (b) The necessary overpotential to reach a current density of 10 mA cm^{-2} for distinc electrocatalysts. (c) Corresponding OER Tafel plots. (d) Time-dependent current density curves of $NiCo_2S_4$ NW/NF, Ni_3S_2/NF, and $NiCo_2S_4$ at a static potential of 1.527 V vs. RHE in 1 M KOH, for 50 h. Reproduced with permission from ref. [62]. Copyright 2016 by Wiley.

Since catalysts containing metal oxide structures show poor properties in stability measurements, researchers have turned to the development of alternative metal sulfide and metal selenide, etc., structures to metal oxide structures. Sivanantham *et al.* [62] developed hierarchical $NiCo_2S_4$ nanowire catalyst systems. These developed sulfur catalysts' low overpotential values and high stability attract attention. The OER results obtained are shown in Figure 10.3. In a study on the Ni_3Se_2 type catalyst, which has properties similar to sulfur structures, Swesi *et al.* [63] examined the OER activities of this catalyst by coating it on different substrates. On different

Table 10.4 Selected Ni-essenced catalysts for OER.

Catalysts	**Electrolyte**	**Electrode/ substrate**	**Overpotential (10 mA cm^{-2}) [mV vs RHE]**	**Tafel slope [mV dec^{-1}]**	**Ref**
$Ni(OH)_2$–NB $Ni(OH)_2$–NF $Ni(OH)_2$–NP $Ni(OH)_2$–NS	1.0 M KOH	GCE	320 310 260 300	92.4 89.3 78.6 77.4	[53]
NCO–HNSs	1.0 M KOH	GCE	340	51	[54]
NiFe LDHs	1.0 M KOH	GCE	220	62.9	[55]
SnNiFe–LDHs	1.0 M KOH	GCE	204	58	[56]
Co-doped Ni–Mn LDH	1.0 M KOH	RDE	310	59	[57]
NiCo–LDH@rGO	1.0 M KOH	RDE	376	74	[58]
$NiMoO_4$	1.0 M KOH	GCE	260	85.7	[59]
$NiMnO_3$	1.0 M KOH	Carbon cloth	256	141	[60]
$NiO/NiFe_2O_4$	1.0 M KOH	Carbon paper	360	40	[61]
$NiCo_2S_4$ NW	1.0 M KOH	Ni foam	260	40.1	[62]
Ni_3Se_2	1.0 M KOH	GCE	310	79.5	[63]
NiP	1.0 M KOH	Carbon cloth	110	64	[64]

NB: nano bud, NF: nanoflower, NP: nanopetal, NS: nanosheet, NCO–HNSs: Ni–Co oxide hierarchical nanosheets, LDHs: layered double hydroxides, rGO: reduced graphene oxide, NW: nanowire, GCE: glassy carbon electrode, RDE: rotating disc electrode.

substrates, such as Au@Glass, Au@Si, GCE, and Ni foam, the lowest Tafel slope value was observed in the GCE electrode, and the lowest overpotential value was observed in the catalyst coated on Ni foam. As with sulfur and selenite-type structures, phosphine-type structures also have an important place in OER studies, especially with their low overpotential and Tafel slope values. Yu *et al.* [64] state that the specifically developed NiP catalyst has a more effective OER activity than both $Ni(OH)_2$ and NiO porous structures. Some of Ni-essenced catalysts for OER are shown in Table 10.4.

10.3.3 Cobalt-Essenced Catalysts for OER

Co-essenced electrocatalysts are found to have more intensive use than Ni- and Mn-essenced catalysts. The main reasons for this include Co's environmental friendliness, low cost, considerable dual-function features, accessibility, economics, and efficiency. In addition, the fact that Co has different oxidation levels (3−/5+), as in Mn, makes it an electrocatalyst that provides different properties. However, Co structures containing 2+ and 3+ oxidation levels are the most stable forms.

Co-metal alloy-type catalysts are widely used as Co-essenced electrocatalysts. Bai *et al.* [65] perused the OER activity of the Co–Fe double-atom catalyst. They investigated the active sites of the Co–Fe alloy-type catalyst they developed using the XAS technique and pointed it out as a catalyst system that could be an alternative to metal oxide-type catalysts. Yu *et al.* [66] combined the NiCo binary metallic catalyst with N-doped carbon nanotube (CNT) structures to increase the low conductibility values of Co–metal alloy structures. They reported that the doped CNT structure provided higher electrocatalytic activity and showed very low overpotential values. They also state that the 1,000-cycle stability of the developed catalyst is higher than that of IrO_2.

As seen in the Ni-essenced catalyst system, metallic LDH catalysts show high OER activity in also Co-essenced structures. Wang *et al.* [67] prepared Co–Ni LDH structures on polypyrrole-coated cotton pads and examined their effects on OER efficiency. It is stated that this LDH structure, especially formed by Ni doping, reduces the deprotonation energy value by DFT calculations. Qiao *et al.* [68] examined the OER activity of Zn–Co LDH structures. They stated that the layered form of this catalyst, synthesized using the microwave-assisted method, is more stable and OER efficient and that these properties are due to the increase in the active site. Kunag *et al.* [69] increased active sites by creating rich surface defects and coordinately unsaturated sites on the cobalt–vanadium–iron (oxy) hydroxide catalysts. They also brought an innovative approach to the earth's crust-abundant metal-essenced electrocatalyst systems with highly low Tafel slope and overpotential value. Jiang *et al.* [70] developed the Co oxychloride electrocatalyst and reported that Cl^- enhanced the coordinated restructuring of unsaturated regions at the atomic level and increased catalytic performance. The role of Co in active site formation and its relationship with Cl were examined in detail with both XANES measurements and DFT calculations.

In studies on CoO_x electrocatalysts, Xu *et al.* [71] synthesized porous cobalt oxide nanoplate structures by a polyol reduction mechanism. These developed porous nanoplates provided more active sites enriched with oxygen vacancies. Haas *et al.* [72] examined the OER activities of $CoO_x(OH)_y$ catalysts. They developed active sites on the catalysts, especially with the oxide hydroxide strategy they developed, and supported the results with DFT calculations by elucidating the structure of the catalyst with detailed characterization. As an example of Fe-doping studies on CoO_x catalysts, studies carried out by Gong *et al.* [73] and Li *et al.* [74] can be given. In particular, the adsorption of Fe^{3+} onto the CoO_x surface increases OER activity by creating oxygen vacancies and increasing activated sites. Recently, Huang *et al.* [75] synthesized MnSe–CeO_2@Co_3O_4 porous core-shell nanospheres as a catalyst with a loft oxygen vacancy content, wide surface area, great electrical conductibility, and marvelous kinetic performance. It is stated that the CeO_2 and Mn–Se elements in this developed core-shell structure perform a significant role in the oxygen vacancy level and have high 40-h stability.

Binary metal oxide catalyst systems have also been developed to boost the OER activities of cobalt oxide-type catalysts. In the study of Zhuang *et al.* [76] examining the OER activity of $FeCoO_x$–S catalysts, it was stated that S doping increased the OER activity and provided low overpotential and Tafel slope value by increasing the active sites of the Fe–Co binary oxide structure. OER activity, polarization, and mass activity results of the developed $FeCoO_x$–S-essenced photocatalysts are shown in Figure 10.4. Bandal *et al.* [77] observed low overpotential and Tafel slope in their study where they examined the OER activity of bimetallic iron cobalt oxide ($FeCoO_x$) catalyst on Ni foam. However, the previous study observed relatively higher overpotential and Tafel slope values in the S-doped structure.

Studies on cobalt-essenced catalyst systems show that the OER activities of CoP structures are particularly high and have been studied intensively. In the first study on CoP catalyst, Surendranath *et al.* [78] investigated the OER activities of CoP catalysts in a neutral pH environment. Liu *et al.* [79] examined the OER activities of cobalt phosphide hollow polyhedron

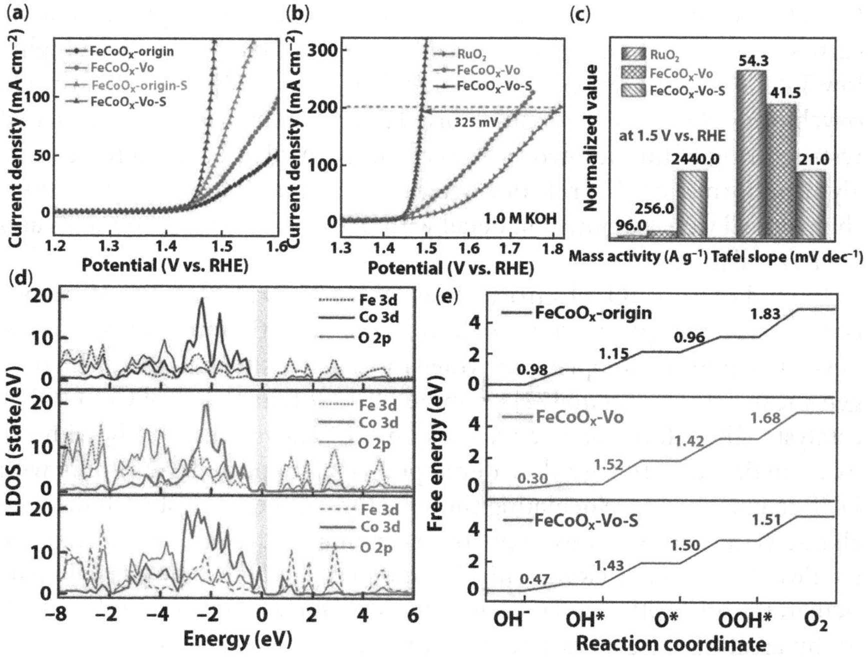

Figure 10.4 Electrochemical oxygen evolution activity assessment. (a) OER polarization curves of $FeCoO_x$-origin, $FeCoO_x$-Vo, $FeCoO_x$-origin-S, and $FeCoO_x$-Vo-S in 1.0 M KOH. (b) OER polarization curves of RuO_2, $FeCoO_x$-Vo, and $FeCoO_x$-Vo-S in 1.0 M KOH to reach a high OER current density. (c) The comparison of mass activity at the potential of 1.5 V vs. RHE and Tafel slope among RuO_2, $FeCoO_x$-Vo, and $FeCoO_x$-Vo-S in 1.0 M KOH. (d) The LDOS of 3d orbitals of Co and Fe atoms, and 2p orbitals of O atoms of $FeCoO_x$-origin, $FeCoO_x$-Vo, and $FeCoO_x$-Vo-S. (e) Energy profiles of the $FeCoO_x$-origin, $FeCoO_x$-Vo, and $FeCoO_x$-Vo-S. Reproduced with permission from ref. [76]. Copyright 2020 by Wiley.

catalysts. In particular, studies on the OER activity of CoP indicate that CoP catalysts are efficient bifunctional electrocatalysts. To increase the OER activities of CoPs, CoFeP-type [80] and CoMnP-type [44] binary phosphine catalysts have been developed, and decreases in the overpotential values of these developed binary catalysts have been observed. Catalyst systems, such as CoS [81], CoSe [82], CoN [83], CoC [84], and CoB [85], which have properties similar to the CoP structure, have also been developed, and effective OER efficiencies have been observed. Some of Co-essenced catalysts for OER are shown in Table 10.5.

Table 10.5 Selected Co-essenced catalysts for OER.

Catalysts	**Electrolyte**	**Electrode/ substrate**	**Overpotential (10 mA cm^{-2}) [mV vs RHE]**	**Tafel slope [mV dec^{-1}]**	**Ref**
Co–Fe	1.0 M KOH	Carbon cloth	495	42	[65]
NiCo@NCNTs	0.1 M KOH	GCE	41	177	[66]
CoNi–LDH	1.0 M KOH	Polypyrrole–cotton pad	350	58	[67]
Zn–Co–LDH	0.1 M KOH	GCE	357	101	[68]
CoV–$Fe_{0.28}$	1.0 M KOH	GCE	215	39.1	[69]
$Co_2(OH)_3Cl$	1.0 M KOH	GCE	300	151	[70]
CoO_x	1.0 M KOH	GCE	306	67	[71]
$CoO_x(OH)_y$	0.1 M KOH	RDE	-	37	[72]
Fe–CoO_x	1.0 M KOH	GCE	309	27.6	[73]
Fe–Co_3O_4	1.0 M KOH	GCE	204	38	[74]
MnSe–CeO_2@ Co_3O_4	1.0 M KOH	GCE	284	94	[75]
$FeCoO_x$–S	1.0 M KOH	GCE	240	21	[76]
$FeCoO_x$	1.0 M KOH	Ni foam	268	57	[77]
Co–Pi	0.1 KPO_4	Ni foam	410	59	[78]
CoP polyhedron	1.0 M KOH	GCE	400	57	[79]
Co–Fe–P	1.0 M KOH	GCE	244	58	[80]
Mn–Co–P	1.0 M KOH	GCE	330	59	[44]

NCNTs: N-doped carbon nanotubes, LDH: layered double hydroxide, GCE: glassy carbon electrode, RDE: rotating disc electrode.

10.4 Other Catalysts for OER

Studies are carried out, especially on carbon-essenced materials, to ensure that the active area is high in the catalysts used in OER. Carbon-essenced materials provide both high surface area and increased conductivity. Researchers have generally used carbon-essenced materials via doping to noble metal/non-noble metal catalysts [86–88]. Perovskite structures have been perused intensively in the last years, especially because of their stable structure and high ionic/electronic conductivity behaviors. Researchers investigated the OER activities of $Ca_2Mn_2O_5$ [89], $LaNiO_3$ [90], $LaFeO_3$ [91], $LaCoO_3$ [92], $ReBaCo_2O_5$ [93], La_2CoMnO_6 [94], $BaTiO_3$ [95], etc., perovskite structures. In recent research, it has been observed that MOF-essenced structures are used as various electrocatalysts [96–104].

10.5 Conclusion

In this chapter, anodic catalysts used in the OER system are investigated. Particular attention was paid to the OER activities, active site, and stability properties of OER catalysts. Catalysts used in OERs were examined in three main groups as follows: noble metal, non-noble metal, and other catalysts. Studies conducted in recent years have been examined, and different examples related to OER catalysts have been evaluated.

Although high current densities and high activities have been observed in noble metal-essenced catalysts, their high cost and difficulties in industrial applications are the biggest obstacles in the development of commercial products. Among non-noble metal-essenced catalyst systems, Ni-essenced, Mn-essenced, and Co-essenced catalyst systems stand out. It is noteworthy that Co-essenced catalysts, which have been developed in recent years, not only show a more effective OER efficiency than noble metal-essenced (especially Ru and Ir) catalysts but also are more stable. One of the biggest disadvantages of OER catalysts is that, because of their loft activity in highly alkaline or acidic environments, the stability of the developed catalyst system cannot withstand these environments.

It seems that it would be beneficial to focus on studies on high current densities with noble/non-noble metal-essenced catalysts that perform well in OER systems. The realization of practical applications in the near future may create a horizon before industrial applications of OER electrodes. Finally, there is a need to focus on further commercial practices and reducing costs.

AI disclosure statement: The author acknowledges the use of AI-based tools for minor language editing or grammatical correction. However, all intellectual contributions, analyses, and conclusions presented in this chapter are entirely the author's own.

References

1. Sen, R., Das, S., Nath, A., Maharana, P., Kar, P., Verpoort, F., Liang, P., Roy, S., Electrocatalytic Water Oxidation: An Overview With an Example of Translation From Lab to Market. *Front. Chem.*, 10, 1–24, 2022.
2. Yang, J., Zhao, W., Xu, F., Hu, X., Weng, B., Nanoelectrocatalysts for Anodic Oxygen Evolution in Acid: A Review. *ACS Appl. Nano Mater.*, 6, 23, 21424–21450, 2023.
3. Gong, M. and Dai, H., A mini review of NiFe-based materials as highly active oxygen evolution reaction electrocatalysts. *Nano Res.*, 8, 1, 23–39, 2015.
4. Reier, T., Oezaslan, M., Strasser, P., Electrocatalytic oxygen evolution reaction (OER) on Ru, Ir, and pt catalysts: A comparative study of nanoparticles and bulk materials. *ACS Catal.*, 2, 8, 1765–1772, 2012.
5. Ravichandran, S., Bhuvanendran, N., Xu, Q., Maiyalagan, T., Xing, L., Su, H., Ordered mesoporous Pt-Ru-Ir nanostructures as superior bifunctional electrocatalyst for oxygen reduction/oxygen evolution reactions. *J. Colloid Interface Sci.*, 608, 207–218, 2022.
6. Nadeem, M., Yasin, G., Bhatti, M.H., Mehmood, M., Arif, M., Dai, L., Pt-M bimetallic nanoparticles (M = Ni, Cu, Er) supported on metal organic framework-derived N-doped nanostructured carbon for hydrogen evolution and oxygen evolution reaction. *J. Power Sources*, 402, 34–42, 2018.
7. Zhu, J., Xie, M., Chen, Z., Lyu, Z., Chi, M., Jin, W., Xia, Y., Pt-Ir-Pd Trimetallic Nanocages as a Dual Catalyst for Efficient Oxygen Reduction and Evolution Reactions in Acidic Media. *Adv. Energy Mater.*, 10, 16, 1904114, 1–7, 2020.
8. Ramírez, A., Hillebrand, P., Stellmach, D., May, M.M., Bogdanoff, P., Sebastian, F., Oxide-supported Ir nanodendrites with high activity and durability for the oxygen evolution reaction in acid PEM water electrolyzers. *J. Phys. Chem. C*, 118, 14073–14081, 2014.
9. Paquin, F., Rivnay, J., Salleo, A., Stingelin, N., Silva, C., Multi-phase semicrystalline microstructures drive exciton dissociation in neat plastic semiconductors. *J. Mater. Chem. C*, 3, 10715–10722, 2015.
10. Wang, C., Sui, Y., Xiao, G., Yang, X., Wei, Y., Zou, G., Zou, B., Synthesis of Cu-Ir nanocages with enhanced electrocatalytic activity for the oxygen evolution reaction. *J. Mater. Chem. A*, 3, 39, 19669–19673, 2015.
11. Zhu, J., Chen, Z., Xie, M., Lyu, Z., Chi, M., Mavrikakis, M., Jin, W., Xia, Y., Iridium-Based Cubic Nanocages with 1.1-nm-Thick Walls: A Highly

Efficient and Durable Electrocatalyst for Water Oxidation in an Acidic Medium. *Angew. Chem. - Int. Ed.*, 58, 22, 7244–7248, 2019.

12. Sun, C., Qin, J., Li, M., Han, G., Song, Y., Ultrafine IrRu Nanoparticles toward Efficient Oxygen Evolution Reaction in Acidic Media. *Inorg. Chem.*, 61, 43, 17362–17369, 2022.
13. Jiménez, P.C., Wiberg, G.K.H., Sievers, G.W., Brüser, V., Arenz, M., Bridging the gap between basic research and application: a half-cell setup for high current density measurements of Ir-based oxygen evolution reaction catalysts on porous transport electrodes. *J. Mater. Chem. A*, 11, 37, 20129–20138, 2023.
14. Luo, X., Wei, X., Zhong, H., Wang, H., Wu, Y., Wang, Q., Gu, W., Gu, M., Beckman, S.P., Zhu, C., Single-Atom Ir-Anchored 3D Amorphous NiFe Nanowire@Nanosheets for Boosted Oxygen Evolution Reaction. *ACS Appl. Mater. Interfaces*, 12, 3, 3539–3546, 2020.
15. Fu, L., Yang, F., Cheng, G., Luo, W., Ultrathin Ir nanowires as high-performance electrocatalysts for efficient water splitting in acidic media. *Nanoscale*, 10, 4, 1892–1897, 2018.
16. Yang, J., Ji, Y., Shao, Q., Zhang, N., Li, Y., Huang, X., A Universal Strategy to Metal Wavy Nanowires for Efficient Electrochemical Water Splitting at pH-Universal Conditions. *Adv. Funct. Mater.*, 28, 41, 1–8, 2018.
17. Xu, Z., Zhou, L., Zhou, G., Wu, S., Wang, P., Li, H., Huang, P., Liu, L., Light-driven Orderly Assembly of Ir-atomic Chains to Integrate a Dynamic Reaction Pathway for Acidic Oxygen Evolution. *Angew. Chem. Int. Ed.*, 62, 21, e202301128, 2023.
18. Cherevko, S., Geiger, S., Kasian, O., Kulyk, N., Grote, J.P., Savan, A., Shrestha, B.R., Merzlikin, S., Breitbach, B., Ludwig, A., Mayrhofer, K.J.J., Oxygen and hydrogen evolution reactions on Ru, RuO2, Ir, and IrO2 thin film electrodes in acidic and alkaline electrolytes: A comparative study on activity and stability. *Catal. Today*, 262, 170–180, 2016.
19. Mattos-Costa, F.I., De Lima-Neto, P., Machado, S.A.S., Avaca, L.A., Characterisation of surfaces modified by sol-gel derived RuxIr1–xO2 coatings for oxygen evolution in acid medium. *Electrochim. Acta*, 44, 8–9, 1515–1523, 1998.
20. Zhang, K., Mai, W., Li, J., Wang, H., Li, G., Hu, W., Highly scattered Ir oxides on TiN as an efficient oxygen evolution reaction electrocatalyst in acidic media. *J. Mater. Sci.*, 55, 8, 3507–3520, 2020.
21. Oakton, E., Lebedev, D., Povia, M., Abbott, D.F., Fabbri, E., Fedorov, A., Nachtegaal, M., Copéret, C., Schmidt, T.J., IrO2-TiO2: A High-Surface-Area, Active, and Stable Electrocatalyst for the Oxygen Evolution Reaction. *ACS Catal.*, 7, 4, 2346–2352, 2017.
22. Xu, J., Liu, G., Li, J., Wang, X., The electrocatalytic properties of an IrO 2/SnO 2 catalyst using SnO 2 as a support and an assisting reagent for the oxygen evolution reaction. *Electrochim. Acta*, 59, 105–112, 2012.

23. Shi, G., Tano, T., Tryk, D.A., Uchiyama, T., Iiyama, A., Uchida, M., Terao, K., Yamaguchi, M., Tamoto, K., Uchimoto, Y., Kakinuma, K., Nanorod Structuring of IrOxon a Unique Microstructure of Sb-Doped Tin Oxide to Dramatically Boost the Oxygen Evolution Reaction Activity for PEM Water Electrolysis. *ACS Catal.*, 13, 18, 12299–12309, 2023.
24. Audichon, T., Mayousse, E., Morisset, S., Morais, C., Comminges, C., Napporn, T.W., Kokoh, K.B., Electroactivity of RuO2-IrO2 mixed nanocatalysts toward the oxygen evolution reaction in a water electrolyzer supplied by a solar profile. *Int. J. Hydrogen Energy*, 39, 30, 16785–16796, 2014.
25. Ma, Z., Zhang, Y., Liu, S., Xu, W., Wu, L., Hsieh, Y.C., Liu, P., Zhu, Y., Sasaki, K., Renner, J.N., Ayers, K.E., Adzic, R.R., Wang, J.X., Reaction mechanism for oxygen evolution on RuO2, IrO2, and RuO2@IrO2 core-shell nanocatalysts. *J. Electroanal. Chem.*, 819, October 2017, 296–305, 2018.
26. Li, H., Pan, Y., Wu, L., He, R., Qin, Z., Luo, S., Yang, L., Zeng, J., IrO2 deposited on RuO2 as core-shell structured RuO2@IrO2 for oxygen evolution reaction in electrochemical water electrolyzer. *Mol. Catal.*, 551, 113619, 2023.
27. Uzgören, IN., Hüner, B., Ylldlrlm, S., Eren, O., Özdoğan, E., Süzen, Y.O., Demir, N., Kaya, M.F., Development of IrO2-WO3 Composite Catalysts from Waste WC-Co Wire Drawing Die for PEM Water Electrolyzers' Oxygen Evolution Reactions. *ACS Sustain. Chem. Eng.*, 10, 39, 13100–13111, 2022.
28. Li, H., Xu, Y., Lv, N., Zhang, Q., Zhang, X., Wei, Z., Wang, Y., Tang, H., Pan, H., Ti-Doped SnO2 Supports IrO2 Electrocatalysts for the Oxygen Evolution Reaction (OER) in PEM Water Electrolysis. *ACS Sustain. Chem. Eng.*, 11, 3, 1121–1132, 2023.
29. Wang, Z., Xiao, B., Lin, Z., Shen, S., Xu, A., Du, Z., Chen, Y., Zhong, W., *In-situ* surface decoration of RuO2 nanoparticles by laser ablation for improved oxygen evolution reaction activity in both acid and alkali solutions. *J. Energy Chem.*, 54, 510–518, 2021.
30. Jiang, R., Tran, D.T., Li, J., Chu, D., Ru@RuO2 Core-Shell Nanorods: A Highly Active and Stable Bifunctional Catalyst for Oxygen Evolution and Hydrogen Evolution Reactions. *Energy Environ. Mater.*, 2, 3, 201–208, 2019.
31. Qin, Y., Yu, T., Deng, S., Zhou, X.Y., Lin, D., Zhang, Q., Jin, Z., Zhang, D., He, Y.B., Qiu, H.J., He, L., Kang, F., Li, K., Zhang, T.Y., RuO2 electronic structure and lattice strain dual engineering for enhanced acidic oxygen evolution reaction performance. *Nat. Commun.*, 13, 1, 3784, 2022.
32. Wu, Y., Tariq, M., Zaman, W.Q., Sun, W., Zhou, Z., Yang, J., Ni-Co Codoped RuO2 with Outstanding Oxygen Evolution Reaction Performance. *ACS Appl. Energy Mater.*, 2, 6, 4105–4110, 2019.
33. Qiu, L., Zheng, G., He, Y., Lei, L., Zhang, X., Ultra-small Sn-RuO2 nanoparticles supported on Ndoped carbon polyhedra for highly active and durable oxygen evolution reaction in acidic media. *Chem. Eng. J.*, 409, 128155, 2021.
34. Tariq, M., Wu, Y., Ma, C., Ali, M., Zaman, W.Q., Abbas, Z., Ayub, K.S., Zhou, J., Wang, G., Cao, L., Yang, J., Boosted up stability and activity of oxygen

vacancy enriched RuO_2/MoO_3 mixed oxide composite for oxygen evolution reaction. *Int. J. Hydrogen Energy*, 45, 35, 17287–17298, 2020.

35. Bai, J., Cheng, L., Liu, S., Zhang, H., Lian, Y., Deng, Y., Zhou, Q., Tang, Y., Su, Y., Unravel the mechanism of d-orbital modulation and oxygen vacancy in cerium-doped RuO_2 catalysts for acidic oxygen evolution reaction. *Appl. Surf. Sci.*, 642, 158613, 2024.
36. Deng, M., Tang, Y., Lu, Z., Wang, Y., Lin, Y., Self–Supporting Mn–RuO_2 Nanoarrays for Stable Oxygen Evolution Reaction in Acid. *Molecules*, 28, 23, 7727, 2023.
37. Zhang, W., Luo, J., Tang, H., Wang, S., Li, W., Zhang, J., Zhou, Y., Co-doped RuO_2 nanoparticles with enhanced catalytic activity and stability for the oxygen evolution reaction. *Dalt. Trans.*, 53, 1031–1039, 2024.
38. Zhang, Y., Dong, J., Sun, T., Zhang, X., Chen, J., Xu, L., Mo-Doped Mesoporous RuO_2 Spheres as High-Performance Acidic Oxygen Evolution Reaction Electrocatalyst. *Small*, 20, 15, 2305889, 20, 15, 2024.
39. Ramírez, A., Hillebrand, P., Stellmach, D., May, M.M., Bogdanoff, P., Sebastian, F., Evaluation of MnO_x, Mn_2O_3, and Mn_3O_4 Electrodeposited Films for the Oxygen Evolution Reaction of Water. *J. Phys. Chem. C*, 118, 14073–14081, 2014.
40. Kölbach, M., Fiechter, S., van de Krol, R., Bogdanoff, P., Evaluation of electrodeposited A-Mn_2O_3 as a catalyst for the oxygen evolution reaction. *Catal. Today*, 290, 2–9, 2017.
41. Guo, C.X., Chen, S., Lu, X., Ethylenediamine-mediated synthesis of Mn_3O_4 nano-octahedrons and their performance as electrocatalysts for the oxygen evolution reaction. *Nanoscale*, 6, 18, 10896–10901, 2014.
42. Gupta, G., Selvakumar, K., Mamlouk, M., Lakshminarasimhan, N., Senthil Kumar, S.M., The effects of morphology, microstructure and mixed-valent states of MnO_2 on the oxygen evolution reaction activity in alkaline anion exchange membrane water electrolysis. *J. Power Sources*, 461, 228131, 2020.
43. Zhuang, Q., Ma, N., Yin, Z., Yang, X., Yin, Z., Gao, J., Xu, Y., Gao, Z., Wang, H., Kang, J., Xiao, D., Li, J., Li, X., Ma, D., Rich Surface Oxygen Vacancies of MnO_2 for Enhancing Electrocatalytic Oxygen Reduction and Oxygen Evolution Reactions. *Adv. Energy Sustain. Res.*, 2, 8, 1–9, 2021.
44. Kaneti, Y.V., Guo, Y., Septiani, N.L.W., Iqbal, M., Jiang, X., Takei, T., Yuliarto, B., Alothman, Z.A., Golberg, D., Yamauchi, Y., Self-templated fabrication of hierarchical hollow manganese-cobalt phosphide yolk-shell spheres for enhanced oxygen evolution reaction. *Chem. Eng. J.*, 405, 126580, 2021.
45. Browne, M.P., Nolan, H., Duesberg, G.S., Colavita, P.E., Lyons, M.E.G., Low-Overpotential High-Activity Mixed Manganese and Ruthenium Oxide Electrocatalysts for Oxygen Evolution Reaction in Alkaline Media. *ACS Catal.*, 6, 4, 2408–2415, 2016.
46. Sumboja, A., Chen, J., Zong, Y., Lee, P.S., Liu, Z., NiMn layered double hydroxides as efficient electrocatalysts for the oxygen evolution reaction and

their application in rechargeable Zn-air batteries. *Nanoscale*, 9, 2, 774–780, 2017.
47. Mosa, I.M., Biswas, S., El-Sawy, A.M., Botu, V., Guild, C., Song, W., Ramprasad, R., Rusling, J.F., Suib, S.L., Tunable mesoporous manganese oxide for high performance oxygen reduction and evolution reactions. *J. Mater. Chem. A*, 4, 2, 620–631, 2015.
48. Frydendal, R., Seitz, L.C., Sokaras, D., Weng, T.C., Nordlund, D., Chorkendorff, I., Stephens, I.E.L., Jaramillo, T.F., Operando investigation of Au-MnOx thin films with improved activity for the oxygen evolution reaction. *Electrochim. Acta*, 230, 22–28, 2017.
49. Sun, W., Fang, Y., Sun, G., Dai, C., Liu, Y., Zhang, J., Zhu, Y., Wang, J., Ruthenium-Manganese Solid Solution Oxide with Enhanced Performance for Acidic and Alkaline Oxygen Evolution Reaction. *Chem. - An Asian J.*, 18, 16, e202300440, 2023.
50. Kakati, U., Elzinga, E.J., Mansley, Z.R., Roe, B., Alimohammadi, F., Schwenk, G., Ning, J., Zhu, Y., Ji, H.-F., Sun, J., Strongin, D.R., Iridium Incorporation into MnO2 for an Enhanced Electrocatalytic Oxygen Evolution Reaction. *ChemCatChem*, 15, 8, e202201549, 2023.
51. Zand, Z., Mohammadi, M.R., Sologubenko, A.S., Handschin, S., Bagheri, R., Chernev, P., Song, Z., Dau, H., Najafpour, M.M., Oxygen Evolution Reaction by Silicate-Stabilized Manganese Oxide. *ACS Appl. Energy Mater.*, 6, 3, 1702–1713, 2023.
52. Luo, Z., Irtem, E., Ibánez, M., Nafria, R., Martĺ-Sánchez, S., Genç, A., De La Mata, M., Liu, Y., Cadavid, D., Llorca, J., Arbiol, J., Andreu, T., Morante, J.R., Cabot, A., Mn3O4@CoMn2O4-CoxOy Nanoparticles: Partial Cation Exchange Synthesis and Electrocatalytic Properties toward the Oxygen Reduction and Evolution Reactions. *ACS Appl. Mater. Interfaces*, 8, 27, 17435–17444, 2016.
53. Luan, C., Liu, G., Liu, Y., Yu, L., Wang, Y., Xiao, Y., Qiao, H., Dai, X., Zhang, X., Structure Effects of 2D Materials on α-Nickel Hydroxide for Oxygen Evolution Reaction. *ACS Nano*, 12, 4, 3875–3885, 2018.
54. Wang, H.Y., Hsu, Y.Y., Chen, R., Chan, T.S., Chen, H.M., Liu, B., Ni3+-induced formation of active NiOOH on the spinel Ni-Co oxide surface for efficient oxygen evolution reaction. *Adv. Energy Mater.*, 5, 10, 1–8, 2015.
55. Wang, Y., Qiao, M., Li, Y., Wang, S., Tuning Surface Electronic Configuration of NiFe LDHs Nanosheets by Introducing Cation Vacancies (Fe or Ni) as Highly Efficient Electrocatalysts for Oxygen Evolution Reaction. *Small*, 14, 17, 1–6, 2018.
56. Bai, L., Liu, Y., Jia, Q., Li, P., Yan, Y., Yuan, N., Guo, S., Quenching modification of NiFe layered double hydroxides as efficient and highly stable electrocatalysts for the oxygen evolution reaction. *J. Colloid Interface Sci.*, 653, PA, 108–116, 2024.

57. Wang, Y., Liu, X., Zhang, N., Qiu, G., Ma, R., Cobalt-doped Ni–Mn layered double hydroxide nanoplates as high-performance electrocatalyst for oxygen evolution reaction. *Appl. Clay Sci.*, 165, 277–283, 2018.
58. Wakamatsu, S., Islam, M.S., Shudo, Y., Fukuda, M., Tagawa, R., Goto, N., Koinuma, M., Sekine, Y., Hayami, S., An efficient oxygen evolution reaction catalyst using Ni-Co layered double hydroxide anchored on reduced graphene oxide. *Energy Adv.*, 2, 9, 1375–1380, 2023.
59. Zhu, J., Qian, J., Peng, X., Xia, B., Gao, D., Etching-Induced Surface Reconstruction of NiMoO4for Oxygen Evolution Reaction. *Nano Micro Lett.*, 15, 1, 1–12, 2023.
60. Dinesh, M., Haldorai, Y., Rajendra Kumar, R.T., Mn–Ni binary metal oxide for high-performance supercapacitor and electro-catalyst for oxygen evolution reaction. *Ceram. Int.*, 46, 18, 28006–28012, 2020.
61. Landon, J., Demeter, E., Inoğlu, N., Keturakis, C., Wachs, I.E., Vasić, R., Frenkel, A.I., Kitchin, J.R., Spectroscopic characterization of mixed Fe-Ni oxide electrocatalysts for the oxygen evolution reaction in alkaline electrolytes. *ACS Catal.*, 2, 8, 1793–1801, 2012.
62. Sivanantham, A., Ganesan, P., Shanmugam, S., Hierarchical NiCo2S4 Nanowire Arrays Supported on Ni Foam: An Efficient and Durable Bifunctional Electrocatalyst for Oxygen and Hydrogen Evolution Reactions. *Adv. Funct. Mater.*, 26, 26, 4661–4672, 2016.
63. Swesi, A.T., Masud, J., Nath, M., Nickel selenide as a high-efficiency catalyst for oxygen evolution reaction. *Energy Environ. Sci.*, 9, 5, 1771–1782, 2016.
64. Yu, X.Y., Feng, Y., Guan, B., Lou, X.W.D., Paik, U., Carbon coated porous nickel phosphides nanoplates for highly efficient oxygen evolution reaction. *Energy Environ. Sci.*, 9, 4, 1246–1250, 2016.
65. Bai, L., Hsu, C.S., Alexander, D.T.L., Chen, H.M., Hu, X., A Cobalt-Iron Double-Atom Catalyst for the Oxygen Evolution Reaction. *J. Am. Chem. Soc.*, 141, 36, 14190–14199, 2019.
66. Yu, J., Zhong, Y., Zhou, W., Shao, Z., Facile synthesis of nitrogen-doped carbon nanotubes encapsulating nickel cobalt alloys 3D networks for oxygen evolution reaction in an alkaline solution. *J. Power Sources*, 338, 26–33, 2017.
67. Wang, W., Lu, Y., Zhao, M., Luo, R., Yang, Y., Peng, T., Yan, H., Liu, X., Luo, Y., Controllable tuning of cobalt nickel-layered double hydroxide arrays as multifunctional electrodes for flexible supercapattery device and oxygen evolution reaction. *ACS Nano*, 13, 10, 12206–12218, 2019.
68. Qiao, C., Zhang, Y., Zhu, Y., Cao, C., Bao, X., Xu, J., One-step synthesis of zinc–cobalt layered double hydroxide (Zn–Co-LDH) nanosheets for high-efficiency oxygen evolution reaction. *J. Mater. Chem. A*, 3, 13, 6878–6883, 2015.
69. Kuang, M., Zhang, J., Liu, D., Tan, H., Dinh, K.N., Yang, L., Ren, H., Huang, W., Fang, W., Yao, J., Hao, X., Xu, J., Liu, C., Song, L., Liu, B., Yan, Q., Amorphous/Crystalline Heterostructured Cobalt-Vanadium-Iron (Oxy)

hydroxides for Highly Efficient Oxygen Evolution Reaction. *Adv. Energy Mater.*, 10, 43, 1–8, 2020.
70. Jiang, H., He, Q., Li, X., Su, X., Zhang, Y., Chen, S., Zhang, S., Zhang, G., Jiang, J., Luo, Y., Ajayan, P.M., Song, L., Tracking Structural Self-Reconstruction and Identifying True Active Sites toward Cobalt Oxychloride Precatalyst of Oxygen Evolution Reaction. *Adv. Mater.*, 31, 8, 1–8, 2019.
71. Xu, W., Lyu, F., Bai, Y., Gao, A., Feng, J., Cai, Z., Yin, Y., Porous cobalt oxide nanoplates enriched with oxygen vacancies for oxygen evolution reaction. *Nano Energy*, 43, 110–116, 2018.
72. Haase, F.T., Bergmann, A., Jones, T.E., Timoshenko, J., Herzog, A., Jeon, H.S., Rettenmaier, C., Cuenya, B.R., Size effects and active state formation of cobalt oxide nanoparticles during the oxygen evolution reaction. *Nat. Energy*, 7, 8, 765–773, 2022.
73. Gong, L., Chng, X.Y.E., Du, Y., Xi, S., Yeo, B.S., Enhanced Catalysis of the Electrochemical Oxygen Evolution Reaction by Iron(III) Ions Adsorbed on Amorphous Cobalt Oxide. *ACS Catal.*, 8, 2, 807–814, 2018.
74. Li, Y., Li, F.M., Meng, X.Y., Wu, X.R., Li, S.N., Chen, Y., Direct chemical synthesis of ultrathin holey iron doped cobalt oxide nanosheets on nickel foam for oxygen evolution reaction. *Nano Energy*, 54, 238–250, 2018.
75. Huang, X., Wang, X., Zhang, M., Jiang, Q., Qin, Z., Liu, Y., Hou, Y., Cao, X., Gu, H., Manganese and selenium co-doped CeO2@Co3O4 porous core–shell nanospheres for enhanced oxygen evolution reactions. *Energy Adv.*, 2, 2, 338–345, 2023.
76. Zhuang, L., Jia, Y., Liu, H., Li, Z., Li, M., Zhang, L., Wang, X., Yang, D., Zhu, Z., Yao, X., Sulfur-Modified Oxygen Vacancies in Iron–Cobalt Oxide Nanosheets: Enabling Extremely High Activity of the Oxygen Evolution Reaction to Achieve the Industrial Water Splitting Benchmark. *Angew. Chem. - Int. Ed.*, 59, 34, 14664–14670, 2020.
77. Bandal, H.A., Jadhav, A.R., Tamboli, A.H., Kim, H., Bimetallic iron cobalt oxide self-supported on Ni-Foam: An efficient bifunctional electrocatalyst for oxygen and hydrogen evolution reaction. *Electrochim. Acta*, 249, 253–262, 2017.
78. Surendranath, Y., Kanan, M.W., Nocera, D.G., Mechanistic studies of the oxygen evolution reaction by a cobalt-phosphate catalyst at neutral pH. *J. Am. Chem. Soc.*, 132, 46, 16501–16509, 2010.
79. Liu, M. and Li, J., Cobalt Phosphide Hollow Polyhedron as Efficient Bifunctional Electrocatalysts for the Evolution Reaction of Hydrogen and Oxygen. *ACS Appl. Mater. Interfaces*, 8, 3, 2158–2165, 2016.
80. Zhang, T., Du, J., Xi, P., Xu, C., Hybrids of Cobalt/Iron Phosphides Derived from Bimetal-Organic Frameworks as Highly Efficient Electrocatalysts for Oxygen Evolution Reaction. *ACS Appl. Mater. Interfaces*, 9, 1, 362–370, 2017.
81. Liu, T., Liang, Y., Liu, Q., Sun, X., He, Y., Asiri, A.M., Electrodeposition of cobalt-sulfide nanosheets film as an efficient electrocatalyst for oxygen evolution reaction. *Electrochem. Commun.*, 60, 92–96, 2015.

82. Liao, M., Zeng, G., Luo, T., Jin, Z., Wang, Y., Kou, X., Xiao, D., Three-dimensional coral-like cobalt selenide as an advanced electrocatalyst for highly efficient oxygen evolution reaction. *Electrochim. Acta*, 194, 59–66, 2016.
83. Chu, H., Zhang, D., Jin, B., Yang, M., Impact of morphology on the oxygen evolution reaction of 3D hollow Cobalt-Molybdenum Nitride. *Appl. Catal. B Environ.*, 255, 117744, 2019.
84. Jiang, J., Liu, Q., Zeng, C., Ai, L., Cobalt/molybdenum carbide@N-doped carbon as a bifunctional electrocatalyst for hydrogen and oxygen evolution reactions. *J. Mater. Chem. A*, 5, 32, 16929–16935, 2017.
85. Ren, X., Ge, R., Zhang, Y., Liu, D., Wu, D., Sun, X., Du, B., Wei, Q., Cobalt-borate nanowire array as a high-performance catalyst for oxygen evolution reaction in near-neutral media. *J. Mater. Chem. A*, 5, 16, 7291–7294, 2017.
86. Song, L., Chang, J., Ma, Y., Tan, X., Xu, Y., Guo, L., Chen, Z., Zhao, T., Li, Y., Liu, Y., Zhang, Y., Chu, W., Cobalt/nitrogen codoped carbon nanosheets derived from catkins as a high performance non-noble metal electrocatalyst for oxygen reduction reaction and hydrogen evolution reaction. *RSC Adv.*, 10, 71, 43248–43255, 2020.
87. Yu, J., Wu, X., Zhang, H., Ni, M., Zhou, W., Shao, Z., Core Effect on the Performance of N/P Codoped Carbon Encapsulating Noble-Metal Phosphide Nanostructures for Hydrogen Evolution Reaction. *ACS Appl. Energy Mater.*, 2, 4, 2645–2653, 2019.
88. Wang, S.-Q., Xia, W.-Y., Liang, Z.-S., Liu, Z.-L., Xu, C.-W., Li, Q.-Y., NiO/C enhanced by noble metal (Pt, Pd, Au) as high-efficient electrocatalyst for oxygen evolution reaction in water oxidation to obtain high purity hydrogen. *Ionics (Kiel).*, 23, 8, 2161–2166, 2017.
89. Kim, J., Yin, X., Tsao, K.-C., Fang, S., Yang, H., Ca2Mn2O5 as Oxygen-Deficient Perovskite Electrocatalyst for Oxygen Evolution Reaction. *J. Am. Chem. Soc.*, 136, 42, 14646–14649, 2014.
90. Liu, J., Jia, E., Stoerzinger, K.A., Wang, L., Wang, Y., Yang, Z., Shen, D., Engelhard, M.H., Bowden, M.E., Zhu, Z., Chambers, S.A., Du, Y., Dynamic Lattice Oxygen Participation on Perovskite LaNiO3 during Oxygen Evolution Reaction. *J. Phys. Chem. C*, 124, 28, 15386–15390, 2020.
91. Peng, X., Zheng, J., Zhang, Y., Wang, Z., Tuning oxygen vacancies on LaFeO3 perovskite as efficient electrocatalysts for oxygen evolution reaction. *Mater. Lett.*, 309, 131317, 2022.
92. Kim, J., Chen, X., Shih, P.-C., Yang, H., Porous Perovskite-Type Lanthanum Cobaltite as Electrocatalysts toward Oxygen Evolution Reaction. *ACS Sustain. Chem. Eng.*, 5, 11, 10910–10917, 2017.
93. Li, L., Dong, Z., Xia, T., Li, X., Li, Q., Sun, L., Zhao, H., Huo, L., A series of bifunctional ReBaCo2O5+δ perovskite catalysts towards intermediate-temperature oxygen reduction reaction and oxygen evolution reaction. *Chem. Eng. J.*, 468, 143762, 2023.

94. Li, S.-F., Zheng, J., Hu, L., Ma, Y., Zhao, S., Zhu, C.-H., Yan, D., Sr-Doped Double Perovskite La2CoMnO6 to Promote the Oxygen Evolution Reaction Activity. *ChemElectroChem*, 9, 15, e202200475, 2022.
95. Tanwar, N., Upadhyay, S., Priya, R., Pundir, S., Sharma, P., Pandey, O.P., Eu-doped BaTiO3 perovskite as an efficient electrocatalyst for oxygen evolution reaction. *J. Solid State Chem.*, 317, 123674, 2023.
96. Jiang, Y., Chen, T.-Y., Chen, J.-L., Liu, Y., Yuan, X., Yan, J., Sun, Q., Xu, Z., Zhang, D., Wang, X., Meng, C., Guo, X., Ren, L., Liu, L., Lin, R.Y.-Y., Heterostructured Bimetallic MOF-on-MOF Architectures for Efficient Oxygen Evolution Reaction. *Adv. Mater.*, 36, 8, 2306910, 2024.
97. Wang, H.-F., Chen, L., Pang, H., Kaskel, S., Xu, Q., MOF-derived electrocatalysts for oxygen reduction, oxygen evolution and hydrogen evolution reactions. *Chem. Soc. Rev.*, 49, 5, 1414–1448, 2020.
98. Tang, Y., Li, J., Lu, Z., Wang, Y., Tao, K., Lin, Y., MOF-Derived CoSe2@ NiFeOOH Arrays for Efficient Oxygen Evolution Reaction. *Nanomaterials*, 13, 19, 2621, 2023.
99. Liu, Y., Jia, G., Wu, Q., Zhang, D., Wu, J., Yin, Y., Sai, S., Guo, Z., Cui, X., MOF-derived FeNiCoOX hierarchical hollow nanocages for oxygen evolution reaction. *Mater. Lett.*, 291, 129564, 2021.
100. Farid, S., Ren, S., Hao, C., MOF-derived metal/carbon materials as oxygen evolution reaction catalysts. *Inorg. Chem. Commun.*, 94, 57–74, 2018.
101. Zhang, J., Zhang, C., Duo, X., Liu, Z., Liu, C., Guo, L., Xie, A., Luo, S., Self-assembled acicular CuCo-MOF for enhancing oxygen evolution reaction. *Ionics (Kiel).*, 26, 10, 5123–5132, 2020.
102. Shi, X., Li, J., Zhang, X., Guo, J., MOF derived NiCo-LDH hollow structure for oxygen evolution reaction. *J. Phys. Chem. Solids*, 172, 111051, 2023.
103. Du, J., Zhang, F., Jiang, L., Guo, Z., Song, H., Enhanced cobalt MOF electrocatalyst for oxygen evolution reaction via morphology regulation. *Inorg. Chem. Commun.*, 158, 111661, 2023.
104. Fiaz, M., Kashif, M., Fatima, M., Batool, S.R., Asghar, M.A., Shakeel, M., Athar, M., Synthesis of Efficient TMS@MOF-5 Catalysts for Oxygen Evolution Reaction. *Catal. Lett.*, 150, 9, 2648–2659, 2020.

11

Performance Characterization and Analysis of Electrolyzers

Mehmet Fatih Kaya[1,2,3,4*], Bulut Hüner[2,3,5], Murat Kıstı[1,2,3,6] and Nesrin Demir[1,2,3]

[1]Heat Engineering Division, Energy Systems Engineering Department, Erciyes University, Engineering Faculty, Kayseri, Türkiye
[2]H2FC Hydrogen Energy Research Group, Erciyes University, Kayseri, Türkiye
[3]ArGePark Research Building, Erciyes University, Kayseri, Türkiye
[4]Bataryasan Enerji San. ve Tic. A. Ş., Erciyes Teknopark, Kayseri, Türkiye
[5]Directorate of Research and Innovation, Osmaniye Korkut Ata University, Osmaniye, Türkiye
[6]Energy Systems Engineering Department, Graduate School of Natural and Applied Sciences, Erciyes University, Kayseri, Türkiye

Abstract
Electrolyzer systems play a crucial role in energy conversion and are key technologies offering potential for sustainable energy production. These systems come in various configurations, including Alkaline Water Electrolyzers (AWE), Polymer Electrolyte Membrane Electrolyzers (PEMWE), and Solid Oxide Electrolyzers (SOE). Each electrolyzer type has its own advantages and disadvantages depending on specific application requirements such as material selection, efficiency, cost, operating conditions, and hydrogen production scale. To advance the utilization of hydrogen as a sustainable and clean energy carrier, it is critical to disseminate and develop these technologies. The characterization and analysis of electrolyzer systems involve determining several critical parameters. These include factors like current density, Faraday efficiency, energy efficiency, stability, gas production rates, and purity. Characterization studies aim to understand the mechanisms affecting electrolyzer performance, enhance efficiency, and guide material selection.

This book chapter focuses on the fundamental principles of electrolyzer systems and highlights key issues in characterizing AWEs, PEMWEs, and SOEs. This study

**Corresponding author*: kayamehmetfatih@erciyes.edu.tr

Inamuddin, Tariq Altalhi, Mohammad Luqman and Jorddy Neves Cruz (eds.) Water Splitting: Production of Hydrogen, (277–308) © 2025 Scrivener Publishing LLC

focuses on electrochemical analysis, materials engineering, and developments in energy storage systems, and aims to pioneer research in electrolysis technologies.

***Keywords*:** Characterization and analysis, water electrolysis, electrolyzers, AWEs, PEMWEs, SOEs

11.1 Introduction

Technological advancements, economic growth, population increase, and evolving living standards worldwide contribute to a rise in energy consumption. This surge in consumption continuously escalates the demand for energy [1]. Fossil fuels such as oil, coal, and natural gas still provide a significant part of the world's energy needs [2, 3]. The use of these resources increases harmful gas emissions (CO_x, SO_x, NO_x) and environmental pollution [4]. The escalation of environmental pollution worldwide intensifies the greenhouse effect, which in turn indirectly contributes to climate change [5]. Fossil-based resource reserves are also known to be at risk of global depletion. The exhaustion of fossil fuels poses significant environmental, economic, and geopolitical challenges. Environmentally, this includes air and water pollution, habitat destruction, and the exacerbation of climate change due to greenhouse gas emissions released during fuel combustion [6]. Considering the depletion of fossil fuel reserves, it is an undeniable fact that the world requires clean, alternative energy sources to combat climate change and mitigate environmental pollution [7, 8]. To meet the increasing energy demand and minimize the damage to the environment, energy production must be increased by using renewable energy sources. Simultaneously, it is crucial that this generated energy is readily accessible, storable, and easily transportable to needed locations. Hydrogen is increasingly viewed as the energy source of the future owing to its high energy density, eco-friendliness, and the capability for safe storage in fuel tanks [9, 10]. As an energy carrier, hydrogen emerges as a better option since it can be chemically and electrochemically produced from biomass and water. Although hydrogen production through water electrolysis has been used for many years, it still requires a significant scientific effort before electrolyzers can be widely used [11]. Currently, water electrolysis, hindered by the high cost and low efficiency of electrolyzer systems, contributes merely 4% to hydrogen production. To facilitate broader adoption of water electrolysis, it is important to not only reduce energy consumption and system costs but also enhance the durability and performance of these systems [12]. Electrolyzers can be divided into three classes according to electrolyte type and working principles. Whereas AWE is commercially available at the most economical rate, PEMWE is less widely available than AWEs. SOE is currently in the experimental phase in laboratory environments [13]. AWE is recognized as one of the easiest and simplest methods. AWEs have been widely used in commercial and industrial applications for many years, and detailed research has generally increased after the 1990s. AWEs use an aqueous solution containing potassium hydroxide (KOH) or sodium hydroxide (NaOH) to increase the conductivity of ions and split water molecules into oxygen and hydrogen [14]. With the increasing use of AWEs technology, a significant potential for hydrogen production in industrial applications has emerged. AWE is central to developing sustainable energy generation and storage [15]. On the other hand, PEMWE has many advantages over AWE and SOE technology such as high voltage efficiency, hydrogen production with 99.999% purity, the ability to switch between electrolyzer and fuel cell operating modes, fast dynamic response time when combined with renewable energy sources, and the ability to operate at high pressure [16]. SOE represents one of the evolving water electrolysis technologies presently undergoing developmental stages. Although they have not yet achieved commercialization, various companies are actively exploring ways to bring them into the market. The ceramics used as electrolytes in SOE technology exhibit high-performance processes for cost-effective hydrogen production. This promising technology also facilitates the co-electrolysis of CO_2 and steam using a specific type of SOE [17]. In contrast to other electrolysis technologies, SOE systems operate at high temperatures between 500°C and 1,000°C. But it is important to point out that the high temperatures also present challenges like material durability, thermal management, and energy consumption required to maintain the elevated temperature. Recently, there has been a surge in interest in Solid Oxide Electrolyzers (SOE) driven by the growing demand for water electrolysis technology and its lower electrical energy consumption compared to traditional electrolysis units. In addition, the heat required for SOE operation can come from external sources like industrial waste heat, solar, and wind [18]. Performance characterization and analysis are critical to the development of electrolysis technologies. The importance of characterization and analysis stems from the need to fully understand, evaluate, and optimize the operational efficiency, durability, and economic sustainability of electrolysis systems. One aspect of this importance stems from the need for researchers and industry professionals to have a detailed understanding of performance characteristics as electrolyzer technologies continue to evolve. Performance characterization is directly related to the optimization of key parameters such as catalysts, electrodes, and electrolytes [19]. Understanding the interactions of these components and how they affect the overall efficiency of the electrolysis process is critical to achieving technological breakthroughs. This information allows researchers to design and implement strategies to increase catalytic activity, reduce overpotentials, and improve the overall energy conversion efficiency of electrolyzers [20]. Analysis of the performance of electrolyzer technologies is not limited to improving efficiency; it also has a direct impact on economic considerations. Cost effectiveness is a critical criterion for the widespread adoption of electrolysis in various industries. Therefore, detailed analysis helps to optimize resource utilization and ultimately lower the overall production cost of hydrogen. As a result, the impact of performance characterization and analysis of electrolyzer technologies is central to the advancement of sustainable energy solutions. It is important to comprehensively understand the performance parameters from solving the complexities of existing technologies to facilitating transitions from laboratory environments to industrially scalable applications. As the world transitions to a hydrogen-based economy, ongoing research in performance analysis will contribute to keeping electrolyzer technologies at the forefront of clean energy innovation, thereby contributing to a more sustainable and environmentally friendly future [21].

11.2 Fundamentals of Electrolysis

The technologies developed for electrolysis include AWE, PEMWE, and SOE technologies. Despite using different materials and operating under different conditions, these systems work on similar principles. A detailed comparison of electrolysis technology parameters is presented in Table 11.1.

SOE has the highest electrolysis efficiency (>89%) among various electrolysis technologies. However, it is limited due to material constraints, short operational lifespan, and high production costs. PEMWE has an efficiency ranging from 50% to 83%, a comparatively compact design, and a quick response. On the other hand, the use of precious noble metals, such as Ir and Ru, as catalysts for PEMWE makes it more expensive compared to other electrolysis technologies [24]. AWEs are considered the oldest and most mature technology, and they do not require the use of precious metals as catalysts. With an average efficiency of 50%–78% and a simple structure, it holds the largest proportion of the water electrolysis market. Therefore, AWEs are the best choice for commercial applications considering system

Table 11.1 Comparison of the technologies for water electrolysis [22, 23].

	AWE	SOE	PEMWE
Anode reaction	$2OH^- \rightarrow H_2O + \frac{1}{2}O_2 + 2e^-$	$O^{-2} \rightarrow \frac{1}{2}O_2 + 2e^-$	$H_2O \rightarrow 2H^+ + \frac{1}{2}O_2 + 2e^-$
Cathode reaction	$2H_2O + 2e^- \rightarrow H_2 + 2OH^-$	$H_2O + 2e^- \rightarrow H_2 + O^{-2}$	$2H^+ + 2e^- \rightarrow H_2$
Moving ion	OH^-	O^{-2}	H^+
Cell efficiency (%)	50–78	80–89	50–83
Purity (%)	99.5–99.9998	99.9	99.999
Operating temperature (°C)	30–90	500–850	30–80
Cell voltage (V)	1.4–2.5	0.7–1.5	1.4–2.5
Current density (A/cm^2)	0.2–0.8	0.3–1	1–3
Stack energy consumption (kWh/Nm^3)	4.2–5.9	<3,5	4.2–5.5
Outlet pressure (bar)	≤ 30	≤ 70	≤ 60
Lifespan (kh)	60–120	8–20	50–100

durability, size, and investment cost [25]. Table 11.2 shows the main techno-economic properties of various electrolysis technologies proposed by the International Energy Agency (IEA).

Tables 11.1 and 11.2 show that SOE technologies are more efficient and consume less electricity than AWE and PEM technologies. However, they operate at higher temperatures, which could complicate integration with other systems. Today, the cost of stack systems for PEM technologies is higher than that of AWE technologies. This state also affects both the CAPEX of the system and the price of replacing the stack. The stack lifetime, a critical factor influencing electrolyzer technologies' overall cost

Table 11.2 Techno-economic properties of various electrolyzer system technologies [26].

Properties	AWE technologies			SOE technologies			PEM technologies		
	Today	2030	Long term	Today	2030	Long term	Today	2030	Long term
Electrical efficiency (%)	63–70	65–71	70–80	74–81	77–84	77–90	56–60	63–68	67–74
Load range (%)	10–110	-	-	20–110	-	-	0–160	-	-
Stack lifetime (kh)	60–90	90–100	100–150	10–30	40–60	75–100	30–90	60–90	100–150
Plant footprint (m^2/kWe)	0.095	-	-	-	-	-	0.048	-	-
CAPEX (USD/kWe)	500–1,400	400–850	200–700	2,800–5,600	800–2,800	500–1,000	1,100–1,800	650–1,500	200–900

effectiveness and maintenance requirements, varies widely among different technologies. According to Table 11.2, the stack lifetime of electrolyzer technologies varies greatly, including 60–90, 30–90, and 10–30 kh for AWE, PEM, and SOE technologies, respectively. This change in stack lifetime directly impacts the total cost of ownership and operational efficiency of electrolyzer technology. A longer stack lifetime generally results in lower maintenance costs and fewer replacements over the electrolyzer lifetime. This makes longer stack lifetime technologies potentially more economically attractive, despite their higher initial costs. The higher costs associated with PEM stack systems, combined with their variable lifespan, require further research and development to enhance their durability and reduce costs. Similarly, due to the shorter lifespan of SOE technologies, advances in materials science and engineering are required to increase their lifespan and cost efficiency [27].

11.2.1 AWEs

AWEs are among the most preferred technologies for hydrogen production compared to other electrolysis methods. When compared to PEMWEs and SOEs, the AWE technology is considered a robust solution that operates at high power density, utilizes less expensive cell components, offers a longer system lifespan (20–30 years), has lower system costs, and consumes less energy employing more cost-effective base metals. The AWE system occurs in anode and cathode electrodes immersed in an alkaline solution. Generally, potassium hydroxide (KOH) or sodium hydroxide (NaOH) is used at a concentration of 25%–30%. A diaphragm is also employed to separate the generated oxygen (O_2) and hydrogen (H_2). The electrolyte, divided by the diaphragm, allows the separation of gases (O_2 and H_2) produced at the electrodes on either side of the diaphragm, which penetrate the OH^- ions [28]. The electrochemical reactions occurring at the anode electrode and cathode electrode of the AWE, as well as the total reaction, are given in Equations 11.1–11.3.

Anode Reaction:

$$2OH^- \rightarrow \frac{1}{2}O_2(g) + H_2O + 2e^- \tag{11.1}$$

Cathode Reaction:

$$2H_2O(l) + 2e^- \rightarrow H_2(g) + 2OH^- \tag{11.2}$$

Total Reaction:

$$H_2O(l) \rightarrow H_2(g) + \frac{1}{2}O_2\,(g) \tag{11.3}$$

AWE has a maximum current density of less than 400 mA/cm^2. The power consumption for hydrogen generation is approximately 4.5–5.5 kWh.Nm^3 with a 60% efficiency. Despite the apparent simplicity of the process, conventional AWEs face various challenges. Efficiency and safety can be compromised by unwanted gas crossover between electrodes and gas bubble formation on the active surface of the electrode. During the reaction, H_2 and O_2 gas bubbles form sequentially on the surfaces of the cathode and anode electrodes. Gas bubbles formed on the electrode surface can prevent the electron transfer between the electrode and the electrolyte increasing the resistance of the electrolyte and leading to increased losses. AWEs also face the challenge of achieving low maximum power density with liquid electrolytes. Bubble formation in electrolysis cells is inevitable due to the negative effects of hydroxyl consumption, low ionic conductivity, and clogging of gas diffusion layer pores, which disrupt the alkali electrolyte (usually KOH) during the reaction. To overcome this major obstacle, it is necessary to adapt and optimize materials and configurations. One advantage of AWEs is their cost effectiveness due to the use of metal electrodes and non-noble metals as catalysts in the electrolyte solution. The performance of electrolysis is heavily influenced by the electrodes or catalysts where oxygen evolution reactions (OER) and hydrogen evolution reactions (HER) occur. Therefore, the electrode materials used in AWEs must possess high corrosion resistance, electrical conductivity, and catalytic activity. It is important to note that electrochemical applications require objective evaluations of electrode materials [29, 30]. While noble metal-based catalysts are recognized for their low initial overpotential and high kinetic activity in HER and OER catalysis, their large-scale application is constrained by their scarcity and high cost. To make AWEs more cost effective in industrial applications, there is a critical need to employ alternative catalysts that are not only scalable and affordable but also demonstrate superior catalytic performance [31, 32].

11.2.2 SOEs

The first SOE technology was reported by Donitz and Erdle in the 1980s. However, SOEs have been intensively developed over the last 15 years, although this technology is still considered immature. This technology has emerged as having high hydrogen production efficiency and potential for commercial application [33, 34]. Operating at high temperatures, typically between 700°C and 800°C, this technology can significantly reduce the energy needed for splitting H_2O into H_2. The efficiency of power–hydrogen conversion can be further increased by supplying external heat to the system through the generation of steam at temperatures between 150°C and 180°C. This improvement in efficiency has an opportunity to significantly decrease the cost of hydrogen [35]. High temperatures can cause material degradation and affect the operating time of SOE. The hydrogen mixture produced from SOE with water vapor may require additional processing to obtain high-purity hydrogen [36, 37]. The working principle of SOEs is generally the reverse mode of the solid oxide fuel cell. An SOE generally comprises four layers: the H_2 (or vapor) electrode, a dense ionic conductive electrolyte, the O_2 (or air) electrode, and the interlayer between the O_2 electrode and the electrolyte. In SOE, water sent by the cathode is reduced to hydrogen and oxygen ions. These ions then pass through the solid electrolyte and form O_2 through oxidation at the anode side [38]. These reactions are given in Equations 11.4–11.6 [39].

Anode Reaction:

$$2O^{-2} \rightarrow O_2(g) + 4e^- \tag{11.4}$$

Cathode Reaction:

$$2H_2O(g) + 4e^- \rightarrow 2H_2(g) + 2O^{-2} \tag{11.5}$$

Overall Reaction:

$$H_2O(g) \rightarrow H_2(g) + \frac{1}{2}O_2(g) \tag{11.6}$$

SOEs based on oxygen ion-conducting electrolytes typically operate at high temperatures due to the low oxygen ion conductivity of ceramic electrolytes. This depends on both the electrode kinetic activity and the oxygen ion conductivity of the electrolyte. However, low-temperature operation of SOEs is possible by increasing ion conductivity with high ion transfer [40].

11.2.3 PEMWEs

PEMWEs have been in existence for over 60 years. General Electric (GE) developed them using a solid sulfonated polystyrene membrane as the electrolyte [41]. PEMWE technology is like PEM fuel cell technology. Both use solid poly-sulfonated membranes, such as Nafion, as the electrolyte, which is proton conducting. PEMWEs offer several advantages, including low gas permeability, high current density, low operating temperatures (20°C–80°C), and high current density. Additionally, PEMWE is a promising hydrogen production technology due to its compact design, fast response, high output pressure, and efficiency. However, it also has significant disadvantages, such as dependence on precious and rare noble metal catalysts, which results in high costs [42]. PEMWEs are electrochemical energy converters that use a polymer electrolyte membrane (or proton exchange membrane) as the ionic conductor between electrodes. Molecular water and ion species are transported through the membrane from the anode side to the cathode side, where they are dissociated into H^+, O_2, and e^-s. Protons are then transferred to the cathode electrode side through the membrane. Electrons leave the anode electrode through an external power circuit that obtains cell voltage for the reaction. Hydrogen is produced at the cathode electrode side through recombination of electrons and protons. The reactions occurring on the anode and cathode sides, as well as the total reaction, are given by Equations 11.7–11.9 [43].

Anode Reaction:

$$H_2O(l) \rightarrow 2H^+ + \frac{1}{2}O_2(g) + 2e^- \tag{11.7}$$

Cathode Reaction:

$$2H^+ + 2e^- \rightarrow H_2(g) \tag{11.8}$$

Total Reaction:

$$H_2O(l) \rightarrow H_2(g) + \frac{1}{2}O_2(g) \tag{11.9}$$

Equations 11.7 and 11.8 represent the OER occurring at the anode electrode and the HER occurring at the cathode electrode, respectively. Equation 11.9 represents the total reaction arising from the combination of two electrochemical half-reactions that occur on the surface of the electrodes requiring a direct current (DC) power source for their formation.

11.3 Performance Metrics

Several performance measures determine the capital and operating costs of electrolyzer systems, including Faradaic efficiency, current density, stability, energy efficiency, and product concentration [44].

11.3.1 Current Density

The efficiency of electrolyzer systems is significantly influenced by the current density, which refers to the amount of electrical current drawn per unit active surface area of the electrode. Higher current density leads to faster consumption of reactants within an electrochemical reactor, especially increasing production rates in reactions with higher output speeds.

The importance of current density is highlighted by its direct influence on the capital cost of an installation, as it determines the electrolyzer size needed to achieve a specific output capacity. In practical situations, a relatively high current density is often required to optimize capital investment, although specific density requirements may vary for each unique system. The current density of an electrolyzer is determined by several factors, including the reactants' mass transportation onto the surface of the electrode, the surface morphology of the catalyst, the electrons' transfer to the reaction-facilitating active sites, and the cell's overall resistance.

11.3.2 Faradaic Efficiency

Faradaic efficiency is defined as the fraction of the overall current density in an electrochemical process that is used to generate a specific product. This measurement evaluates the selectivity of the electrical current applied to a specific transformation product. High Faradaic efficiency for desired products indicates minimal electrical energy distribution in undesired by-products, resulting in low energy losses and optimizing the overall system performance [45]. Moreover, superior Faradaic efficiency for a specific product has the potential to reduce downstream separation costs. The effectiveness of an electrochemical process relies on the careful allocation of the provided current to the intended product. This parameter is influenced by various factors, including catalytic performance, mass transfer dynamics of the system, electrolyte properties, and current density considerations [46].

11.3.3 Energy Efficiency

Energy efficiency assessment is of great importance as a key performance metric for a specific system. This measurement compares the electrical input to the stored energy in the bonds of the resulting products. The operational efficiency of the electrolyzer is primarily determined by the operating voltage, which is affected by both excessive potential demands on the electrodes and ohmic losses within the system. The pursuit of high energy efficiency involves a combination of factors, including achieving notable Faradaic efficiency for the desired product, minimizing cathodic and anodic overpotentials, mitigating ohmic losses, and ultimately reducing the cell potential. The comprehensive energy efficiency of the system is intricately linked to the consideration of both upstream and downstream separation requirements.

11.3.4 Stability

To reduce electrolysis downtime, repair expenses, and replacement costs, it is imperative to establish a highly reliable and stable electrolyzer system. A practical electrolyzer configuration should maintain consistent production of the targeted product while also achieving a commercially viable Faraday efficiency and high current density over extended operational durations often spanning thousands of hours. The stability of electrolyzer systems depends on several factors, including the enduring stability of catalysts, gas diffusion layers, and membranes located at the cathode and anode interfaces [30]. Any decrease in the stability of electrolyzer systems may be due to catalyst degradation, crossover effects, and systemic errors associated with the applied potential to the cell.

11.3.5 Gas Production Rates and Purity

When compared with other parameters, such as Faradaic efficiency, current density, stability, and energy efficiency, less emphasis is placed on the evaluation of reactant and product concentrations. However, optimizing the concentrations of products and reactants has significant techno-economic implications for the overall system, primarily due to the potential impact on recycling and separation costs associated with products and reactants. The objective of this measurement is to increase the concentration of the product in wastewater while minimizing the need for reactant recycling. Challenges may arise depending on the electrolyzer architecture, such as crosslinking or difficulties in achieving the desired outcome. The presence of impurities in the produced hydrogen gas is an undesirable condition in system design. In an ideal system, the gas produced should be completely pure to maximize overall system performance, and removing impurities directly enhances system efficiency.

11.4 Performance Analysis

Water electrolyzer is considered a pivotal technology in facilitating the transition toward a sustainable and clean energy system. However, there are still challenges to overcome, including cost considerations, efficiency improvements, and the development of a dedicated infrastructure for H_2 production, distribution, and utilization [47]. To overcome these challenges, it is crucial to realize water electrolysis as a sustainable and feasible method of hydrogen production. The performance of a water electrolyzer, which is a method that uses electrical energy to carry out a non-spontaneous electrochemical reaction, such as the decomposition of H_2O into O_2 and H_2, can be affected by different factors, including the selection of electrode materials, electrolyte type, temperature, pressure, membrane structure, and cell design [41, 48, 49]. Optimizing these factors in the design of electrolyzers is critical for achieving high efficiency, high durability, and high performance in hydrogen production or other electrochemical processes. The performance of an electrolyzer system is directly influenced by operational conditions, including temperature and pressure [50]. Particularly, pressure operation is critical to the storage of H_2 and has a considerable impact on both material selection and stack design. Electrolyzer system efficiency can be enhanced by operating an electrolyzer in differential pressure mode, whereby at the cathode, H_2 undergoes electrochemical compression. At the same time, it is important to ensure that the electrolyzer system operates at equal pressure for both the anode and cathode sides, especially in stack designs. PEMWE is an electrolyzer technology that offers an advantage over AWE by generating high-pressure hydrogen ranging from 3,000 to 80,000 kPa. The performance of electrolyzers has been previously studied by many researchers, particularly the impact of the clamp pressure [51–53]. This pressure is a crucial parameter that significantly affects the performance and safety of electrolyzers. High performance is provided when a uniform distribution of compression pressure is applied to all bipolar plates (BPs). When designing or operating electrolyzers, it is important to consider and control compression pressure to optimize performance, efficiency, and safety because uniform and appropriate compression pressure contributes to the long-term stability and reliability of the electrolyzer. Compression pressure also helps create effective seals between various components preventing leakage of gases, such as hydrogen and oxygen, produced during the electrolysis process. In this case, it helps prevent mechanical deformation or damage of BPPs, seals, and other components to ensure a longer operating life [54, 55]. For example, Frensch *et al.* [56] examined the impact of compression pressure, ranging from 0.77 to 3.45 MPa, on the electrolyzer system at different temperatures. They commented that increasing the compression pressure provided a lower internal contact resistance (ICR) value at higher compression pressure, which in turn increased the electrolyzer performance. A lower ICR value is beneficial because it means that there is less electrical resistance at the interfaces facilitating better electron and proton transport within the cell. Moreover, compression pressure can also accelerate gas passage through the membrane [57]. Higher compression pressures are thought to increase the rate of gas flow across the membrane, since higher compression pressures increase gas penetration. Stahler *et al.* [58] reported the relationship between compression pressure and membrane permeation. Changes in the physical properties of the membrane material under pressure can be caused by factors, such as increased stress, leading to the generation of defects or pores on the membrane, or changes in the interactions between gas molecules and the membrane [59]. Moreover, the effect of membrane thickness on the performance of the electrolyzer cell has been studied in the literature [57, 60–62]. For example, alkaline electrolyzers use alkaline electrolytes to facilitate the electrochemical reaction that splits water into H_2 and O_2 gases. The impact of membrane thickness on electrolyzer cell performance specifically involves a balance between ohmic losses, voltage efficiency, mass transport limitations, and mechanical stability. Ohmic loss is directly linked to both membrane conductivity and membrane thickness, and diffusion overpotential is influenced by variations in membrane thickness. This overpotential is associated with the transport of reactants to, and products from, the electrochemical reaction sites within the cell. Therefore, optimization of membrane thickness is necessary to achieve the desired balance for efficient and economically viable hydrogen production [61]. Stahler *et al.* [63] achieved a current density of 11 A/cm^2@2 V in a one-cell electrolyzer system with a membrane thickness of approximately 20 mm. They reported that this current density corresponds to an electrical energy consumption of approximately 53.6 $kWh.kg_{Hydrogen}^{-1}$. However, they stated that decreasing the membrane thickness may cause a decrease in both the H_2 permeability through the membrane and its resistance against mechanical stresses in differential pressure mode. The membrane thickness in electrolyzer cells can significantly affect its performance. Because thick membranes cause high ohmic losses, ions have to travel a longer distance and encounter greater resistance. Therefore, high ohmic losses can reduce the overall efficiency of the electrolyzer cell performance. Thinner membranes generally lead to better voltage efficiency. The reason for this is that thinner membranes contribute to lower ohmic losses and have less electrical energy to overcome the resistance within the test cell. It is desirable to increase the voltage efficiency in electrolyzer cells to achieve higher efficiency in hydrogen production [64, 65]. However, new membrane materials with improved barrier properties may prevent the penetration of gases, thus increasing the cost of hydrogen production. The optimal membrane thickness must be determined to maximize the desired hydrogen production. Therefore, it is often necessary to conduct experiments or simulations considering the conditions of the hydrogen production system [61]. The materials used in the construction of electrolyzer components, such as membranes and electrodes, can also

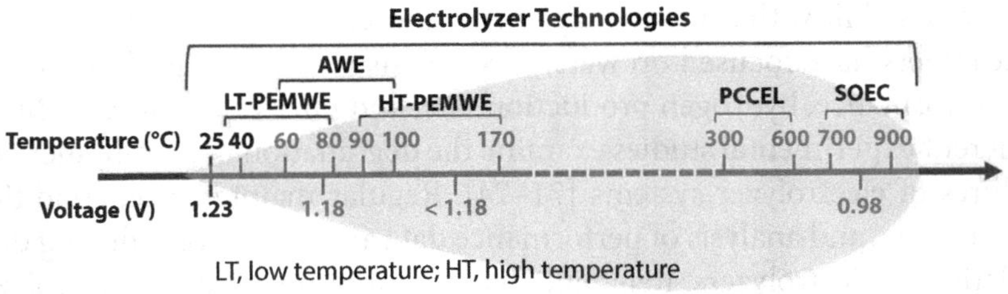

Figure 11.1 Temperature ranges of electrolyzer technology under different operating conditions.

have temperature-dependent properties. Some materials can degrade or lose efficiency at high temperatures affecting the overall performance and lifespan of the electrolyzer systems. Temperature plays an important role in the performance of electrolyzers affecting reaction rates, conductivity, solubility, and overall energy efficiency. Optimum operating temperatures depend on the electrolysis process and the materials used in the electrolysis system. The temperature at which electrolyzers operate can have significant effects on their efficiency, performance, and application [66]. Temperature ranges of electrolyzer technology operating at different temperatures are given in Figure 11.1.

In particular, SOE operating at higher temperatures (between 700°C and 900°C) has a significant potential to reduce electrical energy demand compared to water electrolysis technologies operating at lower temperatures of 50°C to 80°C [67]. At these temperatures, the solid oxide electrolyte (ceramics) used in SOEs allows for efficient ion transport enhancing the overall performance of the cell. Operating SOEs at higher temperatures can lead to increased energy efficiency, and higher temperatures can facilitate faster ion mobility reducing electrical losses associated with the electrolysis process. The performance of SOEs can be impacted by various parameters such as fuel gas inlet, gas flow rates, and operating temperature [68, 69]. Yang *et al.* [70] have conducted long-term tests on SOEs at different operating temperatures (700°C, 750°C, and 800°C) by using impedance analysis. In the analyses, they stated that the best operating temperature of SOEs is 750°C depending on the temperature. They also reported that the cells exhibit different initial performances due to different operating temperatures and that significant degradation is observed within the first 300 h. The high temperatures involved in SOE technology ensure that the resulting electrochemical processes are highly efficient. However, the operation of SOEs at high temperatures also presents challenges such as material stability, thermal management, and energy consumption. Many researchers have focused on water electrolysis technology and improving its performance, hydrogen production rate, and purity of hydrogen. Many different experimental studies examine the degradation of membranes and failures of electrolyzer systems [71–74]. Regular maintenance, condition monitoring, and analysis of performance data are crucial for reducing degradation in electrolyzers. Reliability has an important place in electrolyzer systems to ensure the safe production of hydrogen as fuel. Moreover, the high purity and efficiency of hydrogen production depend on the reliability of electrolytic cells in electrolyzers. It may cause safety problems for electrolyzers due to membrane degradation, especially considering the operating conditions [75, 76].

11.5 Advanced Techniques and Modeling

There are various cases for the analysis of electrolyzer systems, including experimental and analytical methods as well as numerical methods [77–79]. Combining these approaches allows for a comprehensive analysis of electrolyzer systems enabling better design, optimization, and understanding of their performance characteristics. The influence of various parameters, like temperature, pressure, speed, concentration of reactants, and current density, on the electrolyzer can be determined using numerical models. Evaluating the effects of system performance against changes in parameters also helps identify critical factors affecting overall performance. Nonetheless, this approach necessitates the concurrent solution of intricate equations encompassing continuity, momentum, energy, and species conservation, along with electrochemical kinetic equations. By using numerical models, researchers can explore a broad range of operating conditions, evaluate the interaction of different parameters, and ultimately design electrolyzer systems that are both energy efficient and economically viable. Therefore, it is important to use optimization techniques to find the best design parameters and operating conditions that maximize efficiency or minimize costs [80, 81]. These simulations provide valuable information before experimental testing, reduce costs, and accelerate the development of efficient electrolysis technologies. At the same time, each parameter can positively or negatively affect the performance of the electrolyzer system [82]. Another important strategy in electrolyzer technologies is optimizing flow distribution and heat throughout the cell. To achieve optimum mass transport properties, it is important to properly distribute the flow. In addition, the kinetics of the electrochemical processes taking place in the electrocatalyst layer have a critical impact on the homogeneity of the reactant distribution and on the concentrations of both reactants and products [83]. Ma *et al.* [82] investigated the performance of combined thermal fluid, species transport, and electrochemical processes in the PEMWE model using ANSYS/Fluent software. The findings indicate that the assumptions made in the electrochemistry and two-phase flow still offer sufficient accuracy to predict cell performance. They concluded that this modeling could aid in designing components encompassing electrochemical outcomes from electrodes in contact with membranes, porous transport layers, and separate plates with fluid fields contributing to the optimization of PEMWE performance. The study suggests that despite certain assumptions made in the electrochemistry and two-phase flow aspects of the model, the accuracy achieved is sufficient for predicting cell performance. Moreover, numerical calculations of current collector lengths, temperatures, molar fraction distributions, and membrane thicknesses in PEMWEs can be determined with COMSOL Multiphysics software and researchers prefer to model PEMWE components separately [84, 85]. This approach enhances the comprehension of Multiphysics processes within each component, such as the OER/HER catalyst layer and the anode/cathode electrode. Modeling the anode and cathode components through Multiphysics analysis contributes to a comprehensive understanding of the detailed effects on electrode performance. It also facilitates the assessment of specific properties of electrode materials. Zinser *et al.* [86] developed a one-dimensional model for anode electrode mass transport and determined the possible drying phenomenon based on the channel water saturation by studying the influence of electrode material characteristics on the two-phase flow dynamics. This model effectively provides a complete understanding of the effects of key properties of materials on Multiphysics processes under different operational conditions, as well as predicting realistic scenarios for electrolyzer design and operation in scale-up problems. Researchers have extensively investigated the impact of two-phase flows on the efficiency of electrolyzers, with particular emphasis on bubble formations and the dynamics of two-phase flow during water electrolysis. This involves optimizing various parameters, including current density, temperature, pressure, and electrolyte composition, to achieve the best performance under two-phase flow conditions [87, 88]. The existing models predominantly consist of empirical, semi-empirical, and analytical models designed to accurately forecast the performance of PEMWEs. Some models in this category lack validation through experimental data, which limits their applicability in simulating PEMWE performance. Considering the primary application of PEMWEs, incorporating dynamic behaviors is an important consideration in the development of PEMWE models [89]. These strategies are vital for advancing the development of sustainable and cost-effective hydrogen production through electrolysis. Some studies on the optimization of AWEs ignored the influence of gas products on the electrolyte flow. Modeling is an essential tool for ensuring the rapid advancement of AWE, contributing to the understanding of phenomena, analysis of control systems, energy management, and optimization. Furthermore, it facilitates the exploration of design and operational boundary conditions with reduced time. Researchers have investigated the flow performance and distribution of electrolytes in flow channels through a combination of experimental and numerical simulation methods [90–92]. For example, Wang *et al.* [93] developed a numerical model for single-phase flow within a three-dimensional concave–convex BP. Then, they performed visualization flow experiments. In the analysis results, it is determined that there is a lower electrolyte velocity at the boundary compared to the central region and a non-uniform flow within the electrolyzer. They also determined that the mastoid structure did not distribute electrolytes horizontally. The electrochemical model of AWEs involves understanding the various electrochemical processes that occur during electrolysis. Research in this field often aims to establish a comprehensive understanding of the mechanisms involved in the electrochemical reactions within the AWEs [94]. This included investigating the reaction kinetics, the role of different materials used in the electrodes, and the overall efficiency of the process. These models can help optimize the design and operation of AWEs for improved efficiency and durability.

Determining the effect of some parameters on electrolyzer systems also requires the use of different measurement techniques. For example, linear sweep voltammetry (LSV) and cycling voltammetry (CV) characterization tests are powerful tools to characterize the electrochemical behavior of electrolyzers providing information on kinetics, reaction mechanisms, and overall performance. These tests are valuable for optimizing electrolyzer design and operation in the context of hydrogen production and other electrochemical processes [95]. In particular, the type of electrode materials and catalysts used in the electrolyzer (e.g., Pt, Ru, Cu, and Ni-based materials) can significantly affect electrochemical performance, and these analyses help characterize the behavior of materials. It can provide information about the electrolyzer under various operating conditions, such as different temperatures, pressures, flow rates, and electrochemical processes in different electrolyte conditions to understand the performance of the electrolyzer under realistic scenarios [96, 97]. Electrochemical impedance spectroscopy (EIS) analysis of electrolyzers provides a comprehensive understanding of their electrochemical performance, aiding in optimization, and continuous monitoring for improved efficiency and durability. This technique is used to analyze the impedance of the electrolyzer system. EIS provides valuable information about the electrochemical processes occurring within the electrolyzer and can help identify any potential issues [98]. Tendencies toward dissolution in electrolyzer systems can be understood by examining available thermodynamic data. However, predicting the kinetics of a material's stability in real conditions remains challenging and requires experimental measurements, preferably conducted through *in situ* tests. One recently developed technique that addresses this issue is online ICP-MS. Moreover, analysis of electrolyte composition involves techniques, such as ion ICP-MS or titration, to determine the concentration of ions in the electrolyte solution. This helps in understanding the conductivity and overall efficiency of the electrolyte. Gas flow rates of hydrogen or oxygen gases produced in electrolyzers can be measured using flow meters. Gas chromatography is a technique used to analyze the composition of the gases produced ensuring the desired ratio and purity. By employing these measurement techniques, researchers can gain valuable insights into the behavior of electrolyzer systems and optimize their performance based on various parameters [99, 100]. Surface characterization of electrolyzers is crucial for understanding and optimizing the performance of these devices, which are used in electrochemical processes such as the electrolysis of water to produce hydrogen and oxygen. A variety of analytical techniques can be used to examine the surface properties of electrolyzer components, including electrodes and membranes [101]. For example, field emission scanning electron microscope (FE-SEM) is a high-resolution imaging technique that offers detailed surface morphology information, and it can be used to observe the surface structure, morphology, and defects in electrodes and membranes of electrolyzers. On the other hand, the combination of FE-SEM and Energy Dispersive X-ray Spectroscopy (EDX) allows researchers to simultaneously observe the surface morphology of a sample at high resolution and analyze its elemental composition. In particular, X-ray diffraction (XRD) and X-ray photoelectron spectroscopy (XPS) analysis can be used to analyze the chemical composition of material surfaces, obtain information about crystal phase structures, and detect any changes in the crystalline structure that may occur during operation [102]. The use of these techniques together allows researchers to comprehensively analyze the surface properties of electrolyzer components aiding in the design and optimization of efficient and durable electrochemical systems.

11.6 Experimental Procedures for Characterizing Electrolyzer Performance

Performance analysis of electrolyzers in industrial settings requires a thorough assessment covering efficiency, durability, hydrogen purity, operating conditions, maintenance needs, and economic viability. A key aspect for these applications is evaluating the scalability of electrolyzer technology, which involves understanding how effectively the technology can be adapted for larger-scale production demands.

While the basic test systems for electrolyzers are generally similar, they can vary in specific aspects depending on the type of electrolyzer being used. This variance is crucial to account for to ensure accurate assessment and optimal application of the technology in industrial contexts. Figure 11.2 shows a test setup diagram prepared for AWE.

In testing Alkaline Water Electrolysis (AWE) systems, electrical energy is supplied to the electrolyzer to drive the electrochemical reactions. While thermal energy is also necessary to maintain the operating temperature, external heat sources are typically not required due to the system's high irreversibility. The electrolyte, a solution of KOH in water, circulates through the electrolyzer stack facilitating gas evolution. Prior to entering the electrolyzer, this electrolyte is heated via a heat exchanger to reach the required operating temperature.

As the electrolyte and generated hydrogen gas exit the electrolyzer, they enter a phase separator. Here, hydrogen, the primary product, undergoes drying and purification to meet the desired quality standards. Concurrently, the product gas containing oxygen is also processed in the phase separator.

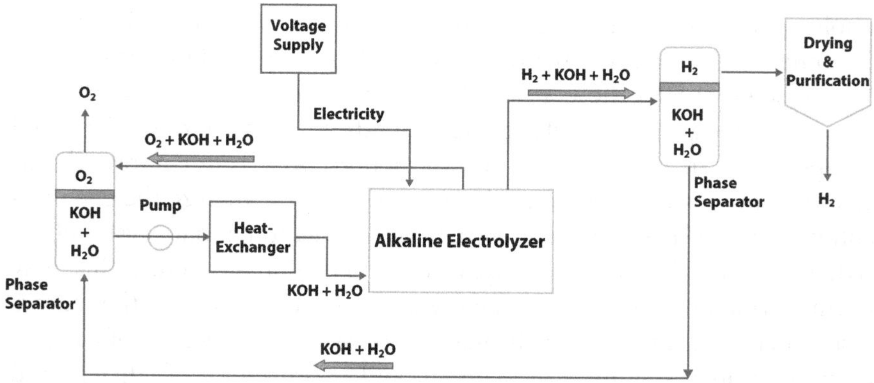

Figure 11.2 Schematic diagram of AWE test systems. Reprinted with permission from ref [68]. Copyright 2024, Elsevier [65].

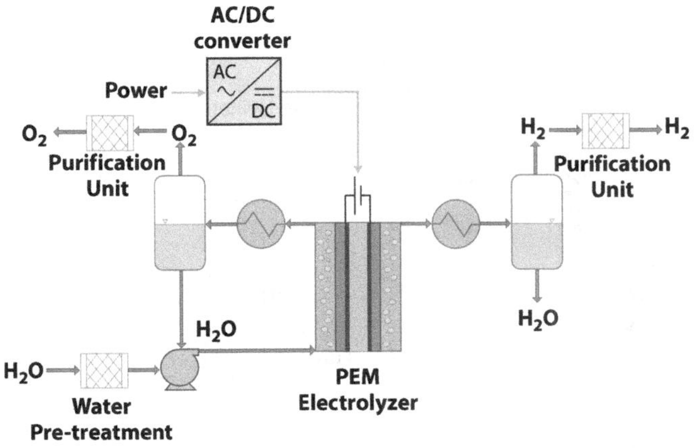

Figure 11.3 Schematic diagram of PEMWE test systems. Reprinted with permission from ref [68]. Copyright 2023, Elsevier [104].

The liquid electrolyte completes its cyclic journey through the system exiting the separator. Moreover, regular monitoring and replenishment of the caustic solution are critical to maintain optimal electrolyte conductivity. This series of operations underscores the intricate and nuanced processes essential for maintaining the efficiency and functionality of an AWE system [103]. PEMWEs require sensors, pumps, power conversion, valves, control equipment, and power conversion to operate the test systems of the electrolyzer. Figure 11.3 shows a simplified schematic view for PEMWE.

Pre-treating feed water is crucial to reducing poisoning of the cell. The flexibility of electrolyzers and their ability to provide ancillary services to the grid depend on high performance power electronics and appropriate controllers operating on the AC–DC converter. Additionally, if the electrolyzer is connected to the grid through an AC–DC converter, it can provide ancillary services to the grid. In this case, it is important to pay special attention to the quality of power supplied to the cell to prevent any additional losses [105]. Pumps supply fresh, pure water to the cell, usually at the anode only, under super stoichiometric conditions to prevent membrane dehydration and to remove generated gases from the cell. The extra water also works as a coolant. The cell outlet has two phases, gas and liquid, which are then split. To reduce the steam ratio, the flows can be chilled to room temperature or lower using a refrigeration circuit. Additional treatments of purification may need to be performed to achieve the purity specifications for the H_2 and O_2 streams [106]. The flow scheme of the SOE system is presented in Figure 11.4.

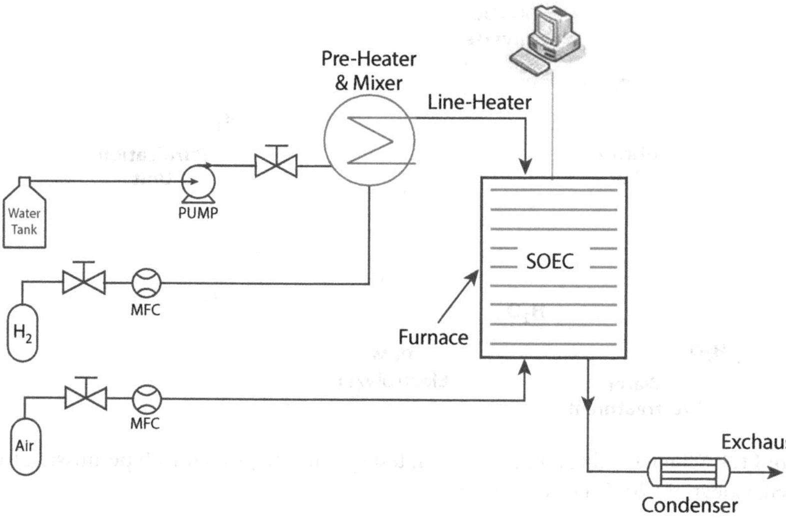

Figure 11.4 Schematic diagram of SOE test systems. Reprinted with permission from ref. [68]. Copyright 2014, Elsevier [107].

The system consists of series-connected SOE stacks and a control system that includes heaters, a steam generator, and a pump. The heat from the SOE outlet gas stream is recovered by heating water in a series of heaters. The heated water is then fed to the steam generator to produce steam, which is superheated in the electric heater. To improve the efficiency of the SOE system and reduce electricity demand, steam is heated in multiple heat exchangers using the exiting H_2 and O_2 streams [108]. Within the electrolyzer, steam undergoes dissociation at temperatures ranging from 650°C to 1,000°C at the cathode resulting in the formation of H_2 and O_2 molecules through a reaction that reduces water. The oxide ions migrate to the anode side from the cathode side releasing electrons in the external circuit and becoming O_2 through an OER. To facilitate electrochemical reactions on both electrodes, oxide ion migration requires high temperatures. The anodic electrode carries the O_2, while the anodic H_2 electrode on the other side of the electrolyte carries the H_2 along with some steam mixture. After the inlet process stream is heated, the downstream H_2-rich product stream is cooled and then passed to a separator to remove H_2 from the condensate stream. The use of experimental methods provides important data for the development of strategies to improve the performance of electrolyzers and the design of materials. These analytical techniques contribute to designing more efficient electrolyzers for sustainable energy production leading to advances in energy storage and conversion technologies.

11.7 Future Trends and Challenges

The water electrolysis technology confronts many challenges. The main focus of current research and development efforts is twofold, the first of which is to increase operating current density. The second is to increase efficiency levels to reduce operational expenditure. While general efforts continue to increase the efficiency of electrolyzers, production and operating costs are also reduced. Competing on cost with other hydrogen production methods is crucial for the widespread adoption of electrolysis. Electrolyzers are increasingly integrated with renewable or alternative energy sources, like wind and solar power, to produce green hydrogen, and this trend is consistent with the growing emphasis on sustainable and environmentally friendly hydrogen production [109]. Moreover, this could lead to the design of modular and scalable electrolyzer systems that enable easier integration into various environments, from small-scale distributed systems to large industrial applications. This approach could help optimize hydrogen production according to demand. Among hydrogen production technologies, PEMWE, which operates at lower temperatures and pressures, is gaining attention due to its fast response times and high-purity hydrogen production potential. Ongoing efforts aim to increase the durability and reduce the cost of PEMWEs. On the other hand, AWEs are known for their simple structure and long-term stability. Innovations in this technology aim to make AWEs more cost effective and competitive by focusing on the use of non-precious metal catalysts. Researchers are investigating high-temperature electrolysis technologies, known as high-temperature electrolysis, to improve overall efficiency. These technologies can potentially reduce energy input and costs compared to conventional electrolysis performed at lower temperatures [110, 111]. SOE, which is one of these technologies, offers the advantage of higher efficiency and the ability to directly use waste heat from industrial processes. To address the evolving demands of the future, it is imperative to advance the development of water electrolyzers. Innovative electrolysis systems must meet a set of key criteria, including compactness, enhanced gas management, effective liquid and thermal management, heightened load flexibility, prolonged durability and ease of maintenance, reduced costs, a forward-thinking stack design employing components conducive to mass production, and the integration of highly efficient power electronics [26]. Moreover, despite the superiority of PEMWE technology over AWE technology, it is estimated that the production costs of PEMWE will be higher in the future. As technology advances and the hydrogen economy develops, it is expected that these factors will continue to evolve potentially leading to further cost reductions in hydrogen production using PEMWE technology [112, 113]. On the other hand, SOE technology operating at high temperatures is not at the commercialization stage compared to other electrolyzers, and the cost of SOECs may vary depending on various factors such as advances in technology and market demand. High operating temperatures lead to rapid deterioration of active materials and system balance components (such as damage to ceramic layers due to heating and cooling at high temperatures). However, especially from a cost perspective, increasing the operational lifetime of SOEs is at the forefront of current research efforts directed toward SOEs [114]. In summary, the future of electrolysis systems relies on the integration of advanced technologies and innovative designs that meet the outlined criteria. Achieving these goals will not only contribute to the sustainability of hydrogen production but also facilitate the integration of electrolyzers into various sectors of the energy landscape.

11.8 Conclusion

The growing interest in electrolysis technologies is driven by green hydrogen production as a promising renewable energy source for achieving global decarbonization goals. H_2 is considered a promising and versatile energy carrier, and its production plays a key role in the development of sustainable and clean energy systems. A cost-effective, high-capacity water electrolysis system is a pivotal technology that can facilitate the widespread adoption of zero-carbon hydrogen across various applications. Cost competitiveness is critical to the broad acceptance of H_2 as an energy carrier. Continuous improvement of the performance of H_2 production technologies also helps in cost reduction efforts. This can involve optimizing resource utilization, streamlining processes, and identifying more cost-effective materials or catalysts. Electrolyzers, which are among the hydrogen production technologies, are often used in combination with renewable energy sources like wind or solar power. It enables the integration of these technologies to be optimized and to obtain greater efficiency from renewable energy with hydrogen production. The intermittency and variability of renewable energy can impact the overall performance of the hydrogen production system. Moreover, a stable and consistent hydrogen supply is important for meeting the demands of various applications, including transportation and industrial processes. Achieving a reliable supply of hydrogen requires overcoming challenges in production cost, scalability, and development of infrastructure. Ongoing R&D, policy support, and private-sector investment are vital to advancing these goals. The selection of electrolyzer technology depends on application, economic conditions, and availability of resources. PEMWE and AWE technologies are currently the most widely deployed technologies. PEMWE is preferred for applications requiring high efficiency and fast response times, while AWEs are chosen for larger-scale, cost-sensitive applications. SOEs and other emerging electrolyzer technologies are still under development. These technologies may bring improved efficiency or cost reductions, but they often require more time and investment before reaching commercial viability. Governments, industries, and research institutions are working collaboratively to develop innovative solutions and promote the transition to a more sustainable energy production environment based on hydrogen production technologies. This approach aims to contribute to broader goals such as the reduction of greenhouse gas emissions, the mitigation of climate change, and the encouragement of sustainable energy.

Acknowledgments

This research is made possible through the support of the Scientific and Technological Research Council of Turkey (TUBITAK) 1001 Research Projects Funding Program under project number 120M234. The authors express sincere gratitude to the Scientific Research Projects Unit of Erciyes University for their financial support (Project No: FDK-2023-13395). Special appreciation is extended to the Erciyes University Dean of Research for providing essential infrastructure and laboratory facilities at the ArGePark research building. Additionally, the authors dedicate this publication to the 100th anniversary of the Republic of Türkiye.

AI disclosure statement: The author acknowledges the use of AI-based tools for minor language editing or grammatical correction. However, all intellectual contributions, analyses, and conclusions presented in this chapter are entirely the author's own.

References

1. Agency, I.E., *Total Final Consumption (TFC) by sector, World 1990-2018*, The International Energy Agency (IEA), Paris, France, 2018.
2. Agency, I.E., *Total energy supply (TES) by source, World 1990-2018*, The International Energy Agency (IEA), Paris, France, 2020.

3. Demirci, Y.C., *et al.*, DSSCs based on unsymmetrical A3B type Zn (II) and TiO (IV) naphthalenephthalocyanine/porphyrin cocktail dyes: A potential alternative for ruthenium based sensitizers. *J. Photochem. Photobiol., A*, 440, 114642, 2023.
4. Johnsson, F., Kjärstad, J., Rootzén, J., The threat to climate change mitigation posed by the abundance of fossil fuels. *Clim. Policy*, 19, 2, 258–274, 2019.
5. Davis, S.J., Caldeira, K., Matthews, H.D., Future CO2 emissions and climate change from existing energy infrastructure. *Science*, 329, 5997, 1330–1333, 2010.
6. Abas, N., Kalair, A., Khan, N., Review of fossil fuels and future energy technologies. *Futures*, 69, 31–49, 2015.
7. Solaun, K. and Cerdá, E., Climate change impacts on renewable energy generation. A review of quantitative projections. *Renew. Sustain. Energy Rev.*, 116, 109415, 2019.
8. Mehmandoust, M., *et al.*, The determination of timolol maleate using silver/tannic acid/titanium oxide nanocomposite as an electrochemical sensor in real samples. *Electroanalysis*, 34, 7, 1150–1162, 2022.
9. Veziroglu, T.N., 21st Century's energy: Hydrogen energy system, in: *Assessment of hydrogen energy for sustainable development*, pp. 9–31, Springer, Dordrecht, Holland, 2007.
10. Filippov, S.P. and Yaroslavtsev, A.B., Hydrogen energy: development prospects and materials. *Russ. Chem. Rev.*, 90, 6, 627, 2021.
11. Züttel, A., *et al.*, Hydrogen: the future energy carrier. *Philos. Trans. R. Soc. A Mathematical Phys. Eng. Sci.*, 368, 1923, 3329–3342, 2010.
12. Naqvi, S.A.H., *et al.*, Hydrogen Production through Alkaline Electrolyzers: A Techno-Economic and Enviro-Economic Analysis. *Chem. Eng. Technol.*, 46, 3, 474–481, 2023.
13. Hnát, J., Paidar, M., Bouzek, K., Hydrogen production by electrolysis, in: *Current Trends and Future Developments on (Bio-) Membranes*, pp. 91–117, Elsevier, Rende, Italy, 2020.
14. Kraglund, M.R., *et al.*, Ion-solvating membranes as a new approach towards high rate alkaline electrolyzers. *Energy Environ. Sci.*, 12, 11, 3313–3318, 2019.
15. Nikolaidis, P. and Poullikkas, A., A comparative overview of hydrogen production processes. *Renew. Sustain. Energy Rev.*, 67, 597–611, 2017.
16. Corti, H.R., *Polymer electrolytes for low and high temperature PEM electrolyzers*, p. 101109, Current Opinion in Electrochemistry, Elsevier, Holland, 2022.
17. Rolo, I., Costa, V.A., Brito, F.P., Hydrogen-Based Energy Systems: Current Technology Development Status, Opportunities and Challenges. *Energies*, 17, 1, 180, 2023.
18. Zhang, L., *et al.*, *Electrochemical water electrolysis: fundamentals and technologies*, CRC Press, USA, 2020.
19. Young, J.L., *et al.*, PEM electrolyzer characterization with carbon-based hardware and material sets. *Electrochem. Commun.*, 124, 106941, 2021.

20. Li, C. and Baek, J.-B., The promise of hydrogen production from alkaline anion exchange membrane electrolyzers. *Nano Energy*, 87, 106162, 2021.
21. Santos, A.L., Cebola, M.-J., Santos, D.M., Towards the hydrogen economy—A review of the parameters that influence the efficiency of alkaline water electrolyzers. *Energies*, 14, 11, 3193, 2021.
22. Sampangi, S.K. and Lim, H., *Recent advances of hydrogen production through proton exchange membrane water electrolysis–A review*, Sustainable Energy & Fuels, RSC, UK, 2023.
23. Kumar, S.S. and Lim, H., An overview of water electrolysis technologies for green hydrogen production. *Energy Rep.*, 8, 13793–13813, 2022.
24. Rashid, M., *et al.*, Hydrogen production by water electrolysis: a review of alkaline water electrolysis, PEM water electrolysis and high temperature water electrolysis. *Int. J. Eng. Adv. Technol.*, 4, 3, 80–93, 2015.
25. Holm, T., *et al.*, Hydrogen costs from water electrolysis at high temperature and pressure. *Energy Convers. Manage.*, 237, 114106, 2021.
26. Grigoriev, S., *et al.*, Current status, research trends, and challenges in water electrolysis science and technology. *Int. J. Hydrogen Energy*, 45, 49, 26036–26058, 2020.
27. Shin, H., *et al.*, Techno-economic evaluation of green hydrogen production with low-temperature water electrolysis technologies directly coupled with renewable power sources. *Energy Convers. Manage.*, 286, 117083, 2023.
28. Bodner, M., Hofer, A., Hacker, V., H2 generation from alkaline electrolyzer. *Wiley Interdiscip. Rev.: Energy Environ.*, 4, 4, 365–381, 2015.
29. Zhou, P., *et al.*, Anodized Steel: The Most Promising Bifunctional Electrocatalyst for Alkaline Water Electrolysis in Industry. *Adv. Funct. Mater.*, 32, 26, 2202068, 2022.
30. Liu, Z., *et al.*, The effect of membrane on an alkaline water electrolyzer. *Int. J. Hydrogen Energy*, 42, 50, 29661–29665, 2017.
31. David, M., Ocampo-Martínez, C., Sánchez-Peña, R., Advances in alkaline water electrolyzers: A review. *J. Energy Storage*, 23, 392–403, 2019.
32. Karacan, C., *et al.*, Fabrication of High Performing and Durable Nickel-Based Catalyst Coated Diaphragms for Alkaline Water Electrolyzers. *J. Electrochem. Soc.*, 169, 5, 054502, 2022.
33. Khan, M.A., *et al.*, Recent progresses in electrocatalysts for water electrolysis. *Electrochem. Energy Rev.*, 1, 4, 483–530, 2018.
34. Kaya, M., *et al.*, Numerically modeling of anode supported tubular SOFC. *J. Appl. Mech. Eng.*, 3, 1, 1–5, 2014.
35. Nechache, A. and Hody, S., Alternative and innovative solid oxide electrolysis cell materials: a short review. *Renew. Sustain. Energy Rev.*, 149, 111322, 2021.
36. Chi, J. and Yu, H., Water electrolysis based on renewable energy for hydrogen production. *Chin. J. Catal.*, 39, 3, 390–394, 2018.
37. Kaya, M.F., Modelling of a solid oxide water electrolyzer and investigation of the cell performance on different working conditions. *J. Fac. Eng. Archit. Gazi Univ.*, 36, 4, 2109–2119, 2021.

38. Pandiyan, A., *et al.*, Review of solid oxide electrolysis cells: a clean energy strategy for hydrogen generation. *Nanomater. Energy*, 8, 1, 2–22, 2019.
39. Kaya, M.F., Bir katı oksit elektrolizörün sayısal modellenmesi ve farklı çalışma koşullarındaki performansının incelenmesi. *Gazi Üniversitesi Mühendislik Mimarlık Fakültesi Dergisi*, 36, 4, 2109–2120, 2021.
40. Ye, L. and Xie, K., High-temperature electrocatalysis and key materials in solid oxide electrolysis cells. *J. Energy Chem.*, 54, 736–745, 2021.
41. Khatib, F.N., *et al.*, Material degradation of components in polymer electrolyte membrane (PEM) electrolytic cell and mitigation mechanisms: A review. *Renew. Sustain. Energy Rev.*, 111, 1–14, 2019.
42. Carmo, M., *et al.*, A comprehensive review on PEM water electrolysis. *Int. J. Hydrogen Energy*, 38, 12, 4901–4934, 2013.
43. Hernández-Gómez, A., Ramirez, V., Guilbert, D., Investigation of PEM electrolyzer modeling: Electrical domain, efficiency, and specific energy consumption. *Int. J. Hydrogen Energy*, 45, 29, 14625–14639, 2020.
44. Gao, G., *et al.*, *Comparative analysis of electrolyzers for electrochemical carbon dioxide conversion*, p. 114284, Catalysis Today, Elsevier, Holland, 2023.
45. Ingelsson, M., Yasri, N., Roberts, E.P., Electrode passivation, faradaic efficiency, and performance enhancement strategies in electrocoagulation—a review. *Water Res.*, 187, 116433, 2020.
46. Gaikwad, P.S., *et al.*, Enhancing the Faradaic efficiency of solid oxide electrolysis cells: progress and perspective. *npj Comput. Mater.*, 9, 1, 149, 2023.
47. Janßen, H., *et al.*, A facile and economical approach to fabricate a single-piece bipolar plate for PEM electrolyzers. *Int. J. Hydrogen Energy*, 49, 816–828, 2024.
48. Miller, H.A., *et al.*, Green hydrogen from anion exchange membrane water electrolysis: a review of recent developments in critical materials and operating conditions. *Sustain. Energy Fuels*, 4, 5, 2114–2133, 2020.
49. Smolinka, T., Ojong, E.T., Garche, J., Hydrogen production from renewable energies—electrolyzer technologies, in: *Electrochemical energy storage for renewable sources and grid balancing*, pp. 103–128, Elsevier, Holland, 2015.
50. Caparros Mancera, J.J., *et al.*, An optimized balance of plant for a medium-size PEM electrolyzer: design, control and physical implementation. *Electronics*, 9, 5, 871, 2020.
51. Al Shakhshir, S., *et al.*, *In-situ* experimental characterization of the clamping pressure effects on low temperature polymer electrolyte membrane electrolysis. *Int. J. Hydrogen Energy*, 42, 34, 21597–21606, 2017.
52. Borgardt, E., *et al.*, Impact of clamping pressure and stress relaxation on the performance of different polymer electrolyte membrane water electrolysis cell designs. *Int. J. Hydrogen Energy*, 44, 42, 23556–23567, 2019.
53. Jiang, W., *et al.*, Influence of clamping pressure on contact pressure uniformity and electrical output performance of proton exchange membrane fuel cell. *Appl. Energy*, 353, 122021, 2024.

54. Zhou, X., *et al.*, Numerical and experimental characterization of gas permeation through membranes with consideration of mechanical degradation in proton exchange membrane fuel cells. *J. Power Sources*, 556, 232489, 2023.
55. Dang, J., *et al.*, Hydrogen crossover measurement and durability assessment of high-pressure proton exchange membrane electrolyzer. *J. Power Sources*, 563, 232776, 2023.
56. Frensch, S.H., *et al.*, Model-supported characterization of a PEM water electrolysis cell for the effect of compression. *Electrochim. Acta*, 263, 228–236, 2018.
57. Scheepers, F., *et al.*, Improving the efficiency of PEM electrolyzers through membrane-specific pressure optimization. *Energies*, 13, 3, 612, 2020.
58. Stähler, M., *et al.*, Impact of porous transport layer compression on hydrogen permeation in PEM water electrolysis. *Int. J. Hydrogen Energy*, 45, 7, 4008–4014, 2020.
59. Bazarah, A., *et al.*, Factors influencing the performance and durability of polymer electrolyte membrane water electrolyzer: A review. *Int. J. Hydrogen Energy*, 47, 85, 35976–35989, 2022.
60. Müller, M., *et al.*, Water management in membrane electrolysis and options for advanced plants. *Int. J. Hydrogen Energy*, 44, 21, 10147–10155, 2019.
61. Li, Y., *et al.*, Optimization of membrane thickness for proton exchange membrane electrolyzer considering hydrogen production efficiency and hydrogen permeation phenomenon. *Appl. Energy*, 355, 122233, 2024.
62. Razmjooei, F., *et al.*, Elucidating the performance limitations of alkaline electrolyte membrane electrolysis: dominance of anion concentration in membrane electrode assembly. *ChemElectroChem*, 7, 19, 3951–3960, 2020.
63. Stähler, M., *et al.*, A completely slot die coated membrane electrode assembly. *Int. J. Hydrogen Energy*, 44, 14, 7053–7058, 2019.
64. Garbe, S., *et al.*, Insight into elevated temperature and thin membrane application for high efficiency in polymer electrolyte water electrolysis. *Electrochim. Acta*, 377, 138046, 2021.
65. Sharshir, S.W., *et al.*, *A review of recent advances in alkaline electrolyzer for green hydrogen production: Performance improvement and applications*, International Journal of Hydrogen Energy, Elsevier, Holland, 2023.
66. Bonanno, M., *et al.*, *Review and Prospects of PEM Water Electrolysis at Elevated Temperature Operation*, p. 2300281, Advanced Materials Technologies, Wiley, Germany, 2023.
67. Wolf, S.E., *et al.*, Solid oxide electrolysis cells–current material development and industrial application. *J. Mater. Chem. A*, 11, 34, 17977–18028, 2023.
68. Hauch, A., *et al.*, Recent advances in solid oxide cell technology for electrolysis. *Science*, 370, 6513, eaba6118, 2020.
69. Chen, H., Wang, J., Xu, X., Parametric Study of Operating Conditions on Performances of a Solid Oxide Electrolysis Cell. *J. Therm. Sci.*, 2023, 1–16, Springer, 2023.

70. Yang, Y., *et al.*, Study of solid oxide electrolysis cells operated in potentiostatic mode: Effect of operating temperature on durability. *Chem. Eng. J.*, 417, 129260, 2021.
71. Phan, T.T., *et al.*, Degradation analysis of polymer electrolyte membrane water electrolyzer with different membrane thicknesses. *Int. J. Hydrogen Energy*, 49, 875–885, 2024.
72. Schäfer, D., *et al.*, Syngas production performance and degradation analysis of a solid oxide electrolyzer stack. *J. Power Sources*, 433, 126666, 2019.
73. Niaz, A.K., *et al.*, Effects of the operation mode on the degradation behavior of anion exchange membrane water electrolyzers. *J. Power Sources*, 481, 229093, 2021.
74. Le, L.Q., *et al.*, Performance degradation in proton-conducting ceramic fuel cell and electrolyzer stacks. *J. Power Sources*, 537, 231356, 2022.
75. SimaPro, *LCA Software for Informed*, SimaPro, Holland, 1990, [cited 2024; Available from: https://simapro.com/.
76. Tomić, A.Z., Pivac, I., Barbir, F., A review of testing procedures for proton exchange membrane electrolyzer degradation. *J. Power Sources*, 557, 232569, 2023.
77. Toghyani, S., *et al.*, Optimization of operating parameters of a polymer exchange membrane electrolyzer. *Int. J. Hydrogen Energy*, 44, 13, 6403–6414, 2019.
78. Nafchi, F.M., Afshari, E., Baniasadi, E., Anion exchange membrane water electrolysis: Numerical modeling and electrochemical performance analysis. *Int. J. Hydrogen Energy*, 52, 306–321, 2024.
79. Zheng, J., *et al.*, Three-Dimensional Numerical Simulation of the Performance and Transport Phenomena of Oxygen Evolution Reactions in a Proton Exchange Membrane Water Electrolyzer. *Materials*, 16, 3, 1310, 2023.
80. Norazahar, N., *et al.*, Degradation modelling and reliability analysis of PEM electrolyzer. *Int. J. Hydrogen Energy*, 50, 842–856, 2024.
81. Fakehi, A.H., Ahmadi, S., Mirghaed, M.R., Optimization of operating parameters in a hybrid wind–hydrogen system using energy and exergy analysis: Modeling and case study. *Energy Convers. Manage.*, 106, 1318–1326, 2015.
82. Ma, Z., *et al.*, A comprehensive modeling method for proton exchange membrane electrolyzer development. *Int. J. Hydrogen Energy*, 46, 34, 17627–17643, 2021.
83. Hassan, A.H., *et al.*, Characteristics of different flow patterns for proton exchange membrane water electrolysis with circular geometry. *Int. J. Hydrogen Energy*, 49, 1060–1078, 2024.
84. Lin, N. and Zausch, J., 1D multiphysics modelling of PEM water electrolysis anodes with porous transport layers and the membrane. *Chem. Eng. Sci.*, 253, 117600, 2022.
85. Kaya, M. and Demir, N., Numerical investigation of PEM water electrolysis performance for different oxygen evolution electrocatalysts. *Fuel Cells*, 17, 1, 37–47, 2017.

86. Zinser, A., Papakonstantinou, G., Sundmacher, K., Analysis of mass transport processes in the anodic porous transport layer in PEM water electrolysers. *Int. J. Hydrogen Energy*, 44, 52, 28077–28087, 2019.
87. Mo, J., *et al.*, *In situ* investigation on ultrafast oxygen evolution reactions of water splitting in proton exchange membrane electrolyzer cells. *J. Mater. Chem. A*, 5, 35, 18469–18475, 2017.
88. Avci, A.C. and Toklu, E., A new analysis of two phase flow on hydrogen production from water electrolysis. *Int. J. Hydrogen Energy*, 47, 11, 6986–6995, 2022.
89. Falcão, D. and Pinto, A., A review on PEM electrolyzer modelling: Guidelines for beginners. *J. Clean. Prod.*, 261, 121184, 2020.
90. Duan, X., *et al.*, Numerical simulation and structural optimization of electrolyzer based on the coupled model of electrochemical and multiphase flow. *Int. J. Hydrogen Energy*, 49, 604–615, 2024.
91. Gao, L.-Y., *et al.*, Three-dimensional two-phase CFD simulation of alkaline electrolyzers. *J. Electrochem.*, 29, 9, 2207081, 2023.
92. Qi, Y., *et al.*, Polymer electrolyte fuel cell system level modelling and simulation of transient behavior. *ETransportation*, 2, 100030, 2019.
93. Wang, T., *et al.*, Non-uniform liquid flow distribution in an alkaline water electrolyzer with concave-convex bipolar plate (CCBP): A numerical study. *Int. J. Hydrogen Energy*, 48, 33, 12200–12214, 2023.
94. Hu, S., *et al.*, A comprehensive review of alkaline water electrolysis mathematical modeling. *Appl. Energy*, 327, 120099, 2022.
95. Sun, H., *et al.*, Electrochemical water splitting: Bridging the gaps between fundamental research and industrial applications. *Energy Environ. Mater.*, 6, 5, e12441, 2023.
96. Kaya, M.F., PEM Elektrolizörlerin Manyetik Alan Altındaki Çalışma Performansının İncelenmesi ve Katalizör Geliştirilmesi, in: *Enerji Sistemleri Mühendisliği*, Erciyes Üniversitesi: Fen Bilimleri Enstitüsü, Türkiye, 2018.
97. Ghising, R.B., *et al.*, Ruthenium single atoms implanted on NiS 2-FeS 2 nanosheet heterostructures for efficacious water electrolysis. *J. Mater. Chem. A*, 12, 3489–3500, 2024.
98. Garcia-Navarro, J.C., Schulze, M., Friedrich, K.A., Measuring and modeling mass transport losses in proton exchange membrane water electrolyzers using electrochemical impedance spectroscopy. *J. Power Sources*, 431, 189–204, 2019.
99. Kasian, O., *et al.*, Electrochemical on-line ICP-MS in electrocatalysis research. *Chem. Record*, 19, 10, 2130–2142, 2019.
100. Lin, H.Y., *et al.*, Oxygen evolution electrocatalysts for the proton exchange membrane electrolyzer: challenges on stability. *Small Methods*, 6, 12, 2201130, 2022.
101. Lu, H., *et al.*, Optimization of pulse bi-directional electrolysis *in-situ* synthesis of tungsten carbide by response surface methodology. *Int. J. Refract. Met. Hard Mater*, 111, 106063, 2023.

102. Raja, P.B., *et al.*, Characterization of nanomaterial used in nanobioremediation, in: *Nano-bioremediation: fundamentals and applications*, pp. 57–83, Elsevier, Holland, 2022.
103. Firtina-Ertis, I., Thermodynamic and electrochemical assessment of an alkaline electrolyzer (AE) at different operating parameters. *J. Environ. Chem. Eng.*, 10, 2, 107225, 2022.
104. Mucci, S., Mitsos, A., Bongartz, D., Power-to-X processes based on PEM water electrolyzers: A review of process integration and flexible operation. *Comput. Chem. Eng.*, 108260, 2023.
105. Ruuskanen, V., *et al.*, Power quality estimation of water electrolyzers based on current and voltage measurements. *J. Power Sources*, 450, 227603, 2020.
106. Bensmann, B., *et al.*, Optimal configuration and pressure levels of electrolyzer plants in context of power-to-gas applications. *Appl. Energy*, 167, 107–124, 2016.
107. Li, Q., *et al.*, Achieving high-efficiency hydrogen production using planar solid-oxide electrolysis stacks. *Int. J. Hydrogen Energy*, 39, 21, 10833–10842, 2014.
108. AlZahrani, A.A. and Dincer, I., Thermodynamic and electrochemical analyses of a solid oxide electrolyzer for hydrogen production. *Int. J. Hydrogen Energy*, 42, 33, 21404–21413, 2017.
109. Chatenet, M., *et al.*, Water electrolysis: from textbook knowledge to the latest scientific strategies and industrial developments. *Chem. Soc. Rev.*, 51, 11, 4583–4762, 2022.
110. AlZahrani, A.A. and Dincer, I., Design and analysis of a solar tower based integrated system using high temperature electrolyzer for hydrogen production. *Int. J. Hydrogen Energy*, 41, 19, 8042–8056, 2016.
111. Sazali, N., Emerging technologies by hydrogen: A review. *Int. J. Hydrogen Energy*, 45, 38, 18753–18771, 2020.
112. Wang, Y., *et al.*, The multi-scenario projection of cost reduction in hydrogen production by proton exchange membrane (PEM) water electrolysis in the near future (2020–2060) of China. *Fuel*, 354, 129409, 2023.
113. Krishnan, S., *et al.*, Present and future cost of alkaline and PEM electrolyser stacks. *Int. J. Hydrogen Energy*, 48, 83, 32313–32330, 2023.
114. Prosser, J.H., *et al.*, Cost analysis of hydrogen production by high-temperature solid oxide electrolysis. *Int. J. Hydrogen Energy*, 49, 207–227, 2024.

Index

Also of Interest

Check out these other forthcoming and published titles from Scrivener Publishing

Books by the same editor from Wiley-Scrivener

Quantum Optics Devices on a Chip, Edited by Inamuddin, Tariq Altalhi, Naif Ahmed Alshehri, & Jorddy Neves Cruz, ISBN: 9781394248575. *Quantum Optics Devices on a Chip* provides a comprehensive understanding of how the integration of advanced quantum technologies and photonics is revolutionizing multiple industries, making it essential for anyone interested in the future of quantum innovation.

Solid-Gaseous Biofuels Production, Edited by Inamuddin and Tariq Altalhi, ISBN: 9781394204403. Written by a team of industry experts and edited by one of the most prolific and well-respected engineering authors in the industry, this exciting new volume covers the latest processes, equipment, and applications for clean biofuel production.

SEWAGE AND BIOMASS FROM WASTEWATER TO ENERGY: Possibilities and Technology, Edited by Inamuddin, Tariq Altalhi, Mohammad Luqman, and Joseph K. Bwapwa, ISBN: 9781394204311. Written and edited by a team of industry experts, this exciting new volume covers clean energy production from sewage and biomass while achieving a zero-waste strategy.

SOLAR ENERGY CONCENTRATORS, Edited by Inamuddin, ISBN: 9781394204328. Discover the latest techniques and applications for solar energy concentrators in this essential guide for academics, researchers, environmentalists, and professionals seeking to harness the power of solar energy while reducing environmental impact and costs.

WIND ENERGY STORAGE AND CONVERSION: From Basics to Utilities, Edited by Inamuddin, Tariq Altalhi, and Mohammad Luqman, ISBN: 9781394204335. This book provides a comprehensive guide to the benefits and developments of wind energy, including energy storage and conversion methods, making it a must-read for those interested in sustainable energy.

ADVANCED REDOX FLOW TECHNOLOGIES, Edited by Inamuddin and Tariq Altalhi, ISBN: 9781119904793. This book serves as a comprehensive guide to redox flow technologies, from their basic principles to their applications. As a result, this book provides a thorough review of craftsmanship in the subject, as well as the potential for future advances.

HANDBOOK OF WATER POLLUTION, Edited by Inamuddin, Tariq Altalhi, and Arwa Alrooqi, ISBN: 9781119904809. This handbook covers a wide range of water pollution issues and remedies. Researchers, students, and other governmental and non-governmental groups will all benefit from this informative guide. In detail, this book will include chapters in the areas of water pollution, analysis, and solutions.

ESSENTIAL OILS: Extraction Methods and Applications, Edited by Inamuddin, Tariq Altalhi, and Jorddy Neves Cruz, ISBN: 9781119829355. Written and edited by some of the world's top experts in the field, this exciting new volume provides up-to-date information on essential oils' chemical structure, therapeutic, and biological actions, and defines their functional uses and evaluates the advantages and disadvantages of their use in a variety of sectors.

PROTON EXCHANGE MEMBRANE FUEL CELLS: Electrochemical Methods and Computational Fluid Dynamics, Edited by Inamuddin, Omid Moradi, and Mohd Imran Ahamed, ISBN: 9781119829331. Edited by one of the most well-respected and prolific engineers in the world and his team, this book provides a comprehensive overview of hydrogen production, conversion, and storage, offering the scientific literature a comprehensive coverage of this important fuel.

MATERIALS FOR HYDROGEN PRODUCTION, CONVERSION, AND STORAGE, Edited by Inamuddin, Tariq Altalhi, Sayed Mohammed Adnan, and Mohammed A. Amin, ISBN: 9781119829348. Edited by one of the most well-respected and prolific engineers in the world and his team, this book provides a comprehensive overview of hydrogen production, conversion, and storage, offering the scientific literature a comprehensive coverage of this important fuel.

FUNDAMENTALS OF SOLAR CELL DESIGN, Edited by Inamuddin, Mohd Imran Ahamed, Rajender Boddula, and Mashallah Rezakazemi, ISBN: 9781119724704. Edited by one of the most well-respected and prolific engineers in the world and his team, this book provides a comprehensive overview of solar cells and explores the history of evolution and present scenarios of solar cell design, classification, properties, various semiconductor materials, thin films, wafer-scale, transparent solar cells, and other fundamentals of solar cell design.

Biofuel Cells, Edited by Inamuddin, Mohd Imran Ahamed, Rajender Boddula, and Mashallah Rezakazemi, ISBN: 9781119724698. This book covers the most recent developments and offers a detailed overview of fundamentals, principles, mechanisms, properties, optimizing parameters, analytical characterization tools, various types of biofuel cells, edited by one of the most well-respected and prolific engineers in the world and his team.

Biodiesel Technology and Applications, Edited by Inamuddin, Mohd Imran Ahamed, Rajender Boddula, and Mashallah Rezakazemi, ISBN: 9781119724643. This outstanding new volume provides a comprehensive overview on biodiesel technologies, covering a broad range of topics and practical applications, edited by one of the most well-respected and prolific engineers in the world and his team.

Applied Water Science Volume 1: Fundamentals and Applications, Edited by Inamuddin, Mohd Imran Ahamed, Rajender Boddula and Tauseef Ahmad Rangreez, ISBN: 9781119724766. Edited by one of the most well-respected and prolific engineers in the world and his team, this is the first volume in a two-volume set that is the most thorough, up-to-date, and comprehensive volume on applied water science available today.

Applied Water Science Volume 2: Remediation Technologies, Edited by Inamuddin, Mohd Imran Ahamed, Rajender Boddula and Tauseef Ahmad Rangreez, ISBN: 9781119724735. The second volume in a new two-volume set on applied water science, this book provides understanding, occurrence, identification, toxic effects and control of water pollutants in aquatic environment using green chemistry protocols.

Potassium-Ion Batteries: Materials and Applications, Edited by Inamuddin, Rajender Boddula, and Abdullah M. Asiri, ISBN: 9781119661399. Edited by one of the most well-respected and prolific engineers in the world and his team, this is the most thorough, up-to-date, and comprehensive volume on potassium-ion batteries available today.

Rechargeable Batteries: History, Progress, and Applications, Edited by Rajender Boddula, Inamuddin, Ramyakrishna Pothu, and Abdullah M. Asiri, ISBN: 9781119661191. Edited by one of the most well-respected and prolific engineers in the world and his team, this is the most thorough, up-to-date, and comprehensive volume on rechargeable batteries available today.

Printed in the USA/Agawam, MA
September 16, 2026

910661.097